Cheesemaking Practice

Cheesemaking Practice

R. SCOTT

Third edition

R.K. ROBINSON
R.A. WILBEY
The University of Reading
Reading, UK

A Chapman & Hall Food Science Book

An Aspen Publication®
Aspen Publishers, Inc.
Gaithersburg, Maryland
1998

Orders: (800) 638-8437
Customer Service: (800) 234-1660

Editorial Resources: Jon Walmsley
Library of Congress Catalog Card Number: 98–70420
ISBN: 0–7514–0417–9

Printed in Great Britain

1 2 3 4 5

Contents

Preface to the third edition

When the late Reg Scott wrote the first edition of this book in 1981, his intention was 'to produce a script generally interesting to those readers requiring more information on cheese'. It was not conceived as a book that covered the most recent developments with respect to lipid or protein chemistry, for example, but rather it was hoped that the text would reveal cheesemaking as a fascinating, and yet technically demanding, branch of dairy science. The fact that the author had some 50 years' experience of cheesemaking gave the book a very special character, in that the 'art' of the traditional cheesemaker emerged as a system that, in reality, had a strong scientific basis.

Today, cheesemaking remains a blend of 'art and science' for, while much cheese is made in computer-controlled factories relying on strict standardization to handle the large volumes of milk involved, the production of top quality cheese still relies on the innate skill of the cheesemaker. It was considered appropriate, therefore, that this revised edition of Cheesemaking Practice should include, at one end of the spectrum, details of the latest technology for curd handling and, at the other, simple recipes for the production of farmhouse cheeses. Obviously a student of dairy science will need to consult other texts in order to complete his/her knowledge of the cheesemaking process, but if this revised edition stimulates its readers to delve more deeply, then the task of updating the original manuscript will have been worthwhile.

Richard K. Robinson and R. Andrew Wilbey
The University of Reading

Acknowledgements

The authors gratefully acknowledge the benefit that they have derived from the wealth of material accumulated over many years by the original author, but as many of the illustrations are new, a special note of appreciation must be extended to the following organizations:

APV Nordic Cheese AS
Multivac UK Ltd
Tetra Tebel BV
Vantage House
Westfalia Separator Ltd
Wincanton Engineering Ltd
Wrapid Packaging Systems Ltd

The authors are also grateful to Drs David Jukes, Alistair Grandison and Mike Lewis of The University of Reading for their specialized advice; and to Ian Maclean and his team for their invaluable assistance in the preparation of photographs.

1 A brief history of cheese

A perusal of the literature concerning cheese reveals almost 2000 names applied to cheese, and periodically more names appear as new varieties are made. In order to discover whether there really are so many varieties, or whether a curd type has a number of different names, it is of interest to consider briefly the history of cheese, and particularly how the movement of populations by free or enforced migration has caused the spread of cheese almost throughout the world.

The early cradle of civilization as far as the staple foods, particularly bread and cheese, are concerned, appears to have been that rich agricultural area known as the 'Fertile Crescent' situated between the rivers Euphrates and Tigris in Iraq (*c.* 6000–7000 BC). After an archaeological survey of the area around Ur, Sir Leonard Woolley, in 1924, concluded that cheese had been made there from the milk of both cows and goats (Woolley and Hall, 1931).

An early Sumerian frieze from El-Ubaid depicts the milking operation and curdling of milk, and remnants of material found in the Tomb of Hories-Aha (*c.* 3000 BC) proved, on examination, to have been cheese; firmer evidence from carvings, ceramics or stoneware and writings confirms some of the earlier discoveries.

A scene on the walls of a Ramesid tomb (100 BC) depicts goats being led to pasture, and also shows skin bags suspended from poles. Since animal skin bags were a convenient way for the nomadic tribes to store liquids, milk surplus to daily needs would have been carried in this way. Fermentation of the milk sugars in these warm climates would have caused the milk to curdle in the bags, and the swaying motion of animals during journeys would have caused the acid curds to break up into curds and whey. While the whey provided a refreshing drink on hot journeys, the curds, preserved by the acid of fermentation and a handful of salt, gave a high protein food that would have supplemented the meagre meat supply.

This possible scenario has given rise to the assumption that cheese evolved from fermented milks, but it is perhaps more probable that the crude fermentations progressed along two parallel paths: one route leading to the production of liquid fermented milks, such as yoghurt, laban, koumiss and kefir, and the other to the formation of a solid curd which, after drainage of whey through a cloth or perforated bowl, was salted to become cheese. Fired earthenware bowls with perforated bases suitable for curd drainage have been found in several locations in Europe and Asia. Baskets woven from reeds and other stems have also been found, and such baskets are used in India today for making curds for both Surati Panir and Dacca. Impressions of baskets at Windmill Hill in Dorset, England (dated about 1800 BC), indicate that cheese may have been made in southern England well before the Roman occupation.

The spread of cheesemaking methods from the 'Fertile Crescent' area probably followed the same pathways as did the spread of wheat for bread. These pathways spread like the spokes of a wheel from the Sumerian plains east towards India and Tibet, north to the Steppes of Russia, northwest via the Caspian and Black seas to the Dniester, Dnieper and Danube rivers, and thence to Central and Northern Europe and west via the Mediterranean, Aegean and Adriatic seas to Southern and Central Europe. The spread of cheesemaking south to Africa was hindered by the hot climate.

There are references to cheese in biblical times: – Job X.10 (*c.* 1520 BC), and Samuel 1.XVII, 2.XVII.29 (*c.* 1017 BC), but any written history is scarce until the periods of the Greek and Roman Empires when various authors have left documented evidence. Homer (*c.* 1184 BC) referred to cheese made from the milk of sheep and goats in caves by the 'Cyclops' Polyphemus (Murray, 1919), and this cheese may have been the ancestor of the Feta cheese made so widely in Greece today. Thus, since the terrain dictates the type of animal that can be kept for milk production, it would have been natural for sheep and goats to be kept in the hilly mountain areas of Greece and the surrounding countries. Cythnos, another sheep's milk cheese, also spread throughout Greece from the island of that name, and this cheese is probably the first to be named after the place of its origin, a practice which expanded as the years went by.

Later, Herodotus (484–408 BC) makes reference to the 'Scythian' cheese made from the milk of mares (Godley, 1921), while Aristotle (322 BC) noted that 'Phrygian' cheese was made from the milk of mares and asses. Again these animals were of the type that might have been kept by wandering tribesmen in the eastern countries of the Mediterranean area.

Although the trade in cheese was expanding, especially in countries around the Mediterranean Sea, Varro (116–27 BC) reported on the 'variability of cheese', and suggested that because of 'the varying climates and the different animal foods, the same practice does not occur in every place' (Storr-Best, 1912). This evidence of variable quality in these early days is not surprising, and Varro also comments on the 'laxative qualities' of cheese. In particular, he noted that 'cheese made of cow's milk is the most nourishing but passes through the body with greatest difficulty, while cheese from goat's milk is least nourishing and most laxative, while cheese from sheep's milk is intermediate'. Unfortunately, there is no indication as to why the nutritional effect was so variable.

Although Varro wrote at length on agriculture, Columella (*c.* AD 50) in his *De Re Rustica* concentrated on cheese and its manufacture. He stressed particularly the need for hygiene in milk production and in cheesemaking (Curtis, 1945). Coagulation of the milk was obtained in many ways at this time, and the early coagulants for milk included rennets from the hare and kid, which were better than that from the lamb, as well as the sap from a branch of the fig tree and vinegar. The flower of the thistle, the seed of the bastard saffron, bruised thyme and extract of green pineapples were also used as coagulants. In order to scald the curd, the coagulum was cut and 'sprinkled with boiling water'. Smoking of cheese with smoke from apple tree wood is also mentioned in the treatise.

Indeed, the accounts by Columella show how far the art of cheesemaking had progressed. The spread of cheesemaking was also indicated by Pliny (23–79 AD) when he wrote of the foreign races 'that they condensed their milk into an agreeable sour milk cheese and butterfat' (Jones, 1963). As mentioned earlier, these sour milk cheeses may have been the ancestors of the 'white brined cheese' found today around the eastern Mediterranean.

The trade in cheese between countries, especially those on sea routes, became so great that the Roman emperor Diocletian (284–305 AD) had to fix maximum prices for cheeses. Amongst these cheeses was Lunar, a special variety with its own trademark, the 'Horns of the Moon'. This variety was later to become popular as Parmesan cheese.

The migration of populations due to the wandering habits of tribes, or due to famines, conflicts and invasions brought settlers, with their old crafts including cheesemaking, into new areas. One example is the movement of the Helveti tribe which settled in the Alpine regions (Switzerland), and from their craft developed a whole family of cheeses of which Emmental is the best known variety. The invasion of Britain by the Romans brought a new art of cheesemaking to the country, for, although cheese may have been made there previously, Palladius, a Roman in the 3rd Century, reported on cheese in Chester which was one of the Roman strongholds of the time.

This geographical migration of cheesemaking led to the emergence of new varieties of cheese being made, for differences in climate and terrain influenced the type of lactating animal that was kept. Thus high mountainous regions favoured sheep and goats, which even today have held their place as milk-producing animals (Table 1.1); the more fertile agricultural conditions in the plains led to the cow being the chief milk animal. The type of local animal determined the type of milk available for cheese and, together with local preferences for different flavours, the craft of cheesemaking underwent modifications and new varieties of cheese emerged. In addition, large volumes of cheese were made from the milk of the buffalo, as well as local cheeses from the milk of the yak, zebu, reindeer, mare, ass and, more rarely, the camel.

Although the displacement of peoples by wars or invasions aided the spread of cheese, the hostilities often caused a decline in local cheesemaking operations. This was particularly so in the 14th–17th Centuries in the

Table 1.1 Production of milk from goats and sheep in some of the areas with large herds of these animals

Country	Production 1994 (thousand tonnes)	
	Sheep	Goats
Algeria	220	140
Albania	58	73
Egypt	15	43
France	205	18
Greece	648	75
Italy	656	158
Morocco	27	34
Spain	306	276
Tunisia	16	13
World Production	8000	10 000

(Source: IDF, 1996)

Bohemian, Moravian and Slovakian areas of central Europe, where the hostilities restricted cheesemaking to shepherds' huts in the least accessible mountainous regions, with sheep as the main milk animal. During this period of decline, a large number of descriptive and local names arose to identify smaller cheeses of the sour milk or Quark type, especially in the area bounded by the rivers Danube and Dniester. For instance, in the Romanian plains and uplands (Carpathians), there are over 40 named varieties of the curdled milk cheeses, designated according to location, type of milk or treatment. Although the craft was kept alive in the monasteries, cheesemaking operations did not fully revive until co-operative organizations established creameries in Moravia (1878), Bohemia (1885) and Slovakia (1894). Schools at Pilsen and Kromeriz assisted in this revival of the cheese industry in these areas.

Just as in the Romanian regions, the plains and uplands of Hungary became home to many varieties of cheese, with about 50 varieties being based on Tarho, an old Hungarian yoghurt-like product. For example, Tarkoturo is a type of Quark, and is one of the many varieties made from goat's, sheep's and cow's milks. As in Romania, the designations indicate the type of milk, as well as whether the product is salted, ripened or fresh or, as in the case of Fustos Turo, that it is a smoked cheese. Further south, through the Macedonian mountains to Montenegro and Yugoslavia, there are also many local names, and over 60 named varieties of cheese, Quark and whey cheese of local origin have been identified. The separate languages in the three regions are obviously responsible for the different names, but even with cow's milk cheese, the shape, size and flavour differ from locality to locality. However, the methods of manufacture of cheese from one type of milk are remarkably similar in the different regions.

Up until the end of the 18th Century, the owners of large estates and, more particularly, the monasteries in Europe, kept alive the methods and recipes for the 'Art' of cheesemaking (Tusser, 1557). Otherwise, the craft was handed down, usually from mother to daughter, by word of mouth or by practical teaching. Sometimes this method led to misunderstandings and difficulties arose. Occasionally, a new variety of cheese was born, but this method was not so satisfactory as the written word kept by monks. The monasteries, by the interchange of monks, spread the methods of cheese-making over many areas. The Dales' cheeses (Cotherstone, Swaledale) in Yorkshire (England) were fostered by monks from Jervaulx Abbey and Fountains Abbey, while Wensleydale cheese was derived from methods used by monks at Rievaulx Abbey in Yorkshire. Other cheeses developed by monks in other monasteries were Fromage de Tamie from the Abbey of Tamie Lac d'Annecy near Geneva, Maroilles cheese from the Abbey of Maroilles near Avesnes, and Port du Salut cheese from the Monastery of Notre Dame du Port du Salut near Laval, France. Munster cheese is reported to have been taken by monks from Ireland to the Low Countries where it became an established variety.

In other regions of Europe, there were similar locally produced and locally consumed cheese varieties, but there were also cheeses which earned international repute and which were traded across state boundaries, for example Cheddar, Emmental, Edam, Gouda, Roquefort and Gorgonzola. However, the names given to these varieties often differ in various countries, so that the popular Edam cheese from The Netherlands is now made in many countries under a variety of names:

Edam	–	The Netherlands, Ireland, France, Japan, USA
Edamer	–	Germany, Austria, Norway, Finland
Edami	–	Hungary
Edamic	–	Bosnia–Hercegovina
Eidamsky Salam	–	Czech Republic
Svensk Edam	–	Sweden
Ser Edamski	–	Poland
Hollanda	–	Argentina
Gollandskij	–	Russia
Arad or Eilath	–	Israel

Apart from the influence of language on cheese names, there are reasons for using names to distinguish the different varieties of cheese. Often the name is not descriptive of the cheese, although some are, e.g. cream cheese, mature cheese or low-fat cheese, but, by common usage, a name can become associated with the appearance, taste and aroma of a particular type of cheese. There are instances where a name can only be used for cheese made in a specific geographical area. For instance, a blue cheese now called 'Royal Stewart' cannot be called Stilton because it is not made in the Midlands area of the UK. Even a trademark does not necessarily identify a cheese unless it also incorporates a description. Agreements, patents and legal regulations, whilst applying mainly to the labelling of cheese rather than the name, are likely to become more prominent in future descriptions of cheese.

There are many references in literature to names of cheese varieties known today, but it is essential to recognize that many cheeses have changed in character over a period of years. A list of some of the names with the date first noted is given below:

Gorgonzola	AD 879
Schabzieger	1000
Roquefort	1070
Marolles	1174
Schwangenkase	1178
Grana	1200
Taleggio	1282
Cheddar	1500
Parmesan	1579
Gouda	1697
Gloucester	1783
Stilton	1785
Camembert	1791
St. Paulin	1816

These cheeses have obviously satisfied the consumers over a long period of time, but there are many cheeses which were once produced and are now extinct because they were unacceptable to the consumer. For example, the East Anglian cheese 'Suffolk' was too hard and indigestible, and has, therefore, disappeared. Changes in agricultural practice, changes in food habits and changes in sociological conditions within a society have been given as reasons for the disappearance of cheese varieties, but consumer reaction, as in the case of Suffolk cheese, is probably the major reason. Thus, in spite of efforts to keep the varieties in production, the following cheeses have disappeared from tables in the UK – Berkshire, Banbury, Cotherstone, Cottenham, Essex, Hereford, Kentish, Peakland, Shropshire, Suffolk, Warwick and Wiltshire. In spite of these deletions from the list of names, new cheeses are being added to the world listing all the time.

Many of these new cheeses are, of course, market-led modifications to existing varieties, and it is not clear to what extent tradition will continue to dictate which types of cheese are acceptable. Obviously travel has made consumers in the industrialized world more cosmopolitan in their tastes, but national preferences are often deep-rooted. The English territorial varieties, for example, are all hard-pressed, because historically they tended to be made in the country and then transported for a hundred kilometres or more to buyers located in large towns. Consequently, the only acceptable varieties were those capable of withstanding the abuses of transport by horse-drawn cart and indifferent conditions of storage. In France, cheeses like Cantal would have been equally capable of withstanding poor conditions during transport, but the majority of French cheeses were manufactured for more immediate consumption, so that numerous soft varieties emerged over the years. A contrast can be observed again in Greece, where the dominance of sheep and goats, together with the high ambient temperatures, led to the cheese market being dominated by the brined group that includes Feta and Telemes. Why hard-pressed varieties like Graviera did not make an appearance until quite recently is not clear, for certainly this variety appears to store and travel just as well as the white-brined types. Similar patterns can be traced elsewhere, and it is likely that national groups will continue to retain, at least to some degree, preferences that have an historical element.

Although many traditional cheeses were undoubtedly of excellent quality, the cheesemaking process did not begin to be studied scientifically until the late 1800s. For many years, cheesemakers had recognized that the milk from different districts behaved differently in the cheese vat, but there had been few investigations into the reasons why this might be so. However, one of the earliest investigations was published in a report by the Bath and West and Southern Counties Society to the Board of Agriculture in 1899. F.J. Lloyd was the presenter of this report, which covered all aspects of cheesemaking, including geographical details of soils, rainfall, farming practices, land management, cheesemaking operations, defects and the systems of cheesemaking employed. It is important to stress that, even without scientific backing, the farmhouse cheesemaking systems in the 1850–1900 period were well advanced. An article in the *RASE Journal* of 1875 by J. Chalmers Morton on 'Cheesemaking in Home Dairies or Factories' is informative in this respect and when larger quantities of milk for cheese became available, as in America, the farmhouse methods were readily translated into factory practice.

Cheese factories were set up in America from 1851, followed by others in Europe, mainly through co-operation between farmers. The first factory in England was established at Derby in 1870 and by 1874 there were six factories in Derbyshire alone; factories in other counties followed quickly afterwards (Scott, 1954).

The emigration of people from the temperate climates of Europe to America, Canada, New Zealand, Australia, Africa and the Far East brought a demand, in the first instance, for cheeses from the 'old home', and later for cheeses made locally in the 'new home'. In Australia in 1970, there were some forty-three varieties of cheese available for sale and, since many of the new immigrants to Australia had come from European countries, it is to be expected that the list included mainly European names. These new immigrants took up the craft of cheesemaking to satisfy not only their own needs but also, as in New Zealand, to provide a commodity for export. In some instances the cheese process had to be modified to meet prevailing conditions, so that the characteristics of New Zealand Cheddar are not quite identical with English (Somerset) Cheddar cheese.

This brief dip into the history of cheese shows that the art and craft of cheesemaking is nearly as old as civilization, and that the range of cheese varieties is still expanding. Consequently, the aim of the cheesemaker should be not only to maintain and improve the quality of existing varieties and to devise new methods of manufacture, including mechanization, but also to acquire basic knowledge in order to devise new varieties to meet changing food habits.

References

Board of Agriculture (1899) *A Report on the Results of Investigations into Cheddar Cheesemaking, carried out on behalf of the Bath and West and Southern Counties Society in the years 1891–1898*, by F. J. Lloyd, FLS, FIC.

Curtis, M.C. (1945) *De Re Rustica*, (trans. from Columella, L.J.M.), Bk 7, Ch. 8, Heinemann Press, London.

Godley, A.D. (1921) *Works* (trans. from Herodotus), Vol. 2, Bk 4, Ch. 22–23, Loeb Library, Heinemann Press, London.

IDF (1996) *Production and Utilisation of Ewe and Goat Milk*. S.I. 9603, International Dairy Federation, Brussels

Jones, W.H.S. (1963) *Natural History* (trans. from Pliny), Bk 28, Ch. 33, Heinemann Press, London.

Murray, A.T. (1919) *The Odyssey* (trans. from Homer), G.P. Putnam, New York.

Scott, R. (1954) *Dairyman*, **1**(11), 229.

Storr-Best, L. (1912) *Libri Rerum Rustiarum* (trans. from Varro), Bk 2, Ch. 11, Loeb Library, Heinemann Press, London.

Tusser, R. (1557) *Five Hundred Points of Good Husbandry*, Company of Stationers, London.

Woolley, Sir Leonard and Hall, H. R. (1931) *Ur of the Chaldees*, Ernest Benn Ltd, London.

2 Importance of cheese as a food

According to the FAO (1994), the worldwide production of all types of cheese rose from 5 934 000 tonnes in 1963 to 14 880 089 tonnes in 1993, and Table 2.1 shows the production of cheese in various areas of the world. The precision of the figures is, of course, open to debate, because there is no doubt that much locally made cheese in countries from South America and elsewhere does not enter the statistics. Similarly, there are cheeses from the industrialized countries that have more in common with dairy desserts than traditional products and, again, their position in the production figures for any given country may be in doubt. Nevertheless, the general view is that the volume of cheese produced across the world will continue to increase and that, with fermented milks, will become the major outlet for liquid milk not required for direct consumption. However, the annual increase in cheese production depends not only on the availability of milk, but also on the ability of the industry to sell the cheese, bearing in mind the impact of such factors as: the prevailing economic conditions, market changes due to eating

Table 2.1 Some guideline figures for the production of cheese (all types) in different regions of the world during 1993 (thousand tonnes)

Region	Total	Country	Production
Africa	495	Egypt	333
		South Africa	38
North America	3861	USA	3385
		Canada	305
		Mexico	116
South America	613	Argentina	330
		Brazil	60
Asia	873	Iran	200
		China	164
		Turkey	139
Oceania	423	Australia	233
		New Zealand	190
Eastern Europe		Russian Fed.	708
Western Europe		France	1562
		Italy	919
		The Netherlands	647
		UK	362
		Denmark	288
		Greece	210
		Ireland	91
		Bel-Lux	7
Total world production	14 880		

(Source: *FAO Production Yearbook, 1994.)*

habits (e.g. rise in the number of vegetarians), availability of alternative high-protein foods (e.g. meat or fish), levels of disposable income and any changes in barriers to trade (e.g. the introduction of Certificates of Origin for certain varieties of cheese). In addition, milk production quotas of the type imposed by the European Union can have a devastating impact, and it is generally agreed that they have adversely affected the volumes of dairy produce, including cheese, manufactured in a number of EU countries.

However, cheese is an important commodity in world trade, so that production figures alone do not convey a satisfactory impression of the market for cheese. Thus, a number of countries like France, The Netherlands and Denmark are nett exporters of cheese, even though they also import cheese, and the export/import trade in cheese is complex. It has been noted also that there is no correlation between *per capita* cheese consumption and either milk production or liquid milk consumption. In Finland and Ireland, liquid milk consumption is high but cheese consumption is low, whilst in Germany and France the opposite occurs; it is often assumed that cheese and wine are consumed together in wine-producing areas.

New food habits have increased cheese consumption in Japan, while rapid increases in personal incomes in the Middle Eastern countries have brought an increased demand for cheese. Thus, imports of cheese into Kuwait, Iran, Saudi Arabia, Qatar, Jordan, Syria and Egypt have all risen dramatically over the last 20 years for, due to climatic conditions, milk, and therefore cheese, production is low. Changes in the more traditional markets have occurred with respect to cheese varieties, and there has been an increased demand for 'cream' cheeses like Quark or *fromage frais*, especially for light meals.

Clearly, cheese consumption can be increased where consumers have access to a wide variety of cheeses that satisfy all tastes. The emergence of a diet-conscious sector in industrialized societies has already brought into being a market for low-fat variants (15–18% fat) of several conventional varieties, but the maintenance of edible quality is an essential requirement for the future of this market. Several imitation cheeses using skim-milk with unsaturated vegetable fats have also been produced in the USA (Mozzarella-type, Rule *et al.*, 1980; low-fat cream cheese, Baker, 1981) and in Sweden (Hansen, 1981).

Although national figures for production or consumption are inevitably rather crude indicators, they do reveal that cheese is a major food item in many societies. Consequently, it is pertinent to consider the potential importance of cheese with respect to its nutritional value.

2.1 Nutritional value of cheese

The *per capita* consumption of all cheeses ranges from below 2 kg/year in Japan and South Africa through to 15–20 kg/year in countries such as

Table 2.2 Typical values for the composition of the major groups of cheese (all figures per 100 g of cheese)

Component	Parmesan	Cheddar	Edam	Feta	Cottage Cheese
Water (g)	18.4	36.0	43.8	58.0	79.9
Protein	39.4	25.2	6.0	20.0	14.0
Fat	32.7	34.4	25.4	21.0	4.0
Cholesterol (mg)	100	100	80	70	13
Energy (kcal)	452	412	333	250	98
Vitamins (μg)					
Vitamin A	345	325	175	–	–
Vitamin D	0.25	0.26	0.19	0.5	0.03
Vitamin E	700	530	480	370	80
Thiamine	30	30	30	40	30
Riboflavin	440	400	350	210	260
Niacin	120	70	70	200	100
Pyridoxine	130	100	90	70	80
Cobalamin	1.9	1.1	2.1	1.1	0.7
Folate	12	33	40	23	27
Pantothenate	430	360	380	360	400
Biotin	3.3	3.0	1.8	2.4	3.0
Minerals (mg)					
Sodium	1090	670	1020	1440	380
Potassium	110	77	97	95	89
Calcium	1200	720	770	360	73
Magnesium	45	25	39	20	9
Phosphorus	810	490	530	280	160
Iron	1.1	0.3	0.4	0.2	0.1
Copper	0.3	0.03	0.04	0.07	0.04
Zinc	5.3	2.3	2.2	0.9	0.6
Sulphur	250	230	–	–	–
Chloride ions	1820	1030	1570	2350	550

(After: Anifantakis, 1991; Holland *et al.*, 1991; Shaw, 1993; Tamime, 1993)

France, Germany and Italy (Tamime, 1993). Consequently, there are wide variations in terms of dietary importance, an aspect that may be exaggerated further by compositional differences between varieties of cheese. Thus, as can be seen in Table 2.2, the moisture content of the varieties is extremely important with respect to the characteristics of the group, but these essential contrasts mean that intakes of fat or protein per weight of cheese will differ significantly.

The tendency today is for industrial management to highlight the need to process raw materials in order to obtain the maximum profit margin. In the present context, the raw material, milk – a short-life commodity – is used by the cheese industry for conversion into cheese which, according to variety and storage, has a shelf-life of from 4–5 days up to 5–10 years. The cheese industry, therefore, provides a useful service in extending the life of a valuable human foodstuff.

However, the practical cheesemaker, in addition to turning milk into cheese, must ensure that the cheese is an edible, acceptable, marketable and nutritious food. It must maintain these qualities during its normal shelf-life, and be non-toxic and free from human-disease organisms. In the past, the nutritional aspects of cheese were rarely considered and, since cheese is made from a 'natural' product (milk), were taken for granted. Nowadays, the consumer has become more aware of food composition and the risks – real or imaginary – that may be associated with excessive intakes of certain components in foods. Nevertheless, there is no controversy over the fact that apart from water, man requires five major groups of nutrients in food, namely fat, protein, carbohydrate, vitamins and mineral salts, and it is worth considering the extent to which cheese can contribute to this demand.

2.1.1 Milk fat

The intake of fat should be such that the body can utilize it satisfactorily and, to avoid arteriosclerosis, the recommended daily intake of fat ranges from 25% (sedentary worker) to 35% (heavy manual worker) of the calorific content of the diet. The suggested calorific (energy) intake of the diet (which includes carbohydrates) is 2000–2750 kcal of energy per day and, since fat provides 9.0 kcal/g, the total daily intake of fat should be in the order of 55 g (for sedentary workers) and 106 g (active workers). It is further suggested that fat must have a proportion of 'essential fatty acids', i.e. linoleic, linolenic and arachidonic acids. These fatty acids are unsaturated, which means they are reactive at points along their structure where double bonds occur.

$$CH_3(CH_2)_4CH{=}CHCH_2CH{=}CH(CH_2)_7COOH \text{(Linolenic acid)}$$

$$CH_3(CH_2)_4CH{=}CHCH_2CH{=}CHCH_2CH{=} \\ {=}H(CH_2)_4COOH \text{(Linolenic acid)}$$

$$CH_3(CH_2)_4CH{=}CHCH_2CH{=}CHCH_2CH{=}CHCH_2CH \\ {=}CH(CH_2)_3COOH \text{(Arachidonic acid)}$$

There is concern also as to the relative nutritional importance of the *cis* and *trans* isomers of these acids.

Cheese made from whole milk is known to contain most of the essential fatty acids, but there have been some attempts to introduce higher levels of the desirable unsaturated fatty acids (linolenic) into the diet of the cow in anticipation that they might augment those in normal milk. Although successful, these extra unsaturated fats in the milk quickly cause rancidity problems unless antioxidants are added immediately after the milk is collected, and hence this approach to nutritional manipulation is unlikely to be acceptable to the dairy industry. Most of the fatty acids are present in the form of triglycerides, but up to 5 g/kg of cheese can be in the form of free fatty acids (Renner, 1983). These latter acids are derived from the hydrolysis of the milk fat by microbial lipases – unless rennet paste or the enzyme, lipase, is added during manufacture of the cheese – and their presence is essential to the flavour of a number of cheese varieties; they will also be rapidly absorbed during digestion.

The other controversial issue about consuming animal fats centres on the associated intake of cholesterol, but it may be relevant that (a) the maximum cholesterol content of cheese has been reported to be 100 mg/100 g, and it is suggested that cheese contributes no more than 3–4% of total cholesterol intake (Renner, 1993), and (b) cholesterol in the diet has only a limited effect on the level in serum, for there is a natural mechanism that lowers cholesterol synthesis by the body as the amount of ingested cholesterol increases (Finegan *et al.*, 1968; Flaim *et al.*, 1981). Consequently, Speckmann (1979) concluded that the contribution of cheese to the cholesterol intake of the average consumer was not significant.

2.1.2 Protein

The protein content of food for man must contain those amino acids essential for nutrition, and animal proteins (e.g. cheese) are more likely to contain these essential amino acids than vegetable proteins.

The daily protein requirement for adults is approximately 1 g/kg of body weight so that, for a person weighing 70 kg (154 lb), the daily requirement would be about 70 g of protein. Cheese is a suitable source of protein since it normally contains all the essential amino acids (Table 2.3), and a 100 g portion of soft cheese will provide 30–40% of the daily protein requirements of an adult; an equivalent portion of hard cheese will supply 40–50% (Renner, 1993). Casein is the main protein in cheese, and the

Table 2.3 The essential amino acids in milk protein and casein (%)

	Milk	Casein
Arginine	3.7	3.9
Histidine	2.2	3.0
Threonine	4.6	4.5
Valine	7.1	7.4
Leucine	12.1	10.0
Isoleucine	6.7	6.4
Lysine	7.4	8.1
Methionine	2.8	3.3
Phenylalanine	5.5	5.4
Tryptophan	1.4	9.6

Note: Arginine may not be essential for children, while histidine may be essential for children but not for adults.

differences in the analytical figures between milk and cheese are due to the loss of the whey proteins during cheesemaking. As these whey proteins are nutritionally superior to casein, which has a lower content of sulphur amino acids, the biological value of proteins in cheese is marginally lower than that of whole milk proteins; if the essential amino acid index for whole milk protein is rated as 100, then the corresponding value for cheese ranges from 91 to 97 depending upon the variety. None of the normal cheesemaking operations, e.g. renneting, ripening or maturation, affects the biological value of the protein (Staub, 1978). In some soft cheeses where the milk is concentrated by ultrafiltration to give a total solids level equivalent to that of the finished cheese, all the whey proteins are retained and the nutritional value of the product is enhanced accordingly.

It is relevant also that many cheeses undergo a period of maturation in which a portion of the casein is converted into water-soluble peptides and free amino acids, an action that tends to increase the digestibility of the proteins. Indeed, the digestibility of protein in cheese is higher than that of whole milk (Dillon, 1984), and the mean extent of utilization of essential amino acids is 89.1% compared with the value for egg protein – often regarded as the ideal – of 89.6%. Maturation may also result in the release of variable quantities of D-amino acids which are considered of no nutritional value for humans (Renner, 1993), and also amines, such as histamine, tyramine, putrescine and phenylethylamine. The concentrations of these amines in matured cheeses usually reaches a maximum of 30 mg/100 g for histamine and 40 mg/100 g for tyramine, but tyramine values of 420 mg/100 g have been reported (Evans *et al.*, 1988). Although these amines can affect blood pressure, it is usually agreed that a healthy person can consume quite large quantities of cheese without any adverse reaction, because oxidases in the body should rapidly convert the amines to aldehydes and thence onto carboxylic acids. However, some people do suffer from attacks of migraine after eating cheese, and it is believed that this affect is linked with an inherent deficiency of monoamine oxidase (Davidson *et al.*, 1979).

2.1.3 Lactose

Although milk does contain milk sugar (lactose – a carbohydrate providing energy in the diet), cheeses which have been ripened (and some soft varieties) do not contain appreciable amounts of lactose, since it is either lost into the whey during manufacture, or is converted into lactic acid or lactates during processing. People rated lactose-maldigestors, i.e. those adults that have a deficiency of lactase in their digestive secretions, can normally eat cheese, except for those very soft, fresh cheeses which may still contain appreciable amounts of unfermented lactose. Where some of the lactose has been converted to lactic acid, the lactic acid contents range from 0.2% (Camembert) to 1.3% (Cheddar) (Florence *et al.*, 1984), with the balance between the L(+)

and D(−) isomers varying, at least initially, with the type of starter culture. However, in cheeses subject to maturation, non-starter bacterial activity can transform as much as 50% of the acid into the D(−) isomer (Thomas and Crow, 1983), a form that can only be metabolized to a limited extent by humans. In practice, no toxicity of D(−) lactic acid to adult humans has ever been detected, and the only advisory limit from the WHO applies to children under 1 year old.

2.1.4 *Minerals and vitamins*

As shown in Table 2.2, cheese contains appreciable levels of minerals, of which calcium, iron and phosphorus are the most important. Indeed, 100 g of hard cheese can supply the daily calcium requirement of an average adult, plus nearly 50% of the phosphorus requirement, while the bioavailability is good compared with some other foods (Buchowski and Miller, 1990). On the other hand, some criticism has been made of the salt (NaCl) levels in cheese, but it should be noted that, even in countries with high levels of cheese consumption, cheese contributes no more than 5% of the daily intake of sodium/person/day (Edwards *et al.*, 1989). Consequently, it would be reasonable to conclude that the general population can safely eat cheese without any adverse consequences with respect to sodium, and that manufacturers can continue to use salt during production with the organoleptic features of the cheese being the primary concern.

Various trace elements, such as zinc, iron, iodine, selenium and copper are all found in milk to some extent and, although the total dietary intake from cheese may not be high, the regular consumption of cheese is a benefit with respect to these minerals.

The levels of fat-soluble vitamins in a cheese depends upon its fat content; in a full-fat cheese, some 80% of the vitamin A in milk passes into the product. The concentrations of B vitamins are also useful (Table 2.2), but the precise amounts will depend on a number of factors, e.g. type of starter culture and length of maturation period. Thus, a number of starter bacteria have the ability to synthesize B vitamins (Dillon, 1984), and it could be important that some of the propionibacteria in varieties of Swiss cheese can release vitamin B_{12} (cobalamin) into the retail product.

2.1.5 *Miscellaneous considerations*

While the overall nutritional value of cheese depends upon its chemical composition, there are some additional aspects that, while not fully substantiated, merit attention.

2.1.5.1 Defence against dental caries. There have been a number of studies that suggest that cheese may protect humans against dental caries, and a number of reasons have been suggested for these observations. In particular, it has been proposed that: (a) the high levels of calcium and phosphorus affect the deposition and extraction of minerals from the teeth (Jenkins, 1989) and (b) apart from the physical effect from biting through a firm piece of cheese and dislodging undesirable bacteria from the enamel of the tooth, the increased saliva flow will tend to neutralize acids on the tooth surface. Whether such influences are important within the general population is difficult to predict, but the clinical evidence deserves further attention.

2.1.5.2 Anti-carcinogenic effects. It is widely believed that milk can have a protective effect against several types of cancer (Gurr, 1992), and some fermented milks have been assigned a similar role (Sellars, 1991). The mechanism(s) involved in protection are largely unknown, but in the case of cancer of the colon, it is suggested that bile acids may be an important factor. If this association is crucial, then the action of the microflora in fermented milks may well involve a breakdown of the acids before they can have an adverse effect, while the calcium phosphate from milk or cheese could act to neutralize the acids.

As with all such studies, other dietary and life-style influences cannot be controlled, and hence positive claims must always be treated with due caution. Nevertheless, if some people are being saved the misery of cancer by the ingestion of dairy products on a regular basis, it would seem sensible for both the dairy industry and the medical profession to take the claims seriously.

2.1.5.3 Nitrites and nitrosamines. In some types of cheese, species of anaerobic, spore-forming clostridia, e.g. *Clostridium tyrobutyricum*, can initiate a butyric acid fermentation during maturation, and the resultant carbon dioxide and off-flavour render the cheese unfit to eat. The easiest method of control is the addition of up to 15 g of $NaNO_3$ or KNO_3 per 100 litres of cheese milk because, during maturation, the nitrate is reduced to nitrite which inhibits the growth of clostridia. The knowledge that nitrite is toxic and, perhaps more importantly, that it can react with secondary amines to produce carcinogenic nitosamines, raised questions about the advisability of adding nitrate during cheesemaking.

However, many studies have come to the conclusion that: (a) some of the nitrate is lost into whey, and cheeses made from milk with added nitrate rarely contain residues above 50 mg/kg of cheese, (b) residual nitrite is only present in a small fraction of mature cheeses from any batch (Munksgaard and Werner, 1987) and (c) if the maximum daily intake of nitrite suggested by the WHO is 0.2 mg/kg body weight/day, then the maximum input from milk and cheese represents 0.1% of this figure (Renner, 1993). Given that the main amines in cheese – histamine and tyramine – cannot be converted to nitrosamines anyway, it must be concluded that dairy products pose little risk to the consumer with respect to nitrite ingestion and/or conversion.

2.1.5.4 Mycotoxins. In mould-ripened cheeses, the conditions are unfavourable for toxin production (Chapman and Sharpe, 1990), so that, although *Penicillium roqueforti* can produce roquefortin and the so-called PR-toxin in the laboratory, only trace amounts of roquefortin have ever been detected in blue cheese. Allied with available toxicity data, it has been concluded that mould-ripened cheeses, such as Stilton, Roquefort or Danish Blue (or Danablu) cheese, pose no health risk to man from the presence of mycotoxins (Schoch *et al.*, 1984); the same conclusion can be applied to surface-ripened varieties like Brie and Camembert (Krusch *et al.*, 1977).

However, this confidence does not extend to casual contaminants from cheese rinds, and Bullerman (1979) found that many isolates of *Penicillium* and *Aspergillus* from commercial hard cheeses were capable of toxin production in the laboratory. The low carbohydrate levels in most cheeses are not, of course, conducive to toxin formation (Ahmed *et al.*, 1995), even though Zerfiridis (1985) did isolate one strain of *Aspergillus* that was capable of producing aflatoxin during growth on the Greek variety, Telemes. It is likely, however, that such results are atypical, and that mould growth that might lead to toxin formation can be readily avoided by appropriate manufacturing practices.

References

Ahmed, I.A., Ahmed, A.K. and Robinson, R.K. (1995) *Mycotoxin Research*, **11**, 75.

Anifantakis, E.M. (1991) Traditional Feta Cheese. In *Feta and Related Cheeses* (Eds R.K. Robinson and A.Y. Tamime), Woodhead Publishing Limited, Cambridge, pp. 49–69.

Baker, D.B. (1981). *Low fat cream cheese*. US Patent 4 214 983.

Buchowski, M.S and Miller, D.S. (1990) *Journal of Food Science*, **55**, 1293.

Bullerman, L.B. (1979) *Journal of Food Protection*, **42**, 65.

Chapman, H.R. and Sharpe, M.E. (1990) Microbiology of Cheese. In *Dairy Microbiology*, 2nd edition (Ed. R.K. Robinson), Chapman & Hall, London, pp. 203–290.

Davidson, S., Passmore, R., Brock, J. F. and Truswell, A. S. (1979) *Human Nutrition and Dietetics*. Churchill Livingstone, Edinburgh and London.

Dillon, J.C. (1984) Nutrition. In *Le Fromage* (Ed. A. Eck), Lavoisier, Paris, pp. 497–519.

Edwards, D.G., Kaya, A.E. and Druce, E. (1989) *European Journal of Clinical Nutrition*, **43**, 855.

Evans, C.S., Gray, S. and Kazim, N.O. (1988) *Analyst*, **113**, 1605.

FAO (1994) *FAO Production Yearbook*. Vol. 48, FAO, Rome.

Finegan, A., Hickey, N., Maurer, B and Mulcahy, R. (1968) *American Journal of Clinical Nutrition*, **21**, 143.

Flaim, E., Ferreri, L.F., Thye, F.W., Hill, J.E. and Ritchey, S.J. (1981) *American Journal of Clinical Nutrition*, **34**, 1103.

Florence, E., Milner, D.F. and Harris, W.M. (1984) *Journal of the Society of Dairy Technology*, **37**(3), 13.

Gurr, M.I. (1992) *Journal of the Society of Dairy Technology*, **45**(3), 61.

Hansen, R. (1981) *Nordeuropaeisk Mejeritidskrift*, **47**(2), 36.

Holland, B, Welch, A.A., Unwin, I.D., Buss, D.H., Paul, A.A. and Southgate, D.A.T. (1991) *McCance and Widdowson's The Composition of Foods*. The Royal Society of Chemistry, Cambridge.
Jenkins, G.N. (1989) *Nutrition Quarterly*, **13**, 33.
Krusch, U., Lompe, A., Engel, G. and von Milczewski, K.E. (1977) *Milchwissenschaft*, **32**, 713.
Munksgaard, L. and Werner, H. (1987) *Milchwissenschaft*, **42**, 216–220.
Renner, E. (1983) *Milk and Dairy Products in Human Nutrition*. Volkswirtsch. Verlag., Munchen.
Renner, E. (1993) Nutritional Aspects of Cheese. In *Cheese: Chemistry, Physics and Microbiology*, 2nd Edition (Ed. P.F. Fox), Chapman & Hall, London, pp. 557–580.
Rule, C.E., Manning, R.G. and Zielinski, R.J. (1980) *Mozzarella type cheese*, US Patent 4.232.050.
Schoch, U., Luethy, J. and Schlatter, C. (1984) *Milchwissenschaft*, **39**, 583.
Sellars, R.L. (1991) Acidophilus Products. In *Therapeutic Properties of Fermented Milks* (Ed. R.K. Robinson), Chapman & Hall, London, pp. 81–116.
Shaw, M.B. (1993) Modern Cheesemaking: Soft Cheeses. In *Modern Dairy Technology*, 2nd edition (Ed R.K. Robinson), Chapman & Hall, London, pp. 221–280.
Speckmann, E.W. (1979) In: *Proc. 1st. Marschall International Cheese Conference*, Wisconsin, USA, 339.
Staub, H.W. (1978) *Food Technology*, **32**(12), 5.
Tamime, A.Y. (1993) Modern Cheesemaking: Hard Cheeses. In *Modern Dairy Technology*, 2nd edition (Ed. R.K. Robinson), Chapman & Hall, London, pp. 49–220.
Thomas, T.D and Crow, V.L. (1983) *New Zealand Journal of Dairy Technology*, **37**, 131.
Zerfiridis, G.K. (1985) *Journal of Dairy Science*, **68**, 2184.

3 Cheese varieties

3.1 Definition of cheese

The present word 'cheese' is derived through the Old English words 'cese' and 'chiese' from the Latin 'caseus'. The equivalent words in German, French, Spanish and Italian are 'kase', 'fromage', 'queso' and 'formaggio', respectively.

The expansion in the numbers of types of cheese makes a simple definition of cheese difficult, and a description, such as 'the curd of milk produced by enzyme activity and subsequent separation of the whey from the coagulum to give a more solid curd which is cheese' does not cover whey cheese, lactic cheese, cream cheese and some of the cheeses produced by the newer techniques (e.g. ultra-filtration or reverse osmosis). The definition is not, therefore, universally acceptable.

The Food and Agriculture Organization of the United Nations (FAO) devised a 'Code of Principles' which gave the following definition: 'Cheese is the fresh or matured product obtained by the drainage (of liquid) after the coagulation of milk, cream, skimmed or partly skimmed milk, butter milk or a combination thereof'. This definition does not allow for whey cheeses nor does it allow for cheeses made by the newer processes. A second definition was added for whey cheese: 'Whey cheese is the product obtained by concentration or coagulation of whey with or without the addition of milk or milk fat'. This second definition covered the German Ziger cheese, the Scandinavian whey cheeses like Mysost and Gjetost, the Romanian Urda cheese and a variety of Ricotta-type cheeses.

Although international bodies are beginning to list the names of cheeses, and indeed their origin, composition and other details, these definitions and details are not entirely satisfactory for the practical cheesemaker, who needs to know much more about individual varieties with respect to the type of milk, method of manufacture and scope for marketing.

3.2 International agreements

In 1951, the 'Stresa Convention' concluded an agreement on the use of the names of the better known cheeses. This agreement between Austria, Belgium, France, Denmark, Italy, The Netherlands and Switzerland gave international protection for cheese names in two groups as follows:

Group A: names to be used only for cheese produced in:

France	–	Roquefort
Italy	–	Pecorino Romano, Gorgonzola, Parmigiano Reggiano

Group B: names of 29 varieties, which could be used for cheeses imitated in other countries, provided that the name of the producing country was also included:

Austria	–	Pinzgauer Bergkase
Denmark	–	Samsoe, Maribo, Danbo, Fynbo, Elbo, Tybo, Havarti, Danablu, Marmora
France	–	Camembert, Brie, St. Paulin
Italy	–	Fontina, Fiore Sardo, Asiago, Provolone, Caciocavallo
Norway	–	Gudbrandsdalost, Kokkelost
Sweden	–	Svecia, Herregaardsost
Switzerland	–	Emmental, Sbrinz, Gruyère
Netherlands	–	Gouda, Edam, Friese, Leyden

In 1961, the Code of Principles covering milk and milk products was published by the FAO defining cheese and other dairy products, but not quality standards. Following this publication, representatives from 33 countries were gathered together jointly by the FAO and WHO to fix a Code of Quality Standards for cheese. Some 25 FAO/WHO standards for cheese were agreed between 1966 and 1972. The list of cheeses in the agreed Standards was as follows: Cheddar, Cheshire, Blue Stilton, Danablu, Danbo, Samsoe, Havarti, Maribo, Fynbo, Norvegia, Gudbrandsost, Herregaardsost, Hushallost, Svecia, Edam, Gouda, Emmental, Gruyère,Tilsiter, Butterkase, Harzer, Limburger, St. Paulin, Provolone and Cottage cheese. This FAO/WHO list of cheeses are those cheeses which are made frequently in countries other than the country of origin, and which should bear the names of both the cheese variety and the country where produced. Further lists have been issued from time to time giving not only the names of cheese, but also the characteristics of the cheese including weight, shape, moisture content, dry matter, fat in dry matter, salt content and other relevant features.

In addition, the system of Appellation d'Origine began to be applied more widely, and many Italian cheeses became identified with specific regions. Thus, Fontina is now linked with the Valle d'Aosta in Italy, and the Spanish cheeses, Roncal and Mahon, have been similarly identified with specific locations. To some extent, this activity has served to ease the problems of classification for the major varieties, but it will take time to extend the system throughout Europe and, even then, there will remain the question of whether or not to attempt to incorporate the numerous local variants into the scheme.

While these forms of information will still be produced from time to time by the IDF or by national bodies, the formation of the European Economic Community (EEC), and the regulations made by the EU Commissions, may affect the Cheese Industry in the future, both economically, and by the regulation of manufacturing processes and/or the supply and specifications of raw materials. The final impact of these changes cannot be forecast at the present time, but it may be relevant to cite two examples. Thus, the current quota system for milk production and the compulsory termination of the Milk Marketing Board has led to drastic changes in the dairy industry in the UK. Creameries have been closed with the loss of valuable sources of raw materials; elsewhere production targets have had to be adjusted downwards, while the subsequent decline in commercial activity has led to an inevitable decline in funding for research and development. Exactly how these changes will affect the cheese industry in the long-term remains to be seen, but the impact of Council Regulation (EEC, 1996) No. 2081/92 could be more immediate.

Thus, this Directive means that certain cheeses can only be made at specific locations, so that Stilton Cheese can be produced only in the three counties, Leicestershire, Nottinghamshire and Derbyshire in the UK, and Feta cheese can only be manufactured in Greece. The use of traditional techniques and raw materials is specified also, so that varieties like Danish Feta will soon have to be renamed. In the short-term, Danish White-brined cheese is likely to be less expensive than Greek Feta, so that its market should remain intact. However, it will be interesting to see whether factories for Feta cheese will be established in North America and other non-EU countries, and whether protection of the name can be extended worldwide.

3.3 Classification of cheese

The great range of cheese varieties, excluding minor local variants, makes classification of cheese extremely complicated. It is obvious that a well-known variety has certain distinctive characteristics, such as size, shape, weight, colour, external appearance and analytical data for fat-in-dry-matter, salt content or water-in-fat-free substance, but it is more difficult to determine flavour and aroma, especially if milk from the cow, sheep, goat or buffalo has been used singly or mixed for the manufacture of the cheese.

A very broad classification is shown in Table 3.1 and, although limited in application, the categories are widely accepted.

However, this classification does not define the cheese as ripe or unripe, it does not define size, shape or external appearance, nor does it differentiate between smear or surface mould-ripened varieties or those matured with internal mould growth. The National Committees of the International Dairy Federation (IDF) attempted to improve on this situation, and through their

Table 3.1 A simple classification of cheese based on composition

Cheese type	Water-in-fat-free substance (%)	Fat-in-dry-matter (%)
Extra hard	< 51	< 60
Hard	49–55	40–60
Semi-hard	53–63	25–50
Semi-soft	61–68	10–50
Soft	> 61	10–50

reporters made available information with respect to cheese varieties produced in their territories. The first lists of cheese were completed between 1962 and 1964, but further names were added between 1965 and 1967. These lists of cheese names formed the basis of a report in 1971 (IDF, 1971).

This report listed the characteristics of cheese varieties under the following headings:

1. Country of origin.
2. Source of raw milk – cow, sheep, goat, buffalo.
3. Type of cheese – hard, semi-hard, soft, fresh, acid coagulated or whey cheese.
4. Internal characters – close or open texture, large, medium or small eye-holes, slit openings in curd, ripened with blue or white moulds, colour of curd, presence of herbs or spices.
5. External characters – rind that is hard or soft, smooth or rough, smear or mould-coated, dusted with spices, herbs or ash, type of final coating (plastic, wax, leaves).
6. Weight of cheese – shapes and sizes.
7. Fat-in-dry-matter – percentage maximum.
8. Water content – percentage maximum.
9. Water-in-fat-free substance (WFF).

There were 395 cheeses listed from 27 countries, but it was notable that the same variety was entered by several countries. For example, Cheddar is listed by 20 countries, Camembert by 15, Gouda by 15, Edam by 13, Emmental by 12, Tilsiter by 9, Gruyère by 8, Parmesan by 5 and Brie by 6. The remaining cheese names were mainly of local character and usage but, even so, the list was by no means exhaustive. Thus, FAO (1994) has compiled details of numerous cheeses made in developing countries and, although these products are extremely important at village level, their names and features never appear on any international listing. It is true, of course, that similar omissions can be cited from the industrialized countries and, for instance, the traditional Cornish cheese, Yarg, is little known outside the county of Cornwall, even though it has a history stretching back over 100 years and is still making a contribution to the local economy. However, compiling a descriptive list of all 700–800 cheeses made worldwide

Table 3.2 Classification of cheese in respect of composition

Cheese type	Country	Moisture (% max.)	FDM (% min.)
Extra hard			
Parmesan	Italy	34	32
Romano	Italy	34	38
Sbrinz	Switzerland	28	47–50
Hard (with eyes)			
Emmental	Switzerland	41	43
Gruyère	France	39	45
Hard (without eyes)			
Cheddar	UK	39	48
Cantal	France	42	45
Cheshire	UK	44	48
Leicester	UK	42	48
Provolone	Italy	45	45
Cacciocavallo	Italy	40	42
Semi-hard			
Caerphilly	UK	46	48
Edam	The Netherlands	45	40
Gouda	The Netherlands	45	48
Semi-hard (internally ripened with moulds)			
Stilton	UK	42	48
Roquefort	France	45	50
Blue d'Auvergne	France	50	40
Gorgonzola	Italy	42	48
Blue Cheese	USA	46	50
Semi-soft (surface-ripened with bacteria)			
Limburger	Belgium	52	40
Munster	France	46	45
Port Salut	France	56	42
Tilsiter	Germany	45–55	49–55
Bel Paese	Italy	47	49–55
Brick	USA	44	50
Monterey	USA	44	50
Soft (surface-ripened with moulds)			
Brie	France	56	40
Camembert	France	48	45
Neufchâtel	France	52	51
Soft (unripened)			
Coulommiers	France	58	46
Cream	UK	55	45
Cottage	USA	80	4
Mozzarella	Italy	53	18
Ricotta (whole milk)	Italy	72	12

After: Chapman and Sharpe (1990).

would be a major undertaking indeed and, while the number of cheese names continues to grow (Kosikowski, 1977), the IDF list of cheese names, with country of origin, revised by the International Dairy Federation in 1981 remains the most exhaustive available (IDF, 1981).

Despite this apparent limitation, it is probably true that the practical cheesemaker is more interested in a classification of the major varieties based on moisture content, as an indication of the keeping quality of the cheese, and also the ripening agency of the cheese. Consequently, Table 3.2 lists a number of important varieties according to both ripening agency and range of moisture content, together with the country of origin.

For interest, some of the cheese varieties made in the United Kingdom are given in Table 3.3, and a number of the so-called 'territorial varieties' like Cheddar are now made worldwide. The dominance of the hard-pressed

Table 3.3 Some of the cheese varieties manufactured in the UK

Variety	Approximate weight (kg)	Description
Cheddar	18–28	Pale/deep creamy yellow, firm body, close texture.
Cheshire	20–22	Coloured or white, firm body, open crumbly texture.
Leicester	13–18	Highly coloured, firm to soft buttery body, fairly close texture.
Derby	14–16	Mainly white, fairly firm body, close texture.
Sage Derby	14–16	White curd with layers of coloured, ground-up sage leaves in the body.
Lancashire	22	White, soft body, friable texture.
Dunlop	20–27	Slightly soft cheese of the Cheddar type.
Double Gloucester	22–28	Lightly coloured curd, firm body but texture more open than Cheddar.
Single Gloucester	10–12	Lightly coloured curd, firm body, slightly open short texture (rarely made today).
Caerphilly	3–6	White, soft to firm body, mild flavour, sold at an early age (1–2 months old).
Wensleydale	3–5	White curd similar to white Cheshire, but smaller size
White Stilton	4–8	White cheese, soft body, close texture, not intended for internal mould ripening, often salty
Internally mould-ripened (blue-veined) cheeses		
Blue Stilton	6–8	Creamy white curd, open texture with blue–green well-distributed veins, rough brown coat.
Blue Wensleydale	3–5	Creamy white, firm body, open texture, irregular blue veins, coat usually cloth covered.
Blue Cheshire	10–20	White or coloured Cheshire-type cheese, but more acid curds, open texture with blue veining.
Blue Vinney	5–7	Rare farmhouse cheese made from hand-skimmed milk, irregular blue veins in a hard cheese body, very piquant.
Soft cheeses		
Colwich	0.25–0.50	Soured milk curds, flat round cheese with turned up edges, unsalted.
Cambridge	0.25–1.0	Rennet cheese with coloured layers, rectangular shape on straw base, unsalted, sometimes piped with whipped cream.
Melbury	2.5	Factory made cheese, similar to Brie with white coat.
Crowdie		Skim milk curd, crumbly, sometimes with added cream and/or salt, buttermilk may be used as well.

cheeses is evident, and it is likely that the blue-veined and soft varieties are of more recent origin. Most of these cheese varieties are available in smaller sizes than those quoted, so that while Cheddar cheeses are normally 18–28 kg, other sizes (e.g. 0.5 kg, 2.2 kg, 4.5 kg and 9 kg), usually packed in air-tight plastic wrappings, are available. Indeed, there is a tendency to pack most cheeses, after maturation, into 25–225 g plastic or foil packs for the convenience of the retail outlets.

The popular cheeses presently available in the UK have been made for many years (Cheddar, Cheshire, Gloucester, Caerphilly), but there has been a proliferation of 'mixed' cheeses in more recent times. These cheeses are usually wheel-shaped and flat rather than tall, with one or two layers of flavouring (vegetable-based) ingredients to provide a complex flavour, rather after the manner in which sage is used in Sage Derby (shire) cheese; sometimes the flavouring material(s) is scattered throughout the cheese. Some of these more recently developed dairy and farmhouse cheeses have been made by taking curd or partly ripe cheese and mixing it with other cheeses or non-dairy products (e.g. vegetable or savoury mixes), sometimes in layers, then pressing the cheese into shape (e.g. Walton cheese). The old name 'Cotswold' is now given to a cheese made from a mixture of Double Gloucester and chives. A blue-veined cheese, 'Lymeswold', was made by the English Milk Marketing Board in response to a perceived niche in the market. The blue veins were layered to give an attractive finish, and the cheese was not quite as salty as some other blue cheeses. The fact that this seemingly attractive product failed to hold the interest of consumers has been the subject of much debate, but its demise does highlight the risks of new product development. 'Melbury' is another cheese of recent introduction and is similar to Brie.

Low-fat cheeses – made in response to the current trend for low-fat foods – have appeared under trade or brand names like 'Tendale' or 'Shape', and these cheeses are Cheddar or Cheshire types with only 15–18% fat.

In contrast to the large number of varieties of cheese made on farms in France (180 according to Androuet, 1976), there are less than 50 named varieties currently made on farms in the UK, including cheeses made from goat's and sheep's milk. Some farmhouse cheesemakers have supplies of milk allocated to them for the manufacture of mainly traditional types of cheese, but others use their own milk to make special varieties. As milk produced and used on the farm does not carry a subsidy, the cheese must sell at an economic price and possess very special characteristics if it is to succeed commercially. The following list gives a few of these specialist farmhouse cheeses:

Applewood or Charnwood – a smoked Cheddar-type curd with a paprika coating

Cheviot – Cheddar-type curd with chopped chives

Costwold	–	Double-Gloucester curd with chopped chives and onions
Huntsman	–	cheese made up of layers of Stilton and Double Gloucester curds
Ilchester	–	Double Gloucester curds blended with mustard pickle
Nutwood	–	Cheddar curds blended with cider, raisins, and hazelnuts
Rutland	–	Cheddar curds blended with beer, garlic and parsley
Sherwood	–	Double Gloucester curds blended with mixed sweet pickle
Walton	–	cheese made of Stilton and Cheddar curds with a coating of walnuts
Windsor Red	–	cheese made from Cheddar curds mixed with elderberry wine or commercial concentrated wine (Tamime, 1984).

As mentioned earlier, some of the English cheeses are now manufactured worldwide, but in some countries the basic technology has been modified to produce a variety that has become distinctive in its own right. The situation in North America is typical of this trend for, as shown in Table 3.4, there are a number of varieties that are truly American as against being European cheeses that happen to be manufactured in the USA. Many of the cheeses that are popular in North America today reflect the influx of immigrants from Europe, and factories producing Mozzarella, Ricotta or Bel Paese tend to employ procedures that are Italian in origin. Technical innovations, such as the use of cow's milk for Mozzarella instead of buffalo milk, arose out of necessity, but others like the introduction of Pizza cheese reflect the needs of the market. Thus, Pizza cheese, which is really a drier form of Mozzarella,

Table 3.4 Some varieties of cheese that originated in North America

Baker's Cheese	An unripened skim-milk cheese similar to quark
Brick	Around 2.5 kg and 'brick-shaped', firm body and mild/moderate flavour derived from smear coat of *Brevibacterium linens*
Colby	A typical stirred-curd cheese in which the curd is washed before salting, firm body and mild flavour
Cottage Cheese	Rennet-assisted, acid coagulation of skim-milk curd, granular in texture and sold in pots and tubs, may be plain, creamed, flavoured or with vegetable or spice additions
Monterey	Similar to Cheddar, but lightly matured with a smear coat to give a mild flavour
Liederkranz	A soft, high moisture cheese with a strong/moderate flavour derived from smear coat of *B. linens*
Pineapple	Scalded at 50°C after pressing; Cheddar-style curd in the shape of a pineapple, but the low moisture content makes it good for grating

After: Davis (1976).

needs a structure that is amenable to shredding and a protein matrix that will provide 'strings' of a desired length after cooking. Consequently, both the process and the choice of starter culture are critical, the more so as the product must, as far as possible, be identical from batch to batch.

In other countries, variability is expected by the consumer, and personal preference determines which variety or brand is purchased. Consequently in the major dairy countries of Europe like France or Italy, the range of cheeses on the market is extensive, as are variations within each variety, and classification into even broad categories becomes difficult. Stobbs (1984) suggested the groupings shown in Table 3.5 to identify the major types of French cheese, and Fox and Guinee (1993) proposed the categories shown in Table 3.6 as a preliminary organization for the discussion of Italian cheeses. As mentioned earlier, such groupings are useful as a starting point, and certainly any cheeses located in one of these groupings will have at least some important features in common. In some cases it may be the chemical composition, in others the method of maturation, but so long as the limitations of the schemes are recognized, then the cheesemaker will find them useful guides.

In countries with more limited access to milk, more specific trends are evident. In Greece, for example, reliance on sheep or goat milk, together with the warm climate, has led to the dominance of white-brined cheese

Table 3.5 A suggested classification of French cheeses

Type	Example	Moisture (g per 100 g of cheese)	Fat (g per 100 g of cheese)
Fresh	Fontainebleau Lune Rousse	70–80	10–20
Soft (bloomy rind)	Brie, Camembert	55–60	16–20
Soft (washed rind)	Livarot, Maroilles, Epoisses	40–50	20–33
Soft (goat milk)	Saint Maure	40–60	16–27
Internal moulds	Roquefort, Bleu d' Auvergne	40–45	25–30
Pressed (non-scald)	Tomme de Savoie	35–45	20–26
Pressed (scalded)	Beaufort, Cantal	35–40	26–30

After: Stobbs (1984).

Table 3.6 A suggested classification of Italian cheeses

Type	Examples
Extra hard/hard	Asiago, Parmesan, Grana Reggiano, Romano, Sardo, Fiori Sardo
Pasta filata (mature)	Caciocavallo, Provolone, Incanestrato, Cartonese, Foggiano
(fresh)	Mozzarella, Provatura, Scamorze, Trecce
Hard with 'eyes'	Asin, Fontina, Montasio
Internal mould	Castelmagno, Gorgonzola, Moncenisio, Stracchino di Gorgonzola
Smear-coated	Bel Paese, Caciotta, Crescenza, Raviggiolo, Taleggio
Soft (fresh)	Formaggini, Fresa, Mascarpone, Nostrale, Ricotta

After: Fox and Guinee (1993).

Table 3.7 Some varieties of cheese manufactured in Greece; the nature of the cheeses is a reflection of both the climate and the dominance of sheep and goat milk as a raw material

Hard-pressed cheeses	
Kefalotiri	A firm, close texured cheese for table use or grating depending on the moisture content
Graviera	Often named after place of manufacture, e.g. Graviera of Crete, a firm, Cheddar-style cheese but often with large mechanical openings and/or 'eyes'
Kefalograviera	As the name suggests, a hybrid of the above varieties with a firm body, many 'eyes' and a mild/medium piquant flavour
Soft cheeses	
Galotiri	A traditional Greek cheese made from sheep milk, but only made in limited quantities.
Kopanisti	A soft, spreadable cheese with a piquant, salty flavour
Cheeses in brine	
Feta	The cheese that is identified with Greece; a semi-hard, sheep milk cheese that is matured in brine.
Telemes	A cheese similar to Feta, but more recent in origin; mainly produced in Thrace and Macedonia

After: Anifantakis (1991a).

(Anifantakis, 1991b), and the hard cheeses shown in Table 3.7, have only been developed within the last 100 years. Why such cheeses did not evolve earlier is a matter for speculation, for their in-store stability at ambient temperature is excellent, and other Mediterranean communities, such as those on Sicily, have a long tradition of making hard cheeses from sheep milk (Robinson, 1995).

Nevertheless, in spite of the seemingly vast differences between the cheeses, there are many features of the process of manufacture that are common, and it is these basic similarities that make it possible to discuss the essential aspects of cheesemaking within one text.

References

Androuet, P. (1976) *Encyclopaedia of Cheese*, Omnibus Press, London.

Anifantakis, E.M. (1991a) *Greek Cheeses*. National Dairy Committee of Greece, Athens.

Anifantakis, E.M. (1991b) Traditional Feta Cheese. In *Feta and Related Cheeses* (Eds R.K. Robinson and A.Y. Tamime), Woodhead Publishing Ltd., Cambridge, pp. 49–69.

Chapman, H.R. and Sharpe, M.E. (1990) Microbiology of Cheese. In *Dairy Microbiology* (Ed. R.K. Robinson), Chapman & Hall, London, pp. 203–290.

Davis, J.G. (1976) *Cheese*, Vol. 1, J. & A. Churchill Ltd., London.

EEC (1996) Council Regulation No. 2081/92, Brussels.

FAO (1994) *FAO Production Yearbook*, Vol. 48, Food and Agriculture Organization, Rome.

Fox, P.F. and Guinee, T.P. (1993) Italian Cheeses. In *Cheese – Chemistry, Physics and Microbiology* (Ed. P.F. Fox), Chapman & Hall, London, pp. 221–256.

IDF (1971) Cheese Names. Ann. Bull., International Dairy Federation, Brussels.

IDF (1981). Catalogue of Cheeses, Bull. Doc. 141, International Dairy Federation, Brussels.

Kosikowski, F.V. (1977) *Cheese and Fermented Milk Foods*. F.V. Kosikowski and Associates, Box 139, Brooktondale, New York.
Robinson, R.K. (1995) *A Colour Guide to Cheese and Fermented Milks*. Chapman & Hall, London.
Stobbs, W. (1984) *Guide to the Cheeses of France*. Apple Press Ltd., London.
Tamime, A.Y. (1984) *Dairy Industries International*, **49**(7), 30.

4 Introduction to cheesemaking

4.1 Principles of cheesemaking

Although cheesemaking survived for 6000–7000 years as an 'Art' practised on the homestead or farm, the advance of scientific knowledge, while still incomplete, has enabled cheesemaking to expand from small-scale farm operations to the large-scale factory production of consistently sound cheese.

The transmission of cheesemaking methods by spoken recipes and, later, written recipes was doomed to failure, because there was an absence of suitable quantitative terms to define the stages in the process. A lack of accurate knowledge of the composition of ingredients, and of the physical and chemical reactions upon which the derivation of a satisfactory end-product depended, made the process an 'Art' rather than a 'Science' up to the middle of the 19th Century. Indeed, the approach was little different from the 'Art' of Tusser's day – 1557, but this situation changed as a result of the research of Pasteur, Conn, Storch, Hansen and Lloyd. Thus, from this time, the recipe began to become a meaningful guide to the process, and some changes of particular note were as follows.

(a) The use of heat to destroy bacteria harmful to a process, first reported by Pasteur in 1857, was developed as a treatment of milk that would destroy bacteria damaging to the cheese process, as well as pathogenic micro-organisms.

Obviously the boiling of milk for yoghurt and soft cheesemaking had been practised for centuries, but proper control over the system was a much later development. The earliest factory heat treatment of cheese milk was probably the batch or Low Temperature–Long Hold (LTLH) method of heating and cooling, which involved holding the milk at around 60°C for 30 min. Later, the High Temperature–Short Time (HTST) process became standard practice, with the milk being heated to around 70°C for 15 s. Some cheesemakers stipulate that lower temperatures are essential for good quality, but the advantages of pasteurization with respect to public health are so obvious that most major companies now insist that all cheese must be made from HTST milk or its equivalent. Of course not all cheesemakers agree with this view, but where raw milk is used for manufacture, it is becoming increasingly common for Local Authorities to insist that such cheeses are clearly labelled as to the possible risks for the consumer (Keceli and Robinson, 1997).

(b) The introduction of pure cultures of micro-organisms (starter cultures) replaced the sour milk and whey starters used previously (Storch, 1890; Orla-Jensen, 1919).

(c) The refinement of the extraction of rennet from calf vells and its subsequent standardization by Hansen in 1870 in Denmark resulted in a more standard quality of curds, and less contamination of the curds by micro-organisms associated with the vells.

(d) The development of the acidimeter test by Lloyd immediately provided the recipes with measurements of acidity for each stage of the process (Lloyd, 1899).

Given these developments, the skilled cheesemaker now had measurements to guide the process in line with the recipe and, with more standard materials, the factory cheesemakers were able to improve the average quality of cheese and, equally important, less waste cheese was produced. For example, in the late 1800s Stilton cheesemakers would have expected to manufacture 30% first quality cheese, 40% second quality and 15% that was just about 'marketable' for food use; only the remaining 15% would have been sent for pig feed along with the cheese whey.

4.2 Summary of cheesemaking procedures

Prior to the middle of the 19th Century, recipes were largely handed down from generation to generation by word of mouth, and although Fahrenheit had developed a thermometer in the 1730s, the temperature of cheese milks and whey was often assessed by the dairymaid dipping an elbow into the liquid. Therefore, temperatures were often ignored in recipes.

However, as a more scientific approach became the 'norm', so the recipes became more precise, but it is still important to note that the cheesemaking method/recipe is simply a guide for the manufacture of a specific variety of cheese. It is not a guarantee of the quality of the end-product. Variations in the chemical, bacteriological and enzymic quality of the raw milk, starter or rennet may necessitate deviations from the guidelines laid down in the recipe. For example, in order to accommodate seasonal variations of milk quality, some recipes include a range of specific adjustments, and the recipes for Cheshire cheese are divided into three to accommodate spring, summer and autumn variations in milk quality.

The recipe for a specific cheese should define the ingredients, their proportions and their quality. It should define the acidities and temperatures at each stage of the process, and describe the operations to be performed in the manufacturing process. Finally, the recipe should define the storage conditions for the maturation period. A more detailed treatment is given in the following chapters but, at this stage, it may be helpful to enumerate the principal stages in the process, the ingredients and the operations.

Before commencing operations, the cheesemaker must have obtained a mental or written description of the end-product, including shape, size and nature of the cheese, together with the necessary plant and equipment with which to perform the process.

4.2.1 Ingredients for cheesemaking and their handling

(a) Milk of an agreed quality and type (cow, sheep, goat) must be available, along with an indication of composition (i.e. percentage total solids, fat, protein). The recipe must also stipulate, if appropriate, the in-silo conditions of storage of the raw milk, e.g. temperature and maximum holding time, for excessive lipolysis or proteolysis following the growth of pseudomonads can adversely affect cheese yield and quality. There will be an agreed heat treatment for the destruction of pathogens and other unwanted micro-organisms and, in many cases, this stage will be preceded by standardization of the milk with respect to the fat/casein ratio. Homogenization may be applied as well, though temperatures and pressures must be subject to careful control.

(b) A specific type of starter of known species composition must be selected, and details of usage rate and optimum temperature for growth and acid production should be to hand. The time taken to achieve the desired acidity should be recorded for each starter culture employed, because deviations from the 'norm' can be useful indicators of phage activity or other culture problems.

(c) Colouring agents – the use of permitted colours may be demanded by trade requirements.

(d) Added chemicals – some recipes will advocate the use of calcium chloride, calcium hydroxide, phosphates, sodium or potassium nitrate/nitrite or decolorizing (bleaching) agents, and the rates of usage of such materials can be critical.

(e) Coagulants – a variety of rennets from animal and microbial sources are available, and each type will have specific rate of usage, precise optima with respect to temperature and acidity for maximum activity of the constituent enzymes and a characteristic stability in the curd or whey. The conditions for coagulation of the milk, including temperature, acidity and method of application of the coagulant, must be standardized, as must the firmness of the coagulum at cutting. Thus, cutting the coagulum to release the whey must be timed correctly, and the method of cutting, e.g. scooping with a ladle or slicing with knives into small pieces, is specific for each type of cheese. The size of curd particles, in particular, affects the rate of loss of whey, the rate at which scalding can be done and, therefore, the rate of future acid development.

(f) The temperature to which the curds and whey are heated and the rate of heating (scalding) affects the rate of syneresis of whey and the rate of growth of starter organisms. The final temperature will also affect the survival of the starter bacteria, so that some scalding temperatures are selected to reduce the numbers of *Lactococcus lactis* subsp. *lactis* while leaving the population of *Lactococcus lactis* subsp. *cremoris* largely intact. This differential inhibition allows many of the enzymes of *L. lactis* spp. to remain in the curd to assist with ripening, but prevents excess acid production. The dilution of whey with hot water, a step used in the American washed curd process, similarly reduces the rate of acid development.

(g) Stirring the curd, so that the curds are kept as separate particles floating in the whey, assists with effective whey removal. The term 'pitching' the curd is often used to denote the end of stirring.

(h) Removal of the whey is one of the most important stages in the process, for by removing the curd from the whey, the curd particles can begin to coalesce together to form a cheese. In some cases, natural compaction – with or without pressure – is sufficient to form the cheese, but with cheeses like Cheddar, a 'developed' texture is brought about by manipulation of the curds after whey removal. By contrast, the curd of Pasta filata cheeses are kneaded in hot water after the initial compaction, and the success of all these operations depends upon the initial draining of the whey being completed satisfactorily. It can be important also that most of the lactose is lost in the whey, so limiting the substrates for post-production microbial activity.

(i) During the manufacture of hard-pressed cheeses, the 'textured' curd needs to be broken into small pieces so that the dry salt can be evenly distributed throughout the entire mass of curd. This breakdown is achieved by milling, and the size of milled pieces has a dramatic influence on the texture of the finished cheese.

(j) Salt: the recipe will include details as to the rate and time of application (including values for pH/acidity at the time of salting), whether dry salting or immersion in brine is most suitable, the grain size of the dry salt, as well as the temperatures that should be sought during salting. This definition of the conditions for salting or brining the cheese is essential, for the application of salt (a) inhibits the growth of lactic organisms and, therefore, slows down the rate of acid formation, (b) assists with the syneresis of whey during moulding and/or pressing and (c) imparts essential flavour to the final product.

(k) Moulding or hooping the curd: at a prescribed stage, the prepared curd is formed into a shape – often typical of a particular variety – by the use of metal, plastic or wooden moulds or hoops. In some instances, the cheeses are moulded to shape by hand.

(l) Mechanical pressing – either with weights or with screw/hydraulic presses – assists with the final whey removal, as well as providing texture and shape to the cheese. The initial, intermediate and final pressures are

CHEESE LOG

Variety .... **Date** ....

Vat No. .... Volume of Milk ....

Milk -

Analysis: Fat % .... MSNF % ....

Time /temperature of: Storage................ Heat treatment ....

Starter – Type .... Age of Starter ....

pH / Acidity of Starter .... Quantity of Starter ....

Quantity of Rennet ....

Quantity of Salt ....

	Time	*Temperature*	*pH / Acidity*
Milk	....	....	....
Adding Starter	....	....	....
Rennetting	....	....	....
Cutting	....	....	....
End of Scald	....	....	....
Pitching	....	....	....
Drawing Whey	....	....	....
Milling	....	....	....
Salting	....	....	....
Vatting	....	....	....
Pressing	....	....	....
Date out of Press	....		

Whey Analysis: Fat .... MSNF % ....

Acidity / pH Immediately ....At 6 hours .... At 24 hours ....

Remarks:

....

Figure 4.1 A cheese log. A record of measurements and assessments made during cheesemaking.

important to achieve the correct final quality, as is the timing of any increases in pressure during the overall stage. The temperature of the pressing room may be defined as well.

(m) Once the cheese has been removed from the mould, it will undergo a 'finishing' process that varies with the variety. In some cases, the cheese may be allowed to stand in a warm atmosphere to develop a dry rind; with other cheeses, like Edam, a coating of coloured wax is traditional. Other varieties have to be stored under special conditions to develop a bacterial smear or fungal coating over the cheese. The nature of the rinds/coats determines, to some extent, the conditions of storage although, in the case of Emmental, a pre-maturation period is required for development of the 'eyeholes' in the curd. The storage conditions in terms of temperature, humidity and turning of the cheese are defined in the recipe. However, as the temperature determines the rate of maturation of the curd, this factor is sometimes dictated by the market demand for a slow- or quick-ripening cheese.

In spite of the existence of recipes for specific cheeses, it is important that cheesemaking and maturation involve natural biological processes, and it is inevitable that there will be occasions when the procedures deviate from the normal. Hopefully, the final product will still be acceptable and edible, but where a failure has occurred, it is imperative that the cheesemaker has a means of determining the cause.

It is in these circumstances that the use of a Cheese Log (Figure 4.1) is of such great value, for the Cheese Log is a written record of the cheese process made at the time of manufacture. It should be possible, therefore, to analyse the Log in a manner that may both explain any current problems, and ensure that future operations, particularly where changing conditions of a seasonal character exist, can be controlled with more precision. Indeed, a knowledge of the general principles of cheesemaking allows the cheesemaker to adjust a standard recipe to use raw materials of variable quality, or to adapt a process to accommodate new plant or equipment.

4.3 Good manufacturing practice

This general phrase is useful in that it implies that a manufacturer has taken steps to optimize each stage of a process and that, within any constraints imposed by the nature of the plant for example, no reasonable steps could be taken to improve on current practices. An excellent guide covering this topic has been prepared by the Institute of Food Science and Technology (UK) (IFST, 1991), and no doubt comparable organizations in other countries have prepared similar documents. What is essential is that (a) all factors that could affect the quality of the end-product are considered by the producer, from the selection of staff through to maintenance of plant and equipment and specifications for raw materials and (b) that documentation covering all these aspects is available for consultation by anyone with a legitimate interest in the product and it quality.

Aspects of hygiene and food safety tend to be treated separately under a HACCP regime (Chapter 6), but a policy of adopting 'good manufacturing practice' should be regarded as the 'norm' for all cheesemaking operations.

References

IFST (1991) *Food and Drink Manufacture – Good Manufacturing Practice*, 3rd edn, Institute of Food Science and Technology, London.

Keceli, T. and Robinson, R.K. (1997) *Dairy Industries International*, **62**(4), 29–33.

Lloyd, F.J. (1899) *Cheddar Cheesemaking*, Report of the Bath, West and Southern Counties Society to the Board of Agriculture, HMSO, London.

Orla-Jensen, S. (1919) *The Lactic Bacteria*, Einar Munksgaard, Copenhagen, Denmark.

Pasteur, L. (1857) *Report on Lactic Fermentation*, Lille Scientific Society, France.

Storch, W. (1890) Lactic Streptococci, *Beretning fra Forsoegslaboratoriet Koebenhaven*, 18.

5 Milk as a raw material for cheesemaking

As already indicated, milk from a number of species of mammals is used for cheesemaking, but problems can arise because milk varies in composition from species to species, and even from animal to animal, so that indications of probable composition may not be accurate. In addition, the manufacture of cheese does not depend solely on the macro-composition, i.e. fat, protein, lactose and ash, but also upon the nature of the individual components, such

Table 5.1 Composition of cow's milk as a raw material for cheese

Macro-components	Approximate %	Micro components	
Fat	4.0	Some diglycerides but mainly triglycerides ($C_4 - C_{18}$, $C_{18:1}$, $C_{18:2}$, $C_{20:2}$, $C_{20:3}$)	
Phospholipid	0.05	Lecithin, cephalin, sphingomyelin	
Proteins	3.3	Caseins	2.7%
		α-casein	1.62%
		β-casein	0.60%
		γ-casein	0.11%
		κ-casein	0.36%
		Whey proteins	0.60%
		α-lactalbumin	0.13%
		β-lactoglobulin	0.35%
		immunoglobulin	0.08%
		serum albumin	0.04%
		Traces of other nitrogenous substances	
Lactose	4.6		
Salts (minerals)	0.75	Calcium, magnesium, sodium, potassium, phosphates, citrates, chlorides, sulphates (iron, manganese, copper, cobalt)	
Water	87		
Minor constituents			
Pigments	Carotene, riboflavin, xanthophyll		
Enzymes	Lipases, proteases, reductases, phosphatases, lactoperoxidases, catalase, oxidases		
Vitamins	Fat-soluble A, D, E and K Water-soluble C and the B group		
Gases	Oxygen, nitrogen, carbon dioxide (as carbonic acid), ammonia, hydrogen sulphide		
Volatiles	Extraneous volatiles – petrol, paraffins		
Cellular matter	Epithelial cells, leukocytes		
Micro-organisms	Bacteria (normal udder flora), contaminants (i.e. bacteria, yeasts, moulds)		
Contaminants	Seeds, straw, leaves, disinfectants, manure, urea and soil (Note: the presence of these contaminants is a result of carelessness during milk production.)		

as fatty acids, caseins, albumins, globulins and other variables. Table 5.1 presents a general idea of the macro-composition of cow's milk, and similar tables could be constructed for the milk of other species.

5.1 Variations in the composition of milks

The quality of milks used for cheese depends on the species of milk-producing animal, and some typical figures are cited in Table 5.2. However, it is important that the figures are the predicted average values, and variations in composition between milks from individual breeds of one species can be quite large (see Table 5.3). Even these more specific figures may vary according to the breeding policies adopted (i.e. pure or mixed breeding), as well as a result of other influences such as the health of animal, stage of lactation, climate, time of year and herd management routines.

Seasonal variations in the composition of cheese milks may cause difficulties in cheesemaking if strict adherence to a standard process is maintained throughout the year. Thus, as the constituents of the curd, e.g. fat, salts and water, are held within the casein network, it is the ratio of casein to fat which assumes importance in maintaining cheese quality. When this ratio is out of balance, the body of the cheese is either too soft or too hard unless manufacturing procedures are adjusted to include more or less water in the curd.

The standardization of milk for some varieties of cheese is normal practice, and the monthly variation of fat and casein in milks produced in the West Midlands area of the UK (normally a Cheshire cheese area) confirms

Table 5.2 Expected composition of milk from various species of mammal (%)

Animal	Fat	Casein	Lactose	Whey proteins	Ash
Cow	3.9	2.7	4.6	0.6	0.75
Goat	6.0	3.3	4.6	0.7	0.84
Ewe	9.0	4.6	4.7	1.1	1.0
Buffalo	6.0	3.8	4.5	0.7	0.75
Ass	1.4	0.75	6.1	1.2	0.5
Camel	3.0	3.5	5.5	1.7	1.5
Reindeer	17.1	8.4	2.4	1.2	0.3

Table 5.3 Commercial composition (average expected) of cow's milk (%)

Breed	Fat	Protein	Lactose	Ash
Jersey	5.14	3.8	5.0	0.75
Guernsey	4.9	3.85	4.95	0.75
Shorthorn	3.65	3.3	4.8	0.69
Ayrshire	3.85	3.35	4.95	0.69
Friesian	3.4–4.0	3.25	4.6	0.73

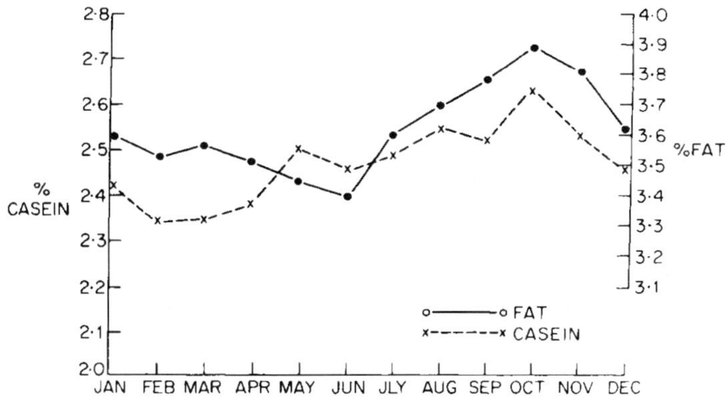

Figure 5.1 Average monthly variation in the percentage of fat and casein in milks from an area in the West Midlands of the UK.

the need for adjustment (see Figure 5.1). There are similar variations in the monthly fat and casein contents of milks produced in other geographical regions, except that the ratios vary from one region to another (see Figure 5.2). As mentioned earlier, breed is important as well, and Table 5.4 gives some fat/casein ratios for milks from some popular breeds of dairy cattle.

When the manufacture of dairy products in a region changes from one type to another, for instance, from butter and cream making to cheesemaking, the composition of the milk in that region may be such that the casein to fat ratio is not ideal for cheese production (Figure 5.2). Recommendations have been given for a fat/casein ratio of 1:0.70 for mechanized Cheddar

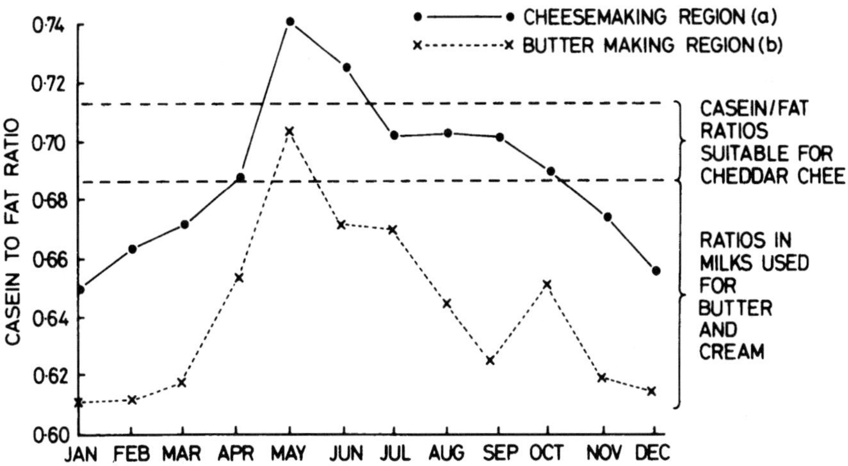

Figure 5.2 Ratios of casein to fat in milks produced in two regions of the UK: (a) a traditional cheesemaking region; (b) a traditional butter and cream producing region.

Table 5.4 Ratio of fat to casein in milk from some dairy breeds

Breed	Fat %	Casein %	Fat/casein ratio
Friesian (1970s)	3.5	2.4	1:0.69
(1990s)	4.0	2.6	1:0.65
Ayrshire	3.85	2.5	1:0.65
Shorthorn	3.65	2.4	1:0.67
Jersey	5.75	3.0	1:0.52
Guernsey	5.0	2.7	1:0.54

cheesemaking and 1:0.68 for manual methods (Australian Society of Dairy Technology, 1970), so that clearly much of the milk from the butter-producing region falls outside these targets. Methods of standardization of milk usually entail the addition of skim milk to the cheese milk (Tamime, 1993; Haddadin *et al.*, 1995), with the cream removed from the skim milk being used as either cream for direct sale on the retail market or for the manufacture of butter.

In connection with the testing of milk prior to standardization, it is important to note that lactose itself does not contribute very much (except in soft cheese) to the weight of the mature cheese. Thus, although lactose contributes a high proportion of the solids-not-fat (SNF) in milk and is itself subject to quantitative variation from time to time, it does not vary from day to day as much as the other SNF constituents. Indeed, if milks from the same source are compared daily, the SNF figure may be used to indicate changes in the protein, although not necessarily the casein, content. Therefore, lactose determination in milk is not meaningful in the context of cheesemaking.

The intervals between milking, especially in the case of high-yielding cows, may affect both the fat/casein ratio as well as the yield of milk, and while the nourishment of the cow affects its well-being as a milk producer, the feed may affect specific components of the milk. This is particularly so with respect to fat, protein and salts, and also in respect of taints derived from feeds (e.g. garlic, weeds, brewers' grains). Similarly, when cows are turned from winter stall feeding onto grass in April or May, there is a change not only in the colour of the milk but also in the composition of the lipids, and a lowering of the fat percentage. Many of these variations in the quality of milk from individual suppliers may be smoothed out when supplies are bulked, but in spite of bulking local supplies, variations due to geographical areas still exist. Obviously the cheesemaker has little control over these variations, but the use of yearly records is helpful in forecasting when difficulties may arise.

One of the results of the introduction of milking machines onto farms has been an increased incidence of mastitis. If mastitic milk is used for cheese, it can cause whey retention in the curds and, later, the development of off-flavours in the cheese. Unfortunately the cheesemaker has to face more serious difficulties when whole herds are affected. Low yields of cheese are expected from mastitic milk due to the low casein content, but a far more serious difficulty arises where the veterinarian or farmer uses medication for the disease without keeping milk from treated cows out of the supply to the cheese factory. Although, in many countries, there are rules governing the sale of milk from cows receiving medication, there are still difficulties in some areas. Antibiotics and chemotherapeutic agents injected into the udder appear in the milk for several milkings after injection and such milk, if used for cheese, prevents the growth of the lactic acid bacteria required for the process; some other medicines (even if given orally) can cause persistent taints in the cheese milk. Large-scale bulking of milk (90 000- to 135 000-litre (20 000- to 30 000-gallon) silos) is usually sufficient to dilute small amounts of antibiotics. The bulking of milk overnight is sometimes used to allow time for antibiotic testing procedures to be applied, e.g. the Delvo test, but more recent tests enable the cheesemaker to test incoming milks within 5–10 min. Consequently, vat failures from antibiotics should become much fewer, and farmers much more cautious in allowing contaminated milk to pass onto the market.

Although the cheesemaker is primarily interested in the macro-components of milk (i.e. fat, protein, salts and water), the success of the cheesemaking operation is often dependent on the micro-components. For this reason, a brief resume of the composition and structure of fats and proteins is given to provide a basis for further discussion on curd formation, and its capacity to develop texture, flavour and aroma; some of the difficulties encountered in processing are also associated with the micro-components.

5.2 Milk fats

While Table 5.1 gives the detailed composition of milk as a raw material for cheesemaking, there are certain components for which even more detailed information is required, especially if seasonal variations in milk quality are to be accommodated and/or new varieties of cheese developed.

For example, the fat in milk is a source of components that are partially responsible for the flavour and aroma, as well as body, of ripe cheese, and these effects will depend not only on the variety of cheese, but also on the composition and physical character of the fat itself. Cheese made without fat usually dries out to give a hard body and, when young, is bland in flavour and does not develop the typical 'cheesy' flavour expected. Indeed, even 1% fat in cheese milk imparts a background flavour that is missing if no fat is used, and it is relevant that hand-skimmed milk still has from 1 to 1.75% of fat remaining; some cheeses are made from this type of milk. Machine-skimmed milk (0.1–0.2% fat) is used for cheese curds where flavours other than lactic flavours may be expected (e.g. Quark), or where other ingredients are incorporated, e.g. Cottage cheese plus herbs or Cottage cheese plus cream dressing.

Fat in milk exists as an aqueous suspension of small globules ranging, in bovine milk, from 0.1 to 22 μm in diameter. The average size depends on the breed of animal, and there is a mean that is individual to each species. The incorporation of fat into the curd is not related to the amount of fat alone, but the composition of the fat and its enclosing membrane is important also. For example, the composition of fat affects its melting point, and hence there may be a release of liquid fat on manipulation of the curds. Large fat globules, molten inside, are easily pressed out of the curds, and this fat is either lost into the whey if the temperature of the vat is above 25–26°C, or fills the open spaces in the curd to give a greasy, mottled body to the cheese. Many cheesemakers prefer to use milks with small fat globules (Ayrshire or Holstein milks) because of the apparent ease of incorporation of fat in the curd; Table 5.5 shows an average, yearly content of selected fatty acids in the milk fats from two herds, as well as an indication of the softening points of such fats. These figures suggest that, because of the differences in melting temperatures, the composition of the fats may be more important in relation to cheese structure than the size of the fat globules.

The milk fat inside the globules is an extremely complex array of fatty acids which are joined to the 'stems' of glycerols – in threes – to give triglycerides. These triglycerides are arranged in a structure, crystalline when cold, which is enclosed within a lipid–protein membrane to form the globule. These complexes of triglycerides melt within the range 28–33°C, but solidify in the range of 19–24°C. As the fatty acids in milk fat can theoretically be arranged in 200 000 different ways in the fat, some of the individual fats are likely to melt or solidify outside these temperature ranges. The temperature history of the fat also affects these ranges, since the heat transfer in fat is so slow that it may take 4 h to reach a steady state; in cheese scalded at high temperature, some of the fat in the pressed cheese may still be molten when the product leaves the mould. This slow heat transfer may also be involved in some of the rennetting difficulties discussed elsewhere (section 11.3).

Table 5.5 Average yearly fatty acid content (%) of the milk fat from a Shorthorn and a Jersey herd. An indication of the softening point of the fats is also given

	Shorthorn herd	Jersey herd
Myristic C_{14}	10.21	8.10
Palmitic C_{16}	21.19	22.14
Stearic C_{18}	9.52	15.25
Oleic $C_{18:1}$	21.84	22.93
Softening temperature range of fat	29.7–30.8°C	37.2–39.9°C

Calculated from Parodi (1973).

While the central core of the fat globule consists mainly of triglycerides, di- and monoglycerides are also present along with free fatty acids in smaller quantities. In chemical terms, the triglycerides are inactive, and it is only when they are acted upon by enzymes that they are broken down – wholly or in part. However, triglycerides do act as solvents for other components of milk, so that compounds like phospholipids, cerebrosides, sterols (esterified cholesterol), pigments (carotenoids) and anti-oxidants (tocopherols) may be present in fat globules along with aldehydes, ketones, lactones and other metabolites. Some of these components contribute to flavour and may be released on lipolysis of the fat.

The electron microscope reveals a layered structure of the fat in the globule. More specifically, it appears that, on cooling, the higher melting point fats solidify first and occupy the centre of the globule, while the lower melting point fats form layers towards the outside. The milk fat is unlikely to form large crystals in cheese, in contrast with the situation in butter or in other bulk fats, due to the small size of the globules, but small crystals may be formed. The forms of the individual fats affect the structure of the crystals and their density within the globule.

The lengths of the fatty acid chains range from four carbon units to 26 carbon units, and since this distribution affects conformation and the structure of the central core of the globule, it can be important with respect to the extent of rancidity and lipolysis. Thus, the exposure of fats to enzyme activity and the release of fatty acids, as well as the oxidation of fats, will affect the flavour and body of the mature cheese.

The distribution of the fatty acids on the glycerol stem is diverse, but nearly 40% of the milk fat is made up of oleic and palmitic acids with a shorter chain acid at position 2 (e.g. oleic–capric–palmitic). Fully saturated fats are found (e.g. palmitin, myristin, stearin), but usually an unsaturated acid is present (e.g. oleic–butyric–palmitic or oleic–caprylic–stearic). Some mixed diglycerides (e.g. palmitic–oleic) may be found, but not triglycerides of the trioleic or tripalmitic types.

Since the fatty acids play an important role in cheese, it is perhaps useful at this stage to look into a little simple chemistry, even though the cheesemaker can only influence minor parts of the reactions. Thus, the fatty acids are mainly straight chains of carbon atoms held together by simple bonds. The hydrogen atoms are also held by single bonds to the carbon chain. For example, butyric acid has only four carbons in its chain with the hydrogen atoms attached as follows:

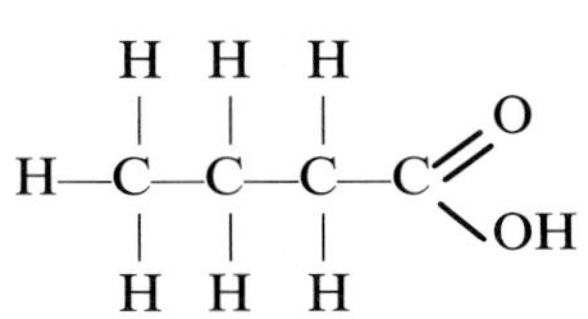

or $CH_3(CH_2)_2\,COOH$

Butyric acid (saturated)

The longer chain acid, stearic acid (also saturated), is formed as follows:

$$CH_3(CH_2)_{16}COOH$$

When the bonds between the carbon atoms are more complex (i.e. double), the acid is said to be unsaturated, and can be reactive at the point of the double bond. The structures of the oleic, linoleic and linolenic acids, all found in milk fat, are unsaturated as shown below:

$$CH_3(CH_2)_7CH{=}CH(CH_2)_7COOH \text{ (Oleic acid; one double bond)}$$

$$CH_3(CH_2)_4CH{=}CHCH_2CH{=}CH(CH_2)_7COOH$$

(Linoleic acid; two double bonds)

$$CH_3(CH_2)CH{=}CHCH_2CH{=}CHCH_2CH{=}CH(CH_2)_7COOH$$

(Linolenic acid; three double bonds)

Branched-chain, as opposed to straight-chain, fatty acids are also found in very small quantities in the milk fat.

Some of the pathways by which the degradation of the fatty acids produces keto acids and ketones are indicated in Figure 5.3, and some of these compounds give rise to flavour and aroma. Thus, although cheese flavour comes from an extremely complex chemical mixture, it is from the shorter chain fatty acids ($C_4 - C_{14}$) that the important ketones are derived.

Rancidity is a term often used to denote the presence of free volatile fatty acids, usually butyric acid. Lipolysis of the naked fats releases the individual

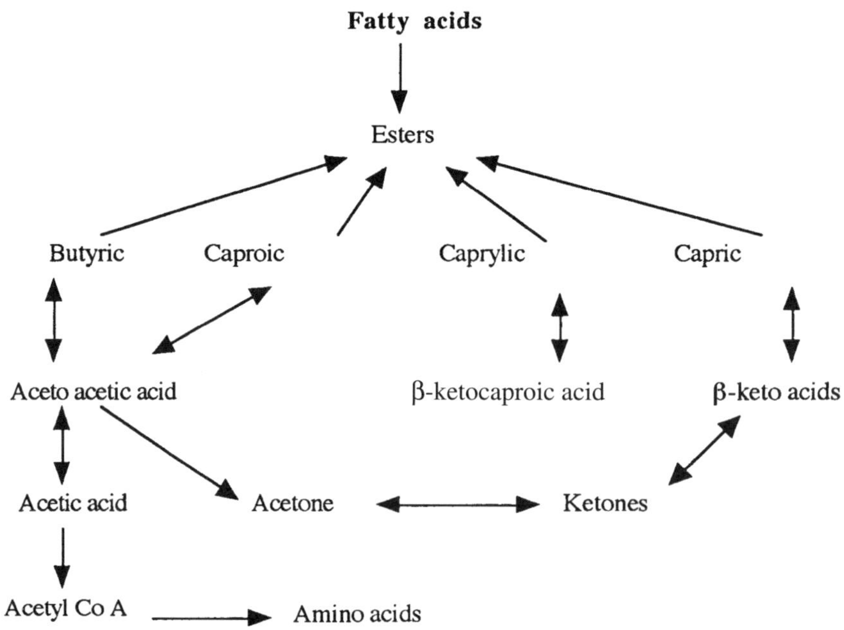

Figure 5.3 Pathways leading to the development of flavour and aroma in ripening cheese.

fatty acids from the glycerol stem, and violent agitation or homogenization may make the free fat available to lipase enzymes. Oxidation of the phospholipids, which are normally adsorbed onto the fat globule membranes, causes an oxidized or cardboard taint in milk, often mistaken for a rancid taint. Agitation or homogenization displaces these lipids from the globule membrane into the serum and so prevents the oxidized flavour from being produced, but the same displacement may allow stronger rancid flavours to develop. In some Italian cheeses, rancid flavour production may be intentional and controlled, and cheesemakers may use extra lipase to produce the typical 'high' flavours.

The onset of rancidity can be caused by the liberation of free fatty acids by active lipolytic enzymes (lipases). Some lipases are normal in milk but, depending on the temperatures employed, their activity is weak compared with lipases produced by micro-organisms. The quantity of microbial lipase depends on the phase of growth of the organism producing the lipase, but it is relevant that the enzymes produced by micro-organisms are often more resistant to heat than the micro-organism itself. Most lipases from psychrotrophic organisms, for example, will survive a normal milk heat treatment of 72°C for 15 s.

The activity of enzymes in breaking down the fat into fatty acids and glycerol depends on the enzyme having access to specific fatty acids, for the position of the fatty acid on the glycerol stem affects the ease with which a lipase can act. For example, butyric acid, often found in the first or α position of a triglyceride, is released by lipases preferentially over the longer chain acids. Nevertheless, the presence of the longer chain acids, palmitic (C_{16}) and stearic (C_{18}), is important in that they tend to smooth or buffer the intensity of other flavours, and hence the proportion of the fatty acids in different milks has a bearing on the flavour of cheese made from those milks.

Table 5.5 indicated the significance of the proportion of the fatty acids, particularly stearic (C_{18}) and oleic ($C_{18:1}$), on the softening point of milk fat, and, since the longer chain fatty acids are mainly derived from feed as opposed to being made in the rumen, the feed controls the composition of fat. For example, the feeding of oil cakes (linseed) produces softer fats, while grass tends to produce the harder fats, and the variations in the values for fatty acids in cow's milk (Table 5.6) may well be indicative of this type of influence; variations between individual animals may be important as well. The levels of fatty acids differ also with species, and Table 5.7 shows a range of fatty acids from some typical samples of milk. The presence of fatty acids with pronounced flavours of their own (e.g. butyric acid) influences the final flavour of cheese. Goat's milk has significant amounts of caproic, caprylic and capric acids, so that cheese made from this milk has the piquant, peppery flavour which is typical of goat's milk. Sheep's milk has an increased amount of capric acid, but the resultant cheese does not have quite so intense a peppery flavour.

Table 5.6 Range of variations of fatty acids in the fats in cow's milk

Fatty acids	Variation from min. % to max. %
Butyric	2.7–4.5
Caproic	1.3–2.2
Caprylic	0.9–2.9
Capric	1.8–3.9
Lauric	2.0–5.0
Myristic	7.0–11.5
Palmitic	22.5–29.5
Stearic	7.0–14.2
Oleic	30.0–41.0
Other acids	2.5–8.2

Table 5.7 Fatty acids present in some samples of cow, goat, sheep and buffalo milks (% by weight of the total fatty acids)

Animal	Fatty acids									
	C_4	C_6	C_8	C_{10}	C_{12}	C_{14}	C_{16}	C_{18}	$C_{18:1}$	Others
Cow	2.9	2.2	1.1	3.0	2.7	9.0	25.0	13.8	33.0	7.3
Goat	3.1	2.8	3.0	10.1	6.0	12.2	27.2	27.5	25.6	3.7
Sheep	4.2	2.0	2.2	6.0	3.1	5.5	16.9	15.8	38.8	5.5
Buffalo	3.1	0.9	1.5	1.8	2.5	9.0	37.5	31.0	11.0	1.7

Note: C_{14}, Butyric; C_6, Caproic; C_8, Caprylic; C_{10}, Capric; C_{12}, Lauric; C_{14}, Myristic; C_{16}, Palmitic; C_{18}, Stearic; $C_{18:1}$, Oleic. Others include linoleic, linolenic.

The central core of the fat globule is surrounded by a lipo-protein membrane – the fat globule membrane – and, if this membrane is weakened by denaturation of the protein moiety by excess acid, the membrane may split or rupture; if molten, the fat then spills out. When this happens the fat is more readily available for hydrolysis and then oxidation. In order to provide a greater opportunity for lipolytic enzymes to break down the fat quickly and produce flavour and aroma precursors in the curd, the fat globules may be broken down by homogenization into small sizes (1–1.5 μm diameter). This process creates the need for a greater area of membrane material to cover the new globules and, although the new membranes may be provided by the proteins in the serum, they may be thinner than the original membranes.

Blue-veined cheese manufacturers take advantage of homogenization to speed up the lipolysis of fats and so accelerate the ripening process with respect to flavour and aroma. Only the cream (25% fat) from the milk is homogenized in this process, and it is then mixed back into its own skim milk. If the high-pressure treatment is applied to the skim-milk fraction as well, the final cheese tends to have a hard, short texture which adversely affects quality.

Apart from homogenization, low pH and high temperature treatments of the milk tend to damage the fat globule membrane by denaturing the protein fraction of the material, so making rupture of the membrane and leakage of fat more likely. Free fatty acids released from this fat by enzymes from the serum may make coagulation of the milk more difficult. The fatty acids are reactive as far as casein is concerned, and may block the sites within the casein moiety required for the formation of a coagulum. Palmer and Tarassuk (1940) and Palmer and Hankinson (1941) found that free lauric, myristic and palmitic fatty acids in milk stored at low temperatures reduced or inhibited coagulation at 35°C.

In connection with fats in milk and dairy products, in recent years there have been efforts to increase the levels of unsaturated fatty acids in human foods, for it has been speculated that saturated fatty acids raise the serum cholesterol levels and increase the risk of coronary heart disease. In spite of the controversy surrounding this view, a method of feeding encapsulated unsaturated oil to cows with the intention of influencing the quality of the fatty acids in milk was devised (Scott *et al.*, 1970). This approach works because the protective coating on the oil means that the oil remains unaffected in the digestive tract and appears in the milk fat. Unfortunately, linoleic acid, being unsaturated, is liable to rapid oxidation, and it was necessary to add an antioxidant (e.g. butylated hydroxyanisole) to the milk immediately it was drawn. In addition, the body of cheese made with such milk tended to be soft and the flavour bland (Czulak *et al.*, 1974). Nevertheless, cheesemakers should be aware of the alleged problems of high levels of saturated fats in the diet, and that it is recommended by nutritionists that the fat intake in the diet should not exceed 25% of the total calorific intake.

5.3 Minor lipids

The group of lipids associated with the fats, though present only in minor amounts, are important during processing with respect to their influence on other milk components. This group of lipids includes the sterols (i.e. cholesterol), the cerebrosides and the phospholipids. This latter group are the most surface active of the lipids and include lecithin and cephalin. The phospholipids are often closely associated with proteins, e.g. form the lipid portion of the lipoproteins in the fat globule membrane. They are almost insoluble in water or fat and tend to form micelles at fat/water interfaces with the polar ends of the molecules in the water phase and the non-polar ends in the fat phase. The fat globule membranes contain between 50 and 90% of the total phospholipids; the proportion depends on various factors, such as agitation, temperature and pH. The association between phospholipids and protein in a lipoprotein layer is a unique feature and, because of the orientation of the phospholipids between fat and serum, they act as an anchor or protective layer against the milk serum. The membrane undergoes changes from time to time according to the conditions, especially temperature and acidity, and, if the membrane is damaged (by agitation), then protein from the serum repairs the damage (Fox, 1995)

Proteins in the outer layers of the globule membrane retain minor amounts of zinc, calcium, iron, magnesium and copper. There are also some enzyme systems associated with the protein, such as esterases and lipases, which are separated from the milk fat by the membrane layer; indeed, these enzymes can be washed away from the intact fat globules. Obviously, disruption of the membrane layer by violent agitation brings the enzymes and fat together to provide an opportunity for lipolysis of the fats.

5.4 Proteins in milk

The proteins in milk may be divided into two main groups: (a) the casein complex which exists mainly in a colloidal state in milk and (b) the serum proteins (i.e. whey proteins) which are mainly in solution in the serum.

The proteins consist of chains of amino acids held together in a formation, often a helix, which gives the protein its character and ability to react. In the helical form, the spirals may be cross-linked to give stability, so that while some proteins are elastic and can shrink, others display a more rigid character. When a protein is denatured by heat or acid, these characteristics change and the protein is less affected by external influences. Obviously little can be done to alter the micro-structure of the components of cheese milk, but a brief review of the nature of these materials may assist in explaining any processing difficulties that may be encountered.

5.4.1 Caseins

Casein, the major milk protein, exists largely in a micellar form in milk. The micelle consists of a complex built up of many casein units which, in turn, are composed of chains of amino acids. The amino-acid sequences of the individual casein moieties have been elucidated by many workers (Mercier *et al.*, 1973). A whole casein unit is composed of approximately 40% α-casein, 35% β-casein, 15% κ-casein and 10% minor components, by weight (Waugh, 1971; Fox, 1992)

The α-casein is found in four variations which are associated genetically with some breeds of animal (Rose *et al.*, 1970; Whitney *et al.*, 1976; IDF, 1980). Fegan *et al.* (1972) showed some of the effects of these genetic variants of milk proteins with regard to curd tension, pH and heat stability, especially when β-lactoglobulin was present.

Variant A of α-casein in cow's milk consists of 186 amino acid units, whereas variants B, C and D each have 199 units. During cheese maturation, the α-caseins can be split-up into smaller units (i.e. α_{s1}, α_{s2}, α_{s3}, α_{s4} and α_{s5}), each having a different flavour depending on the terminal amino acid;

Matoba *et al.* (1970) suggested that phenylalanine as the terminal acid caused bitterness (see also Hodges *et al.*, 1972; Sullivan and Jago, 1972; Visser *et al.*, 1975). Waugh (1961) found that the solubilities of the caseins vary considerably; thus κ-casein is soluble over a range of temperatures and is stable in the presence of calcium ions, while β-casein is soluble at 4°C but only 0.2% soluble at 37°C. In 0.3 M calcium solution, α_s-casein is soluble to 0.03% at 4°C, but soluble to 0.17% at 37°C. At varying temperatures, it is evident that the solubility of β_s-casein may play an important role in the coagulation of milk. When milk is stored at low temperature (4°C), the β-casein may dissociate from the casein complex and, on reheating, form a coat on the outside of the micelle, perhaps hindering enzyme coagulation.

β-casein constitutes 30–35% of the casein complex, and its polypeptide chain has 209 amino acid residues; the variants of β-casein are β-casein A, β-casein B (Jersey), β-casein D (Zebu), and type E (Piedmont). Bitter-tasting peptides resulting from the breakdown of β-casein have been reported by Matoba *et al.* (1970) and by Visser *et al.* (1975).

The γ-caseins are a part of the β-casein chain, and there are at least three variants of these caseins with slightly varying chain lengths. The minor caseins R, TSB, S and TSA^2 have been reported as identical with the γ-caseins (Whitney *et al.*, 1976). Since the component parts of the α_s-casein or β-casein chains are fairly easily removed and broken down, a 'pool' of non-protein nitrogen is available to provide essential nutrients for bacterial activity, as well as to act as possible precursors of flavour and aroma substances.

The κ-casein has only two variants, A and B, and constitutes only 11–15% of the casein complex. In spite of the small proportion of κ-casein in the micelle, it is of the greatest importance in cheese milks because of its ability to stabilize α_s-casein against coagulation; all κ-casein variants stabilize the κ-casein variants. There are 169 amino acid units in the κ-casein chain and, as with other caseins, κ-casein can be divided into a number of smaller chains (Mercier *et al.*, 1973). The important feature of κ-casein in cheesemaking is that the enzyme rennin (chymosin) is able to split the chain of amino acids between units 105 (phenylalanine) and 106 (methionine) (see Figure 5.4). The two parts resulting from the split of κ-casein are an insoluble paracasein

95 100 105 110 115

Met-Ala-Arg-His-Pro-His-Pro-His-Leu-Ser-Phe-Met-Ala-Ile-Pro-Pro-Lys-Lys-Asn-Gln-Asp

↑

Point of cleavage by chymosin

Figure 5.4 Some of the amino acids in a section of the casein chain, showing the point of cleavage by chymosin at the bond between amino acids 105 and 106.

(amino acid residues 1 to 105) which remains associated with the casein micelle and a soluble carbohydrate-containing peptide (glycomacropeptide; amino acid residues 106 to 148).

Wheelock *et al.* (1972) reviewed the role of carbohydrates in connection with bovine κ-casein, and suggested that heating may result in the combination of the non-soluble part of κ-casein with a carbohydrate. This reaction may result in longer rennet clotting times (Zittle *et al.*, 1962). Doi *et al.* (1979) found that α_{s1}-casein and β-casein will complex with different parts of κ-casein, each part having the same amino acids, but different carbohydrate moieties. They suggest that the carbohydrate moiety contributes to the stability of casein.

Although most work has been carried out on cow's milk, goat's and sheep's milk is the major raw material for cheese in many countries, and the differences between the three milks with respect to their caseins has been subjected to serious examination. In particular, the chemistry of the α-caseins within and between breeds of goat has been studied in depth, for the gel formed by the coagulation of goat's milk is always weak, and this feature is ascribed to: (a) higher levels of β-casein than in cow's milk and (b) the nature of the α-caseins (Martin and Addeo, 1996). The situation with sheep's milk has been studied in rather less detail, but a review of current knowledge has been published recently (IDF, 1996).

5.4.2 Whey proteins

Milk serum contains approximately 0.6% of whey proteins, of which β-lactoglobulin accounts for 0.3% and α-lactalbumin for 0.07%; in addition, there are serum albumins and immunoglobulins. The serum proteins are in solution and do not form colloids in the manner of the caseins. The β-lactoglobulins undergo aggregation when subjected to heat and, in milk, may react with the κ-casein causing a longer clotting time (Kannan and Jenness, 1961; Zittle *et al.*, 1962; McKenzie, 1971). Apart from the longer clotting time, the interaction between β-lactoglobulin and the casein tends to cause softer curds which lose moisture more slowly. There are variants of lactoglobulin, namely E, F and G, which have been reported by Bell *et al.* (1981) in connection with Bali (Banteng) cattle.

Whey proteins do not form such elastic or shrinkable curds as casein and tend, therefore, to retain moisture, which favours the growth of bacteria. In cheesemaking, the whey proteins are initially enclosed in the curd, but, being soluble, a proportion are released with the whey when the curd is cut or broken. Those remaining in the curd become part of the body of the cheese and help to form a reservoir of amino acids for flavour or aroma development, as well as metabolites for bacteria. However, when milk is concentrated for cheese by ultrafiltration, most of the whey proteins are incorporated into the curd to give increased yield.

Both colostrum and mastitic milk have a high level of serum protein, and the inclusion of such milks can cause difficulty during the draining of curds.

5.5 Minerals (salts) in milk

Those substances which are normally classed as 'ash' are of the utmost importance in cheese milks. This 'ash' from milk contains a large proportion of metallic components like potassium, sodium, calcium, magnesium, manganese, iron, copper, cobalt, zinc, chromium and nickel, as well as non-metallic elements like sulphur, chlorine, phosphorus and iodine. The latter are usually in the form of the acid radicals, e.g. sulphate, phosphate and chloride, but equally important components like citrate are destroyed during the ashing process. In all, some 25 elements have been reported in cow's milk, although the presence of some depends on the feed.

While many of these components are present in trace amounts in milk, they are very important with respect to enzymes and other structures. Typical examples are: cobalt which forms the centre of the vitamin B_{12} complex; zinc in carbonic anhydrase; magnesium in arginase; molybdenum in xanthine oxidase; iron in both xanthine oxidase and lactoperoxidase. Although these elements are necessary for many of the reactions which occur in milk or curds, they cannot be regarded as possible 'additives' to milk since the amounts required are so low.

However, some salts in milk are of primary importance to the cheese process, and the calcium and magnesium salts of phosphoric and citric acids are of especial note. Although magnesium seems uniquely incapable of assisting micelle formation, it does contribute to the attainment of a stable mineral equilibrium in the milk, but it is the calcium (as phosphate) which enters into the structure of the casein complex (Green, 1982a,b). Approximately two-thirds of the calcium is colloidal and one-third in solution, although the precise balance is influenced by pH and temperature. The reactive calcium is less than one-tenth of the total calcium in milk, so that the bulk of the calcium is complexed with phosphate, citrate and caseins. The quantity of available calcium affects the size of the casein aggregates, so that the addition of calcium chloride before rennetting is likely to increase the size of the casein micelles. On the other hand, dilution of the milk with water prior to rennetting may disperse the micelles into smaller units. The report by Bishov and Mitchell (1956) indicated that calcium ions were more important in the formation of complexes than magnesium, potassium or sodium. Since only 30–35% of calcium is in the ionic or reactive state, any residual detergents – which can sequester calcium – in the milk may cause an increase in the rennet coagulation time, as will any other additive which binds calcium ions (Marshall and Green, 1980).

The speed of coagulation and strength of clot in a rennet coagulum are decreased if the milk is heated to more than 65°C (Jenness and Patton, 1959), and Pyne (1945) showed an effect of calcium phosphate on the hysteresis of rennet curds. Thus, both heat and calcium ion concentration have a marked effect on the rennet coagulation of milk. When cold-stored milk (4°C) is heated to 35°C and held for 30 min, the coagulation time is restored almost to normal (Ramet *et al.*, 1981); Schmutz and Puhan (1980) suggested a temperature of 60°C held for 60 min. However, there is the possibility of interaction between many of the individual components of milk (fats, proteins and salts), and it is important to remember that both proteolytic and lipolytic breakdown of milk components can cause slower than normal rennet reactions, often leading to the production of soft curds.

The aqueous phase in milk is a solution of lactose, soluble proteins, salts, soluble fatty acids, lipids, vitamins and amino acids, while the discontinuous phases are composed of fat globules, calcium–phosphate–citrate–caseinate complexes, calcium phosphate, colloids and cellular matter (leukocytes and bacteria). These two phases are remarkably stable in near neutral conditions, and the buffering capacity is high at around pH 7.

Apart from its ability to combine with salts in the milk, casein is soluble in weakly acidic solutions such as sodium chloride/lactic acid, a condition that exists in some types of raw cheese curd during pressing. This phenomenon affects the 'matting' together of curd particles to give a close texture, or may give rise to dark lines around the curd particles in other types of cheese.

5.6 Enzymes in milk

There are three main sources of enzymes in milk: (a) the milk at the time of its secretion, (b) the micro-organisms present in milk at the time of production (i.e. those micro-organisms present in the teat canal), (c) the micro-organisms which are contaminants gaining entrance to the milk after production from utensils and subsequent handling. Some of these microbial enzymes remain in the milk after death and lysis of the bacterial cells and, although they are present in minute amounts, they show a low level of activity which persists over a long time.

In view of the biochemical activity involved in the digestion and fermentation of feed in the rumen, it is not surprising that some 40 enzymes have been found in milk. The principal enzymes include lactoperoxidase, ribonuclease, xanthine-oxidase, catalase, aldolase and lactase, together with groups of phosphatases, lipases, esterases, proteases, amylases, oxidases and reductases. The lipases associated with membranes are distinct from those in the plasma, and the plasma lipases tend to remain in the serum phase, except for some which become attached to the casein micelles. These plasma enzymes remain inactive in milk until activated by homogenization and the rupture of the fat globule membrane, or heat treatment that may release liquid fat. The membrane lipases are irreversibly bound to the fat globule membrane.

Lipolysis in milk can be hastened by warming cold milk above 32°C, for after subsequent cooling beyond the solidifying point of milk fat, the milk becomes rancid more quickly; this may occur if warm morning's milk is added to cold evening's milk and the whole is then cooled below 20°C. This situation can occur in bulk tanks of farm milks where the cooling equipment is inadequate for the duty required. The incidence of rancidity in milk in some colder northern climates (i.e. Scotland, Scandinavia) has been caused by the rapid conveyance of warm milk (at blood heat) through vacuum pipeline systems from the milking stall to the milk cooler. Aeration and agitation of the milk in the vacuum systems causes activation of the lipolytic enzymes which cause rancidity. In addition, it is relevant that the milk from certain individual cows is naturally susceptible to rancidity, but such milk, if mixed with other non-rancid-susceptible milks, loses the capacity to develop further rancidity.

Since common salt is inhibitory to lipases, some cheesemakers have added salt to stored milk. The overnight dosage of cheese milk with sodium chloride may be a sound method for controlling lipolysis, but the salt can affect rennin activity and, by slowing down coagulation, lead to the production of soft curds which retain moisture.

Some cheesemakers, especially those interested in Italian varieties (e.g. Parmesan), require extra lipolysis to produce the distinctive flavours associated with the variety, and these cheesemakers may avoid the pasteurization of cheese milks in order to minimize the loss of lipases. Some of the lipases are derived from contaminants in the milk, e.g. pseudomonads, micrococci and bacilli – even streptococci have been associated with lipolytic activity, and all these lipases affect the flavour of the final cheese. Factories processing this class of cheese use rennet paste (mixtures of rennet and ground-up glandular tissues from the calf or kid), rather than rennet extract as the coagulant, because the paste has higher lipolytic activity than the extract. Small amounts of esterases, as distinct from lipases, exist in milk as well. These esterases are responsible for some changes in the lipids, and may assume a minor role in cheese maturation.

By contrast, alkaline phosphatase is an esterase which catalyses the hydrolysis of organic phosphates. This enzyme appears to be largely adsorbed onto the fat globule membranes, and since it is destroyed by pasteurization, its presence is used in the phosphatase test for the efficiency of heat treatment. An acid phosphatase (i.e. phosphomonoesterase) is also present in milk. It is heat resistant to at least 96°C, but its effect on cheese milk or curds appears to be minor.

Xanthine oxidase ('Schardinger's enzyme') catalyses the oxidation of aldehydes and is, therefore, of interest in respect of cheese curds. The enzyme is a reductase, active up to 75–80°C and will reduce nitrate to nitrites. Goat's milk does not seem to have a comparable system and is, therefore, weaker in reductases. However, contaminating organisms can inject reductases into the milk, which are then carried over into the curd.

Peroxidase is present in milk in variable amounts and transfers oxygen from peroxides. The enzyme is heat resistant up to 80°C, and has been used in the 'Storch' test for milk which has been heated to temperatures higher than those used for normal pasteurization. However, some milks contain no, or only small amounts of, peroxidase.

Peroxidases transfer active oxygen, but catalase is an enzyme causing decomposition of peroxides to water and inactive oxygen. Colostrum and diseased milks contain large amounts of catalase, which is often accompanied by high counts of bacteria and leukocytes. For this reason, catalase has been used for detecting milk from diseased udders, although the test has been largely replaced by the somatic cell/leukocyte count. Catalase is used for decomposing any peroxide left over after the use of hydrogen peroxide to destroy unwanted micro-organisms, in place of the usual heat treatment of milk. A heat treatment up to 90–95°C is necessary for the complete inactivation of catalase for, although normal pasteurization will cause inactivation, partial reactivation may occur. Catalase is precipitated along with the casein when milk is coagulated.

Proteases in milk catalyse the hydrolysis of bonds or linkages in the chains of amino acids to give peptones, proteases, peptides and amino acids. While the amounts of these substances in cheese are small, they have a profound effect on the flavour, body and even the texture of cheese. Some of these substances are eventually involved in the aroma of cheese, especially when broken down to ammonia as in mould-ripened cheese. While the proteases in milk are progressively destroyed as a heat treatment reaches 75–85°C, the subsequent growth of micro-organisms provides a reserve of proteases and peptidases for continuing the degradation of proteins. While many of these enzymes are precipitated in the curd and are important in cheese maturation, some are lost in the whey attached to the whey proteins.

In bovine milks, lactase is concerned with the breakdown of lactose and, in doing so, forms glucose and galactose which may be more usable by micro-organisms in the milk or curd. Amylase enters milk from the blood and has, therefore, been proposed as a means of detecting damaged udder tissues. In contrast with cow's milk, goat and other mammalian milks contain no, or only traces of, lactase.

The activity of many other enzymes in milk has been regarded as minor in cheese milks, and many are destroyed by normal heat treatment. Organisms growing in the milk after heat treatment (i.e. thermoduric species and starter organisms) produce systems essential to the proper maturation of cheese. Indeed, heat treatment destroys bacterial enzyme systems which might be found useful in providing a fuller flavoured cheese.

5.7 Vitamins in milk

Apart from their potential nutritional role, the level of certain vitamins can influence the metabolic activity of micro-organisms in cheese.

Vitamin A is a fat-soluble vitamin and, although most of the vitamin is absorbed in fat, some remains linked with the globulins and other whey proteins; heat treatment causes little or no loss of this vitamin in milk. Nutritionally, vitamin A activity in the body is derived not only from the vitamin itself but also from the carotenoid pro-vitamins, α, β and γ-carotene, all of which can be converted to vitamin A in the body. This link with the carotenoids is reflected in the fact that the vitamin A content of summer milk is up to 15 times higher than in winter milk. These carotenoids are deep yellow in colour and, while not all will be converted to vitamin A, they do give colour to milk and cheese, which is especially noticeable in summer months. Indeed, the urge to imitate the summer colour of cheese all year round has brought into use various vegetable colours (e.g. annatto, butter yellow, carotene, safranine) for addition to milk and cheese.

The water-soluble vitamins belonging to the B group are well represented in milk, and thiamine (vitamin B_1) is an essential nutrient for human growth. It is present free in solution in milk as well as being bound to proteins, but is progressively destroyed by the heat treatment of milk. Thus, pasteurization destroys 2–3% while sterilization destroys almost all the vitamin, so that the milks for starter cultures prepared on-site and exposed to prolonged high temperature treatments may be deficient in the thiamine needed for bacterial growth.

Vitamin B_2, riboflavin, provides the yellow–green fluorescence in whey. The vitamin is resistant to normal heat treatments but, as with most of the natural vitamins, the content tends to decrease during the maturation of cheese. However, with smear-ripened cheeses, the metabolism of the microflora may increase the riboflavin content of the cheese.

Nicotinic acid (niacin), another member of the vitamin B group, is also resistant to the heat treatment of cheese milk, but is largely lost into the cheese whey. The amide, niacinamide, is involved as a coenzyme in some reactions of importance in the early stages of cheesemaking, but it disappears early in the maturation process.

Vitamin B_6 is present as pyridoxamine or pyridoxal, and is mainly free in the milk serum. It is resistant to pasteurization and is essential for the metabolic processes of micro-organisms. For example, pyridoxal is essential for many enzymatic reactions involving the non-oxidative degradation and interconversion of amino acids including the degradation of tryptophan to niacin, and it plays a role in the amino acid cycles involving alanine, serine, threonine, glycine, tryptophan and the sulphur-containing acids. These reactions are of great importance in the development of cheese flavours.

Pantothenic acid, formed by the union of pantoic acid and β-alanine, is water soluble and resistant to normal heat treatments. It is active in a variety of biochemical reactions, such as the reversible acetylation reactions in carbohydrate, fat or amino acid metabolism. It is important in human nutrition and plays a role in reactions taking place during cheese maturation, as does biotin and folic acid. However, vitamin B_{12}, cyanocobalamin, is perhaps the most important nutritional accessory of the B group vitamins. It is water soluble, and labile to acids, alkalis and light. Vitamin B_{12} is fairly heat resistant, but approximately 10% is lost during the normal processing of liquid milk (71.6°C, 15 s). It is the most complex of all the B vitamins, and is bound to the proteins in milk; it is present in the colostrum, particularly from goats and cows. It is involved, along with coenzymes, in the formation of propionic acid in bacteria, and it affects the metabolism of folic acid. During the maturation of cheese, the growth of lactic bacteria appears to cause a decrease in the vitamin content, as does the growth of moulds on the rind of cheese (Camembert); in some hard cheeses like Gruyère, the vitamin B_{12} content is enriched through synthesis during the internal growth of micro-organisms.

Water-soluble vitamin C, ascorbic acid, is also one of the essential nutrients derived from foods. Unfortunately, this vitamin is sensitive to heat, and long-hold, low-temperature pasteurization (61–65°C, 30 min) destroys from 20 to 30% of the vitamin; HTST pasteurization does not destroy so much. Vitamin C is susceptible to enzymic oxidation, aeration and the presence of metallic contamination (iron and copper), so that agitation (aeration) during storage, especially above 10°C, reduces the vitamin C content. In addition, some 65–90% of the vitamin is lost into the whey, so that matured cheese has little or no vitamin C at the time of consumption.

Vitamin D is present in milk in amounts which depend on the feed of the animals. Sunlight (ultra-violet) is a factor in the conversion of the provitamins (i.e. ergosterol, a plant sterol) to vitamin D (calciferol), and the conversion of 7-dehydrocholesterol in the skin to vitamin D_3; consequently, summer milks have the highest vitamin D content. The vitamin is stable to heat and oxidation and, in humans, is involved in the metabolism of calcium and phosphorus. Since vitamin D is stored in the human body, the precise level obtained from a specific food does not assume great importance nutritionally.

The tocopherols (α, β, γ and δ) all have vitamin E activity and, while they are stable to heat, they are highly susceptible to oxidation. They are, in fact, powerful antioxidants and preserve fats against rancidity and the vitamin A against oxidation. Vitamin E prevents peroxidation of the unsaturated lipid components. Its possible role during cheese maturation remains unresolved, but its antioxidative ability must have some effect on those flavours derived from fats.

Vitamin K activity is present in several compounds, the most potent of which is menadione (vitamin K_3). Vitamin K is heat stable but labile to oxidation and light and, being fat soluble, is not lost in the whey during cheesemaking; its activity, if any, during cheese ripening is undetermined. Nutritionally, vitamin K activity is essential for the synthesis of the prothrombin required to maintain normal blood clotting activity, but as the bacterial flora of the intestinal tract can synthesize vitamin K_2, deficiency should not normally be a problem.

Some of the minor substances in milk are precursors of other essential requirements. For example, *p*-aminobenzoic acid functions as a precursor of folic acid and, as some 20% of the *p*-aminobenzoic acid is present in the bound form, it is likely to be retained in curds. The growth factor required by strains of *Bifidobacterium bifidum* is also present in very small quantities in ruminant milks, and this finding could be relevant to the production of probiotic cheeses (Ghoddusi and Robinson, 1996).

Nutritionally, choline is important to animals and for the growth of micro-organisms, and it exists in milk in several forms; 25–50% of choline is water soluble with a proportion in the free state. The remainder of the choline is combined, possibly as the phosphoric glycerol ester of choline. Insoluble choline may occur in the form of lecithin or sphingomyelin, and pasteurization has little effect on the choline or lecithin content in milk. Lipoic acid is also present in milk and is required by some micro-organisms for satisfactory growth.

The level of vitamins and other accessory factors in milk depends on the feed eaten by the animals and is, therefore, variable from species to species and from season to season. Many of these vitamins are required by micro-organisms for active growth, but some organisms can synthesize vitamins as part of their metabolic processes. However, even these organisms require a minimum vitamin content during the early stages of growth, even though they can manufacture sufficient vitamins later. Cheese containing a large number of micro-organisms is likely to have higher concentrations of some vitamins than the milk itself.

The practical cheesemaker is interested in two things in relation to vitamins: (a) the retention of those that are essential for the growth of cheese flora and (b) the preservation of vitamins as nutritional factors for humans, since one of the reasons for cheesemaking is to prevent the spoilage of a valuable foodstuff – milk. In addition, the promotion of cheese as a 'healthy food' relies on its excellent nutritional value for humans.

5.8 Antibiotics in milk

The direct injection of these agents into the bloodstream of a milk animal or even intake through the mouth, leads to milk being secreted that contains varying amounts of bactericidal or bacteriostatic compounds. The effect of these antibiotics, or indeed other chemotherapeutic agents, in cheese milk is to destroy those organisms necessary for development of acidity, flavour and aroma in the cheese. Volatile carriers (e.g. alcohols) are agents which can appear as taints in the milk, and they may, therefore, be equally harmful.

While remedial treatments are possible, such as the use of the enzyme penicillinase to nullify the activity of penicillin in milk, the cheesemaker is not always aware of the type of antibiotic present. Even if the cheesemaker is aware that penicillin has been employed and makes use of penicillinase, problems can still arise. For example, micrococci and clostridia are susceptible to penicillin but the enterobacteria are not, and hence blown cheese curds can result from a preponderance of coliform organisms in a penicillin-treated milk.

However, more recently, penicillin has been replaced by other natural and synthetic antibiotics and, although their use is controlled by veterinary practice, the antibiotic of choice may vary from time to time. Synthetic penicillins are derivatives of the nucleus of penicillin, 6-amino penicillanic acid, and include benzyl penicillin, methicillin and cloxacillin. Other antibiotics which have been used at some time for the treatment of animals include the following:

aureomycin	benzathine	oxytetracycline	ampicillin
chlortetracycline	bacillin	streptomycin	cepharpirin
neomycin	erythromycin	novobiocin	bacitracin
chloramphenicol (chloromycetin)			

These antibiotics are not all used for direct udder injection, although they can still be secreted in milk. Some favoured treatments have involved two antibiotics together, e.g. penicillin and dehydrostreptomycin, for direct udder injection. A dye marker, Food Blue No. 2, has been used with some injected preparations so that the milk is coloured for the duration of the treatment.

In addition to the problem that not all genera are susceptible to penicillin, it is relevant that different species or strains of susceptible micro-organisms may not be destroyed by the same concentration of antibiotic. Table 5.8 shows the inhibitory levels of penicillin for some micro-organisms used in cheesemaking.

As mentioned earlier, bulking of milk in large silo tanks will, by dilution, reduce the level of antibiotics in the total supply. In other circumstances, cheesemakers have used extra large amounts of starter culture (10–20 times the normal dose rate) on the assumption that, if the inhibitory antibiotic is absorbed by some of the bacterial cells, others will be left to continue the desired acid development. This method worked when penicillin was the only antibiotic being used, but it became less applicable when a wider range of chemotherapeutic chemicals became available. *Lactococcus lactis* sub-sp.

Table 5.8 Levels of penicillin in milk that are inhibitory to the micro-organisms indicated; variations indicate differences between strains

Micro-organism	Inhibitory levels of penicillin (international units)
Lactococcus lactis sub-sp. *lactis*	0.1–0.25
Lactococcus lactis sub-sp. *cremoris*	0.05–0.1
Streptococcus thermophilus	0.01–0.05
Enterococcus faecalis	0.25–0.45
Lactobacillus buglaricus	0.25–0.5
Lactobacillus casei	0.25–0.5
Lactobacillus helveticus	0.25–0.5
Propionibacterium shermani	0.05–0.1
Starter (mixed culture)	0.25–0.5

Note: Since the udders may receive doses of 100 000 units of penicillin or 200 000 to 400 000 units of aureomycin or streptomycin, large amounts of these antibiotics may be secreted along with the milk.

cremoris and *Streptococcus thermophilus* are both susceptible to antibiotics, and the latter, in particular, has been widely used to test for the presence of antibiotics in milk. The more thermoduric *Bacillus stearothermophilus* has now replaced *S. thermophilus* as the organism of choice, because the test can be run at 64°C as against 37°C for *S. thermophilus*; as a result, the test can be competed in two hours rather than four. The use of *B. stearothermophilus* has given rise to a number of practical tests, such as the Delvo Test (Anonymous, 1995) or the Charm Test (Marshall, 1992) and, in general, the procedures require the minimum of basic laboratory facilities. It is worth mentioning that *Propionibacterium shermani* is also susceptible to penicillin, and this is the reason why cheeses such as Emmental have suffered from defective eyeholes when milks treated with penicillin have been used.

An alternative is the Disc Assay method which uses plates of agar heavily seeded with selected test organisms, and various authorities have used *Sarcina lutea*, *Bacillus subtilis*, *Bacillus calidolactis* and *Streptococcus thermophilus*. A disc of filter paper soaked in the milk for testing is laid on the inoculated agar, and the presence of inhibitory substances is indicated by a clear zone around the paper disc. Standardization of the tests for levels of inhibitory substances can be achieved by using control discs with known amounts of compounds present.

More recently, enzyme assays for the β-lactam group of antibiotics have been introduced – e.g. Delvo X-Press and Lac Tek, and they are designed to screen bovine milk for the presence of penicillin G at levels of 2–4 μg/kg. This level of sensitivity is, on average, rather better than that of the other available tests, and the Delvo X-Press or Lac Tek should be capable of detecting around 0.01 international units of a β-lactam antibiotic. Bearing in mind the values for antibiotic sensitivity shown by the starter organisms cited in Table 5.8, it is clear that these enzyme systems for antibiotic detection are as sensitive as the most susceptible species in common use in the

Table 5.9 Antibiotic test regimes as used by some marketing agencies

	Marketing agency			
	Milk Marque (England and Wales)[1]	Northern Ireland	Scottish Schemes Aberdeen	Northern
Frequency of test	Once monthly	Once monthly	Twice monthly	Once monthly
Test used[2]	Bromocresol purple with *Streptococcus thermophilus* or *Bacillus stearothermophilus* as the indicator organism	Disc assay	Disc assay	Disc assay

[1] Since the end of the Milk Marketing Board in 1994, private companies can buy milk direct from farms and institute their own testing schemes.
[2] The EC standard is less than $0.004 \mu g$ (0.007 IU) antibiotic (as penicillin) per ml.
Note: The Lac Tek Test is accepted as an alternative for tanker milk, and a new test – Delvo X-Press – may be accepted as well.

cheese industry. In addition, the speed of these specific tests for antibiotics (10/15 min as against 2 h) means that bulk tanker milk can be tested on reception, so that the transfer of veterinary antibiotics to the cheese vat should be a thing of the past. Confirmatory tests employing the penase treatment of a heat-treated sample should still be employed, and the arrangements for such tests will vary from country to country.

The confidence in these tests does not, of course, ensure that vat failures cannot occur, for cleaning residues of chlorine, iodine or quaternary ammonium compounds can also prove to be inhibitory. According to Neaves (pers. communication), chlorine can prove inhibitory to *Bacillus stearothermophilus* at concentrations >200 mg/litre, iodine at concentrations >50 mg/litre and quaternary ammonium compounds at concentrations between 2 and 10 000, so that inept cleaning procedures could cause problems. However, these levels of compound should be readily detectable by taste, if not by odour, and hence problems of this nature should be infrequent.

The production of acid in milk – by bacteria sensitive to antibiotics – can be used as a test for inhibitory substances, provided that precautions are taken to avoid the inhibition of acid development by phage infection. A more satisfactory test for inhibitory milks is based on the use of reducible dyes, i.e. Methylene blue, Resazurin or Tetrazolium salts, all of which change colour in the presence of actively growing micro-organisms. As the time taken for the dye to change colour is broadly correlated with metabolic activity, the addition of a standard inoculum under standard conditions should induce a colour change in approximately the same time for each batch of milk. As the test can be completed in under 30 min, it provides a simple, routine guide to the acceptability of the milk for cheese production.

There are regulations regarding the use of antibiotics and similar treatments in most countries, especially those interested in manufactured milk products. Financial penalties are used to discourage the sale of milk containing these agents but, in spite of the penalties and regulations, success in cheesemaking requires routine testing of incoming milk supplies. It is also difficult to establish a satisfactory scheme for monitoring, and a typical picture is shown in Table 5.9. Thus, even within the confines of a small country like the UK, different systems apply in the separate regions and, in the absence of effective government control, the situation is likely to grow worse.

5.9 Taints in milk

Taints in milk can be carried over into the cheese and, because of the concentration effect, can in some cases become more noticeable. Some of the more volatile taints can be removed by vacreation (Towler, 1994), but often less volatile substances are fixed to milk components, e.g. fat or protein. Unfortunately, it is very difficult to detect, by smell, even the volatile taints in cold milk as it is received from the farm.

Taints are most likely to occur in milk when a shortage of grass tempts cows to eat other types of herbage, while in the case of sheep and goats, the consumption of seasonally available herbs is a common cause of taints. The following weeds and herbs cause taints in milk which may be transmitted to cheese curds: bitterweed, buttercup, garlic, garlic mustard, wild camomile, tansy, yarrow, ragwort, green hellebore, cowbane, henbane, wormwood, marigold, mint, mayweed, pennyroyal, chicory and landcress. Decaying turnip leaves, kale, cabbage, sugar beet tops and wet brewer's grains also produce taints when fed to stock, as can paraffin, petrol, paints, tars, cresols, carbolic soaps and disinfectants.

Disinfectants of the simple hypochlorite type tend to be neutralized, and should not cause undue difficulty. However, problems can arise with bland cheeses like Cottage cheese, and sterilants based on iodine (iodophores) must be used with care.

5.10 Extraneous substances in milk

Apart from antibiotics, there are other agents which appear as residues in milk. For example, insecticides used for the control of ticks on the animals or insects in cowsheds and fields, and benzene hexachloride, DDT, dieldrin, lindane, methoxychlor, malathion, aldrin and chlordane have been found in milk. Marth and Ellickson (1959) concluded that organophosphate insecticides do not normally appear in milk.

Faulty cleaning practices for utensils may result in contamination of milk by formalin, boric acid or borax, benzoic acid, alkaline salts, salicylic acid, potassium bichromate, hypochlorites, chloramines, quaternary ammonium compounds and, possibly, hydrogen peroxide. Many of these substances affect the growth of starter bacteria. Salts of metals, such as copper, lead, cadmium and zinc may be present in milk from corroded surfaces of utensils or tanks, and poor taste and/or discoloration of curds may result when larger than normal amounts of heavy metals are present.

Although plant toxins may be ingested by cattle feeding under drought conditions, more serious are the mycotoxins derived from moulds on cereals, ground nuts and oil seeds used as cattle feed. The mycotoxin aflatoxin produced by *Aspergillus flavus* and *Aspergillus parasiticus* is a potent liver carcinogen of some concern, and aflatoxin B, the most common in foodstuffs, is converted to aflatoxin M which appears in milk. In the USA, Patterson *et al.* (1980) quoted average values of 0.21 μg of aflatoxin M per litre of bulk milk, but individual cows produced milk with up to 4.82 μg/litre of aflatoxin M. In general, Patterson and Roberts (1980) found that only 4% of dairy feed concentrates had aflatoxin B present, and that toxin levels in milk were likely to be below the permitted level of aflatoxin M of 0.5 μg/litre. This pattern tends to be supported by the data in Table 5.10, but it is notable that winter feeds can give rise to much higher levels than summer grass. Brackett and Marth (1982) found that aflatoxin M was bound to the casein protein, and this location explains the easy transfer to cheese.

Milks from other species are equally liable to contamination (Stoloff, 1980), and Ahmed *et al.* (1995) reported that even camel milk was likely to be contaminated if the animals had been fed on date pits that had been badly stored. Given the stability of aflatoxins to heat (Cole *et al.*, 1993), aflatoxin present in the milk will be transferred to the cheese. However, Fremy and Roiland (1979) found that, in the production of Camembert cheese, aflatoxin M could only be detected in the cheese for 30 days and that 60% of the aflatoxin was lost into the whey.

Insecticides, herbicides, disinfectants and toxins in cheese milks are, of course, of concern to nutritionists, and may add a further burden to the control laboratory. Some reviews of the subject have been produced over the

Table 5.10 Some levels of aflatoxin M_1 found in milk supplies

Country	Positive samples (%)	Range of concentrations (μg/kg)
Germany	45	0.05–0.3
(Winter)	11	1.7–6.5
(Summer)	0	< 0.1
India	14	$\leqslant 13.3$
UK	31	0.03–0.52
USA	63	up to > 0.7

Note: The proposed maximum level is 0.05 μg/kg; the figures are indications only, because the precise values will depend on the methods used.
After: Van Egmond (1989).

years, including Moss (1980) and Turtle (1980), and the IDF has also produced a bulletin on *Aflatoxins in the Dairy Industry* (IDF, 1977). A more extensive coverage of the problems posed by toxins has been produced by Van Egmond (1989).

5.11 Cheese from dried milk powders

While fresh milk remains the predominant source of milk for cheesemaking, dried milks may form a secondary source, either for reinforcing an existing milk or, in some areas where milk is scarce, as the source of milk solids for cheesemaking.

Since cheese in many areas is now made from standardized milk, the standardization could be carried out by adding skim milk instead of extracting the fat. The standardizing process is fairly simple when adding liquid skim milk to whole milk; however, the use of dry skim-milk powder is potentially more complicated. Thus, the heat treatment given to milk before concentrating and drying will result in denaturation of some of the whey proteins. Therefore, for inclusion in cheese milks, skim-milk powder (spray dried) should preferably be of the 'low heat' variety with more than 6 mg of native whey protein per gram of powder, rather than the 'high heat' type with 1.5 mg or less of whey protein per gram of powder. (Milligrams of whey protein per gram of powder = protein index.) Concerns over the microbiological quality of skim-milk powder have led to medium heat powders being those most commonly on offer in the market. Roller-dried powder has been heated too severely to be suitable for cheesemaking.

Reconstituted, spray dried, low heat skim-milk powder can be used for producing skim-milk curds, e.g. Quark, Cottage cheese or Baker's curd cheese, and it can also be used for adding to milk for standardization or to skim or whole liquid milk which is to be used for ultrafiltration of the milk for cheesemaking. When used for Cottage cheese, high rates of powder (20%) of reconstituted liquid can be used and slightly higher acidities and rennet quantities can be usefully employed. Skim-milk powder has been added to high fat whole milks to optimize the fat/protein ratio for Cheddar cheese production. Since the ratio of casein to whey protein and the denaturation of the whey protein varies from sample to sample of dry skim-milk powder, experimental trials are an advantage before major usage.

5.12 Cheese from recombined milks

Whole milk powders are subject to oxidative rancidity, so where cheese milk needs to be created from stable sources, the preferred route is to recombine the reconstituted skim milk with either unsalted butter or preferably anhydrous milk fat. In order to obtain the maximum rennetability, it is desirable to use low heat skim-milk powder in the recombined milk used for the cheese.

Davis (1980) described a process for making cheese made from dry skim milk and butter, and Ali and Robinson (1990) refined the system using skim-milk powder, sodium caseinate/high protein retentate powder and anhydrous milk fat. The cheese was set in moulds and, after salting, the product was almost identical to the white brined cheese Gibna Baida from Sudan. Little whey was expressed from the curd, and it was suggested that the system could prove attractive in less developed countries. Cherry and Lablee (1979) reported on the use of recombined milk for the production of a white cheese using the Hutin-Stenne 13 system, while Ducret (1979) detailed a procedure in which skim-milk and retentate powders were reconstituted with water at 40°C to 45% solids. The coagulum formed with rennet after ripening was cut at pH 4.5–4.7.

Gilles *et al.* (1979) recommended the homogenization of reconstituted milk for the manufacture of Cheddar, Edam and Cottage cheese, and Ghaleb (1979) recommended the use of equal parts of dried and liquid milk for the manufacture of Ras cheese. Poznanski (1978) suggested that recombined milks could be used for cheesemaking in the tropics, e.g. Edam, Tilsit or Camembert. Gilles and Lawrence (1981) reviewed the use of low heat powder and homogenized anhydrous milk fat for Cheddar and soft cheese; a rancid odour of anhydrous fat or a powdery flavour may be defects. A rapid reduction of the pH is alleged to offset the development of rancid flavour in Feta-style cheeses (Ramet *et al.*, 1981), while Shanley and Jameson (1982) recommended the use of low heat powder combined with fresh milk for cheese; Schneid *et al.* (1982) found that the quality of milk powder affected the rate of firming of a rennet coagulum.

Further information on the use of recombined milks is given in Bulletin 142 of the International Dairy Federation (IDF, 1982).

References

Ahmed, I.A, Ahmed, A.K. and Robinson, R.K. (1995) *Mycotoxin Research*, **11**, 75.

Ali, M.Z. and Robinson, R.K. (1990) *Milchwissenschaft*, **45**(11), 699.

Anonymous (1995) Gist-Brocades, Tamworth, Staffs (UK).

Australian Society of Dairy Technology (1970) *Pocket Book of Cheddar Cheese Manufacture*, Highett, Australia.

Bell, K., McKenzie, H.A. and Shaw, D.C. (1981) *Australian Journal of Biological Sciences*, **34**(2), 133.

Bishov, S.J. and Mitchell, J.H. Jr (1956) *Food Technology Campaign*, **10**(7), 312.

Brackett, R.E. and Marth, E.H. (1982) *Zeitschrift fur Lebensmittel Unterschung und Forschung*, **174**(6), 439.

Cherry, G. and Lablee, J. (1979) *International Dairy Federation Bulletin*, No. 126, 36.

Cole, R.J., Dorner, J.W., Bryden, W.L. *et al.* (1993) Occurrence of Mycotoxins in Foods. In *Encyclopaedia of Food Science, Food Technology and Nutrition* (eds R. Macrae, R.K. Robinson and Michelle Sadler), Academic Press, London, pp. 3205–3214.
Czulak, J., Hammond, L.A. and Horwood, J.F. (1974) *Australian Journal of Dairy Technology*, **29**(3), 124.
Davis, J.G. (1980) *Dairy Industries International*, **45**(10), 7.
Doi, H., Ibuki, F. and Kanamori, M. (1979) *Agricultural and Biological Chemistry*, **43**(6), 1301.
Ducret, P. (1979) *International Dairy Federation Bulletin*, Doc. 116.
Fegan, J.T., Bailey, L.F., Hehir, A.F. *et al.* (1972) *Australian Journal of Dairy Technology*, **27**(4), 129.
Fox, P.F. (1992) *Advanced Dairy Chemistry – Volume 1: Proteins*, Chapman & Hall, London.
Fox, P.F. (1995) *Advanced Dairy Chemistry – Volume 2: Lipids*, Chapman & Hall, London.
Fremy, J.M. and Roiland, J.C. (1979) *Annales de la Nutrition et de L'Alimentation*, **33**(5), 619.
Ghaleb, H.M. (1979) *Journal of Agricultural Research, Tanta University, Egypt*, **5**(1), 114.
Ghoddusi, H.B. and Robinson, R.K. (1996) *Dairy Industries International*, **63**, 25.
Gilles, J. and Lawrence, R.C. (1981) *New Zealand Journal of Dairy Science and Technology*, **16**(1), 1.
Gilles, J., Lawrence, R.C., Czulak, J. *et al.* (1979) *International Dairy Federation Bulletin*, No. 116, 33.
Green, M.L. (1982a) *Journal of Dairy Research*, **49**(1), 87.
Green, M.L. (1982b) *Journal of Dairy Research*, **49**(1), 99.
Haddadin, M.S., Shahin, R.M. and Robinson, R.K. (1995) *Journal of the Society of Dairy Technology*, **48**(3), 71.
Hodges, R., Kent, G.B.H. and Richardson, B.C. (1972) *Biochmica et Biophysica Acta*, **257**, 54.
IDF (1977) *Aflatoxins in the Dairy Industry*, Bulletin Doc. 98, International Dairy Federation, Brussels.
IDF (1980) Review, Bulletin 125 (eds J.H. Moore and J.A.F. Rook), International Dairy Federation, Brussels.
IDF (1982) *Recombination of Milk and Milk Products*, Bulletin 142, International Dairy Federation, Brussels.
IDF (1996) *Production and Utilisation of Ewe and Goat Milk*, S.I. 9603 (eds R.K. Robinson, P. Kastanas and F. Vallerand), International Dairy Federation, Brussels.
Jenness, R. and Patton, S. (1959) *Principles of Dairy Chemistry*, Chapman & Hall, London.
Kannan, A. and Jenness, R. (1961) *Journal of Dairy Science*, **44**, 808.
Marshall, R.T. (1992) *Standard Methods for the Examination of Dairy Products*, American Public Health Association, Washington DC, USA.
Marshall, R.J. and Green, M.L. (1980) *Journal of Dairy Research*, **47**(3), 359.
Marth, E.H. and Ellickson, B.E. (1959) *Journal of Milk and Food Technology*, **22**, 112, 145, 179.
Martin, P and Addeo, F. (1996) in *Production and Utilisation of Ewe and Goat Milk*, S.I. 9603 (eds R.K. Robinson, P. Kastanas and F. Vallerand), International Dairy Federation, Brussels.
Matoba, T., Hayashi, R. and Hata, T. (1970) *Agricultural and Biological Chemistry*, **34**(8), 1235.
McKenzie, H.A. (1971) In: *Milk Proteins*, Vol. 2, p. 321 (ed. H.A. McKenzie), Academic Press, London.
Mercier, J.C., Ribadeau-Dumas, B. and Grosclaude, F. (1973) *Netherlands Milk and Dairy Journal*, **27**(2/3), 313.
Moss, M. O. (1980) Mycotoxins in Food. In *Food and Health: Science and Technology*, pp. 115–127 (eds G.G. Birch and K.J. Parker), Applied Science Publishers, London.
Palmer, L.S. and Hankinson, C.L. (1941) *Journal of Dairy Science*, **24**, 429.
Palmer, L.S. and Tarassuk, N.P. (1940) *Journal of Dairy Science*, **23**, 861.
Parodi, P.W. (1973) *Australian Journal of Dairy Technology*, **28**(2), 81.
Patterson, D.S.P. and Roberts, B.A. (1980) *Veterinary Record*, **107**(11), 249.
Patterson, D.S.P., Glancy, E.M. and Roberts, B.A. (1980) *Food and Cosmetics Technology*, **18**(1), 35.
Poznanski, S. (1978) *Deutsch Milchwirtschaft*, **29**(40), 1469.
Pyne, G.T. (1945) *Biochemical Journal*, **39**, 302.
Ramet, J.P., El-Mayda, E. and Weber, F. (1981) *Milchwissenschaft*, **36**(3), 143.
Rose, D., Brunner, J.R., Kalan, E.B. *et al.* (1970) *Journal of Dairy Science*, **53**, 1.

Schmutz, M. and Puhan, Z. (1980) *Schweizerische Milchwirtschaftliche Forschung*, **9**(3), 39.

Schneid, N., Manzanilla, C. and Lenoir, J. (1982) *XXI International Dairy Congress*, Moscow, Vol. 1, Book 1, 446.

Shanley, R.M. and Jameson, G.W. (1982) *XXI International Dairy Congress*, Moscow, Vol. 1, Book 1, 445.

Scott, T.W., Cook, L.J., Ferguson, K.A. *et al.* (1970) *Australian Journal of Science*, **32**(7), 291.

Stoloff, L. (1980) *Journal of Food Protection*, **43**, 226.

Sullivan, J.J. and Jago, G.R. (1972) *Australian Journal of Dairy Technology*, **27**(3), 98.

Tamime, A.Y. (1993) in *Modern Dairy Technology*, Vol. 2 (ed. R.K. Robinson), Chapman & Hall, London, pp. 49–220.

Towler, C. (1994) in *Modern Dairy Technology*, Vol. 1 (ed. R.K. Robinson), Chapman & Hall, London, pp. 61–106.

Turtle, E.E. (1980) Pesticides in Food. In *Food and Health. Science and Technology*, p. 201 (eds G.G. Birch and K.J. Parker), Applied Science Publishers, London.

Van Egmond, H.P. (1989) Aflatoxin M1: Occurrence, Toxicity, Regulation. In *Mycotoxins in Dairy Products* (ed. H.P. Van Egmond), Elsevier, New York, pp. 1–55.

Visser, S., Slangen, K.J. and Hup, J. (1975) *Netherlands Milk and Dairy Journal*, **29**(4), 319.

Waugh, D.F. (1961) *Journal of Physical Chemistry*, **65**(10), 1793.

Waugh, D.E. (1971) In *Milk Proteins*, Vol. 2, p. 21 (ed. H.A. McKenzie), Academic Press, London.

Wheelock, J.V., Sinkinson, G. and Hindle, E.J. (1972) *Dairy Science Abstracts*, No. 171, **34**(9), 655.

Whitney, R.M., Brunner, J.R., Edner, K.E. *et al.* (1976) *Journal of Dairy Science*, **59**(5), 795.

Zittle, C.A., Thompson, M.P., Custer, J.H. and Cerbulis, J. (1962) *Journal of Dairy Science*, **45**(7), 807.

6 Bacteriology in relation to cheesemaking

6.1 The bacterial cell

The fermentation of carbohydrates and nitrogenous compounds in milk or curd by micro-organisms is a basic activity in the cheese process. For this reason it is useful to examine, very briefly, the metabolism of the bacterial cell in order to understand some of the activities of bacteria in the cheese process, although the cheesemaker is concerned not only with the activities of the living cell, but also with the release of cell contents, particularly enzymes, on its death.

There are six phases in the life-span of bacterial cells:

(a) Lag phase, resting condition with no growth;
(b) Growth phase, acceleration of growth from the resting phase;
(c) Logarithmic phase, during which maximum growth is achieved;
(d) Retardation phase, growth slows down;
(e) Stationary phase, resting condition with no growth;
(f) Decline or death phase, cell disintegrates.

The early stage of the cheese process is concerned with the phase of maximum growth (c), but during cheese ripening the retardation phase (d) and death phase (f) become important because of the enzymes and other substances released into the curd. Indeed, large amounts of some enzymes are produced during phase (b), the later stages of phase (c) and during phase (d).

The bacterial cell consists of an outer cell wall composed of three or more layers and enclosing cytoplasm in which the nucleoid body and a number of organelles with specific activities are found. The cell wall is quite complex, since it must transport metabolic requirements from the medium outside the cell to the cytoplasm inside. At the same time, the cell wall must transport organic and inorganic wastes and digestive enzymes from the inner cytoplasm to the outside medium. The cell walls also have the facility to maintain an essential level of metabolic requirements (e.g. amino acids) in the cytoplasm, irrespective of the level in the surrounding environment. Many of the energy-yielding reactions and oxidations also occur in the membranes of the cell walls. The nucleoid body differs from a eucaryotic nucleus in having no defined membrane, but the genetic material exists as a fibrillar network of filaments of DNA (the bacterial chromosome) which are responsible for the genetic replication of the cell.

The cytoplasm comprises all substances and structures inside the cell wall membrane. This semi-fluid matrix envelops the microsomes and other specialized inclusions, and is composed of, *inter alia*, amino acids, proteins, lipocomplexes, peptides, purines, pyrimidines, glucose, ribose, vitamins, nucleotides, coenzymes, disaccharides and water. The denser microsomes, variously designated as episomes, self-generating DNA units or plasmids, are responsible for most of the activity within the cell, and maintain a multitude of enzymes involved in metabolic processes. These plasmids may be integrated with the fibrillar filaments of the main DNA network and confer on the cell the ability to conjugate. As a result of conjugation, genetic material is transferred from one cell to another, and this transfer of DNA may confer certain new specific activities on the receptor cell. Changes can be brought about in bacterial cells to produce new strains of bacteria which have a predetermined activity different from that of the parent cell – popularly called 'genetic engineering'; this subject will be discussed further under 'Starter Cultures', Chapter 9.

When bacteriophage (bacterial viruses) attack the bacterial cell and multiply in the cytoplasm, they destroy the cell. These viruses are often specific for just one or two strains of a bacterial species, so that, introduced into a live, multiple-strain culture of organisms, a phage may decimate just one of the strains of bacteria present.

6.2 Bacteriology of raw cheese milks

The bacteriological quality of milk is the most variable of all factors with which the practical cheesemaker has to contend. Obviously, the farmhouse cheesemaker has, or should have, control of the milk production, especially in respect of the hygiene, and the bacterial flora of farmhouse cheese vats is often quite different from the flora of bulk cheese milk supplies; this is particularly so where the raw milk is collected from a large number of farms.

While penalty schemes for low quality milks are operated in some areas, the schemes are only successful when the price differential between poor and good quality milk is sufficient. In addition, the factory cheesemaker has to contend with extremes of weather, potentially adverse conditions during transport, storage and reception of the supplies, for contamination can occur at any stage. Constant control of critical points in the system is essential, as is routine testing, and it is unfortunate that politicians in some countries are not more aware of the advantages of impartial, nationwide monitoring schemes for vulnerable raw materials.

6.3 Types of microbial contamination

The scientific name of an organism is a means to differentiate it from other organisms, but as the nomenclature and classification of some organisms have been altered from time to time, it is often difficult to correlate the results of work by various authors. In addition, there may be numerous strains of a given species, so that the reported behaviour of a species may, in fact, relate to only one or two strains of that species. In major culture collections, it has become customary to place numbers and/or letters after the species name to indicate the strain, and in most recent papers, it is indicated that the results apply to the activity of an identified strain of bacterium.

Although the cheesemaker is interested in the activity of the micro-organism and the reactions that it brings about in milk and curd rather than in specific names, Table 6.1 lists some of the micro-organisms which have been found in milk supplies. In a survey of bulk tank, farm milks in England and Wales, Panes *et al.* (1979) showed that 86.4% of the samples contained 10 000 to 100 000 cfu/ml organisms, and 13.6% had >100 000 or >1 000 000 cfu/ml organisms. In the same survey, 25% of the samples had psychrotrophic counts of >5000/ml, while in milks from 350 individual farms, there were thermoduric counts of >5000/ml in 16.4% of samples, >10 000/ml in 10.9% and >100 000/ml in 2.7%. The thermoduric flora in cold-stored milk changed very little over 3 days.

Von Brockelmann (1969) concluded that raw milk containing <1 000 000 bacteria/ml could be stored at 5°C for 2–3 days provided that no more than 50% of the population is psychrotrophic, but such figures would now be regarded as excessive. Thus, the pressure is to obtain milks of good quality and, for some purposes, the presence of specific groups like spore-formers should be monitored as well. Thus, Matteuzzi *et al.* (1977) reported that *Clostridium tyrobutyricum* was responsible for 'blowing' (late gas formation) in Grana cheese, and Bergere (1979) showed that the presence of *C. tyrobutyricum* from silage could cause blowing in Dutch Emmental and Gruyère cheeses.

Since spores are very heat resistant, it should be noted that *Bacillus licheniformis*, *B. subtilis*, *B. pumilis*, *B. firmus*, *B. cereus* and, sometimes, *B. circulans* can all be derived direct from teats (Underwood *et al.*, 1974), and spore counts of 3000/ml from teat swabs have been reported. In 1980, the International Dairy Federation issued Doc. 120 (IDF, 1980) about factors which influence the quality of raw milk, and a further review of the situation was published recently (IDF, 1996a); some aspects of the situation with respect to goat and sheep milk are discussed elsewhere (IDF, 1996b).

Pathologically abnormal milks (e.g. mastitic milk) are often disturbed in their protein and salt contents, but they may also contain large numbers of organisms other than those associated with disease. These milks also contain

Table 6.1 Micro-organisms found in raw milk supplies

Micro-organism		Comments (mainly contaminants, except N = Normal)	
Streptococcus bovis			
Enterococcus faecalis		Thermoduric	
Streptococcus thermophilus			
Lactococcus lactis sub-sp. *lactis*			N
Lactococcus lactis sub-sp. *cremoris*			N
Leuconostoc mesenteroides sub-sp. *cremoris*			N
Streptococcus agalactiae		Diseased udder	
Streptococcus dysgalactiae		Diseased udder	
Micrococcus luteus		Thermoduric	
Micrococcus varlans			
Other *Micrococcus* spp.		Various from air contamination	
Staphylococcus aureus		Diseased udder	
Pseudomonas fluorescens			
Pseudomonas putida		Psychrotrophic	
Pseudomonas fragi			
Pseudomonas cepacia			
Pseudomonas aeruginosa			
Pseudomonas maltophilia			
Pseudomonas alcaligenes			
Pseudomonas pseudoalcaligenes			
Corynebacterium lacticum			
Corynebacterium bovis		Thermoduric	
Corynebacterium pyrogenes			
Escherichia coli		Coliforms	
Klebsiella freundii (Enterobacter freundii)			
Klebsiella cloacae			
Klebsiella aerogenes		Psychrotrophic	
Enterobacter liquifaciens			
Bacillus cereus		Thermoduric, aerobic, spore-forming rods	
Bacillus subtilis			
Bacillus licheniformis			
Bacillus circulans			
Aeromonas hydrophila		Psychrotrophic	
Alcaligenes viscosus			
Acinetobacter spp.			
Achromobacter spp.			
Brevibacterium spp.			
Microbacterium lacticum			
Arthrobacter spp.			
Lactobacillus spp.			
Streptomyces spp.		From cereal food and decaying matter	
Actinomyces bovis		From diseased udder lesions	
Yeasts	Various	Occasional	
Moulds	Various	Occasional	

a high number of leukocytes. Veterinary medication used to treat the disease (e.g. penicillin) may produce selective effects on the remaining types of bacteria in the milk, and indeed in other milks mixed with it.

Mastitic milks may have *Streptococcus agalactiae*, *S. dysgalactiae* or *S. uberis* as well as *Staphylococcus aureus*, and may also contain Enterococci, *Pseudomonas pyocyanae*, *Clostridium pyogenes* and even *Mycobacterium tuberculosis*. In advanced conditions, *Actinomyces bovis* and some yeasts may also be present; Bramley and McKinnon (1990) reviewed the importance of pathogens in raw milk. Since many bacteria and viruses will survive in hard cheese for long periods, and in soft cheese until consumed, the use of milk containing pathogens for cheese must involve a heat treatment sufficient to destroy them. Even in areas where diseases such as tuberculosis, anthrax and brucellosis have been eradicated, the consumer is still at risk unless the milk is correctly pasteurized.

Some bacteria can pass through the blood into the milk of the lactating animal, but many disease-causing organisms gain entrance to the milk by contamination from soil, hands, utensils and other contact sources. Kaplan *et al.* (1962) suggested that the following diseases caused by bacteria or their toxins could be transmitted through milk: anthrax, botulism, brucellosis, *Campylobacter* infections, cholera, enteritis caused by *Clostridium perfringens*, coliform infections (gastroenteritis, especially in children), diphtheria, listeriosis (spasmodic outbreaks), shigellosis (dysentery), staphylococcal enterotoxic gastroenteritis, Streptococcal infections (Lancefield A), tuberculosis, Q fever (*Coxiella burnetii*), typhoid and paratyphoid fevers and other salmonelloses, and hence few would doubt the wisdom of using pasteurized milk for cheesemaking (Keceli and Robinson, 1997).

Although avoiding contamination of raw milk is clearly impossible, cleaning routines on the farm can exert a major influence on the types of bacteria likely to be found in milk; in general, cold detergent/disinfecting routines are found to be inferior to either hot detergent/disinfecting or hot detergent/steam sterilizing methods. When the plant is seen to be clean before disinfection, the total numbers of bacteria tend to be low, but when the surfaces of the steel are rough, high counts of streptococci may build up and exceed the numbers of psychrotrophs; milk-stone on surfaces may cause the numbers of psychrotrophs to exceed the numbers of streptococci. These crude rules of thumb should not be overlooked by the cheesemaker.

In North America, Australasia and most parts of Europe, milk collection has changed from the traditional churn system to the collection of cold (2–4°C) bulk milk in road tankers, which then discharge their loads into 70 000- to 150 000-litre silos at the factory as illustrated in Figure 6.1; whereas churn milk would have been processed soon after reception, tanker milks may be stored cold for up to 4 days. The capacity of the silos depends on the location, for some silos may have to hold milk for 1–3 days at a temperature of between 3 and 7°C to allow for a shorter working week. This change in milk-handling practices has led to major alterations in the dominant microflora of milk arriving in the cheese vat. Where a farm can store its milk in cold bulk tanks, experience has shown that, if the tanks are well cleaned, the

Table 6.2 Percentage incidence of some groups of bacteria in cold-stored milk on the farm and the same milks stored at the cheese factory for a further period

	Range of numbers per group (%)	
	Ex-farm tanks	Ex-silo tanks
Streptococci	9–21	5–12
Micrococci	15–68	10–15
Psychrotrophs (Gram-negative rod forms)	6–41	53–74
Other groups of bacteria including spore-formers	11–18	10–19

population of Gram-negative rods, mainly pseudomonads, forms only 20–30% of the total surface count in the tank. When the tanks are not effectively cleaned, the total surface counts increase greatly, and the Gram-negative rods account for 70–80% of the total surface count; this large increase is at the expense of the streptococci and micrococci. Where the milk is collected from the farm only once per day (i.e. the morning and evening milks are mixed), the addition of the later milk can produce very high counts of psychrotrophic organisms in farm-stored milks over 18 h of storage at between 4 and 10°C. The effect of such storage is increased if the same milk is further stored at the cheese plant, and some typical figures are shown in Table 6.2 and Figure 6.2. The spore-forming bacteria do not tend to multiply during storage, and this restriction is typical of chill-stored foods (Prentice and Neaves, 1993).

Some typical reactions brought about by various groups of micro-organisms in cheese milk are illustrated in Table 6.3, and this table shows that the number of groups capable of causing lipolysis is greater than cheesemakers may expect. The genera listed below are psychrotrophs which grow at the temperatures of cold-stored milks:

Pseudomonas spp.
Alcaligenes spp.
Klebsiella spp.
Aeromonas spp.

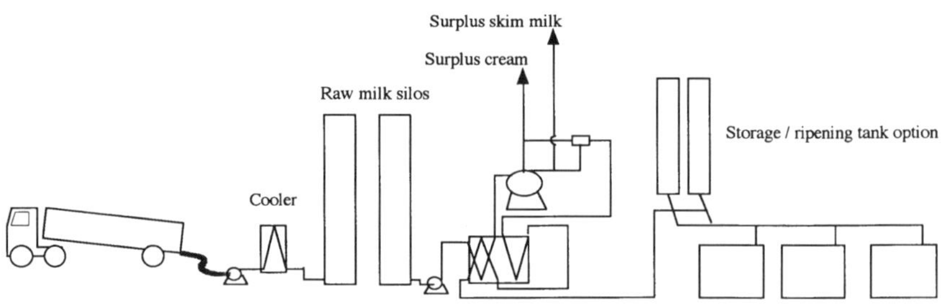

Figure 6.1 Simplified flow diagram of a plant for receiving and processing cheese milks. The throughput will determine the number and size of units.

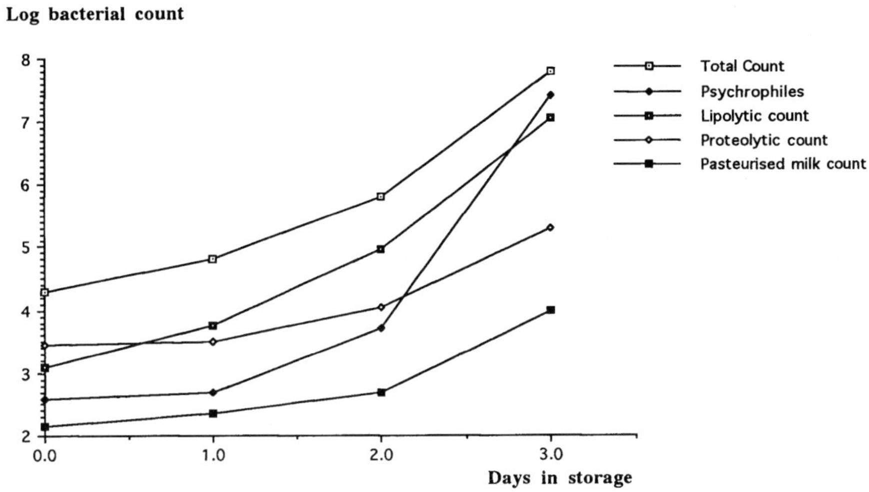

Figure 6.2 The numbers of different types of bacteria which may be found in untreated milk stored cold for a number of days. A total count for pasteurized milk is included for comparison.

and it is notable that the actively proteolytic and lipolytic pseudomonads operate under most storage conditions. Some other Gram-negative rods tend to grow only at higher temperatures, e.g.:

Escherichia spp.
Flavobacterium spp.
Serratia spp.
Aerobacter spp.
Proteus spp.

Strains of *Pseudomonas*, of which *P. fluorescens* is the most prolific, grow well in cold-stored milk, but it is important that almost all groups of bacteria can affect the quality of milk, however slightly. These changes are brought

Table 6.3 Reactions that may be caused during warm weather (in summertime) in cheese milk by the groups of bacteria indicated

Bacterial group	% of reactions by each group growing in warm milk				
	Acidity	Acid and clot	Alkali	Proteolysis	Lipolysis
Lactococci	12	88	nil	9	8
Micrococci	15	3	2	14	18
Coliforms	60	40	1	nil	10
Pseudomonads	nil	nil	22	85	70
Corynebacteria	2	4	1	30	10
Lactobacilli	0	85	nil	18	22
Bacilli	4	8	2	55	44

about mainly by enzymes secreted by the bacteria, or left behind after the death of the organism, and these bacterial enzymes complement the natural enzyme content of the milk. As many of the Gram-negative organisms are psychrotrophs that will grow in milk at 2–7°C, they are able – in a short time (1–3 days) – to initiate reactions in the cold milk, which may be to the disadvantage of the cheesemaker. The suggestion has been made that milk for cheesemaking should be heat treated at low temperature (63°C for 15 s) before cold storage, and a study of this system – usually called 'Thermization' – was published by Griffiths *et al.* (1986). To be effective in destroying the psychrotrophs, this heat treatment would have to be carried out on the farm, especially if the milk is collected every second or third day. It is important also that this type of heat treatment should only be sufficient to destroy the target bacteria, e.g. pseudomonads. For example, the feeding of silage to animals is likely to introduce, for example, *Clostridium butyricum* into the milk supply and, since the spore-formers are largely dormant in cold milks, a severe heat treatment would be a disadvantage in that the heat shock could stimulate such spores to germinate; the subsequent growth of the vegetative forms could cause problems during the later stages of cheese production.

As far as farmhouse cheese milks are concerned, the view that milk with a bacterial population of 1×10^6 cfu/ml produces cheese with a better flavour and aroma in a shorter time than cheese made from very clean milk with a population of 2×10^3 cfu/ml still tends to circulate. Indeed there are even studies that claim to show that, if in addition to the starter culture, pasteurized milk is inoculated with a selected group(s) of organisms, a cheese with a more satisfactory flavour and aroma can be made than would be the case if the starter alone was used. Obviously there may well be differences that can be perceived by experts, but bearing in mind that the majority of consumers are not experts, it is not surprising that most high-volume cheesemakers employ heat-treated milk and a well-defined starter culture with known flavour-producing characteristics.

6.4 Bacteriological testing of milk

Unless there is a change in the cleaning and milking routines, in the climate, in animal health or in the feeding routines on the farm, the average milk supply varies little from day to day with respect to its microbiology. Nevertheless, the cheesemaker must be aware of any changes which may affect the cheese process, and hence the routine testing of milk supplies is essential.

As the majority of contaminants in milk come from plant and equipment (Bramley and McKinnon, 1990), the routine testing of milk contact surfaces on farms or in transport and storage tanks has much to recommend it. Routine laboratory-based methods have been described by Gavron and

Luck (1990), but it is important that ATP-bioluminescent systems are now available for use on-site. The principle is simple – if ATP is present on a surface above the usual background level, then the surface is not clean, so that performing the test on a routine basis provides a degree of confidence that the contact surfaces are hygienically acceptable (Pettifer, 1993).

However, more attention is usually paid to the bacteriological testing of the milk, and these tests can be divided into two categories: (a) the 'broad blanket' type of test, e.g. dye reduction tests or routine plating to give a total colony count, and (b) specific tests for mastitic milk, pathogens of major importance or bacteria that are likely to cause taints or blown cheese. Tests which require the extensive use of selective media may have to be carried out at larger or central laboratories.

This non-routine testing includes special tests for the following:

Micro-organism	Cheese faults/problem
Pathogens	Diseases
Clostridia, i.e. *C. tyrobutyricum*	Blown curds
Coliforms	Blown curds
Pseudomonads	Proteolytic and lipolytic activity
Spores (*B. cereus* and *B. subtilis*)	Proteolytic activity
Psychrotrophic bacteria	Growth in cold storage
Thermoduric bacteria, e.g. *Enterococcus faecalis*	Taints
Antibiotic-producing organisms (nisin and penicillin)	Starter problems
Yeasts	Blown curds

Some recommended methods of testing are described by Gavron and Luck (1990), and more specific details about the composition of media are given in Marshall (1992) and Anonymous (1986).

Although a detailed examination of the milk or a cheese may be able to predict problems, e.g. the possible development of taints, or explain the reasons for a fault, more rapid methods for estimating the numbers of bacteria in milk for cheesemaking are necessary at reception.

In the Breed Smear Method, a typical sample of milk (0.01 ml) is spread over a pre-marked area (1 cm^2) on a slide and, after staining, the bacteria are counted through a microscope. A standard table stipulates how many fields within the square centimetre have to be counted to obtain a reliable count, and a formula is available to turn the figures into counts/ml of the original milk. Although the method has the advantage of speed, operator fatigue can be a problem if the number of samples is high (Harrigan and McCance, 1976).

It was against this background that the Direct Epifluorescent Filter Technique (DEFT) method for estimating the numbers of bacteria in raw milk was developed. It is a membrane technique and, like the Breed Smear, takes about 30 min to complete. Graves and Schipper (1966) also used a membrane technique for counting the numbers of bacteria in milk, but Pettifer (1982, 1983) improved the method by using a polycarbonate membrane. In the DEFT method, the milk is treated with a proteolytic enzyme and then with a detergent (Triton X-100), and this treatment, which breaks down the casein and disperses the fat globules, enables the milk to be filtered more easily. The bacteria retained on the membrane are stained with Acridine Orange and Tinopal AN, and can be counted using an epifluorescent microscope; active bacteria are stained and fluoresce orange–red, while inactive cells fluoresce green. The method allows the counting of numbers of bacteria from 10^3 up to 10^8/ml, and the apparatus can be coupled to an image analyser that allows automatic counting (Pettifer, 1983); it can also be adapted for counting somatic cells (Pettifer, 1982).

As the automatic recording of test results becomes more widespread, so the often-irksome task of keeping records during busy seasons becomes less tedious. This advance is important, for microbiological data can give advanced warning of difficulties, either with milk supplies or with processes, and as such can be a useful tool for the cheesemaker. Perhaps one of the major advantages of accurate records is that it enables a manufacturer to pinpoint milk supplies that are likely to need individual testing.

Thus, while one supply of milk which is defective may be diluted in a large (45 000–112 500 litre) silo, several such supplies arriving together may not be diluted sufficiently, and can jointly cause trouble. Routine 'blanket' testing of supplies using dye-reduction tests is useful to pick out the defective supplies. The dyes used most frequently are resazurin, 2-3-5 triphenyl tetrazolium chloride or methylene blue and, when 1 ml of dye is added to 9 ml of milk and incubated at 37°C, the change in colour of the dye over a set period of time (10 min–2 h) indicates the activity of bacteria in the milk. Unfortunately, such tests are often too slow to monitor all supplies during a period of rapid intake of milk, for the 10-min Resazurin Test will only detect milk of extremely poor quality. Nevertheless, the routine application of such tests can serve as a warning to the cheesemaker that a supplier is falling below the desired standards of microbiological quality.

An indication of the 'condition' of the milk can also be obtained from an acidity test. This test is usually carried out by titration of an aliquot of milk to a standard pH using phenolphthalein as the indicator (pH 8.3). The acidity test is not very useful for individual milk samples, since the buffering capacity of milks differs widely according to the chemical composition of the milk. Thus, a perfectly fresh Jersey milk sample may show an acidity of 0.21% lactic acid, whereas a milk of poor bacteriological quality from malnourished cattle may only be 0.16% lactic acid. In general, however, an increase in acidity is an indication of bacterial activity and, with bulked milks of 13 500 litres and over, acidity can be used as a measure of likely bacterial quality.

Mastitic milks can, in the absence of sophisticated laboratory facilities, be detected by the use of the 'Resazurin–Rennet' test. The resazurin indicates bacterial cell activity in the milk by a reduction of the colour, while rennet clotting of the milk (fast or slow) shows up any chemical abnormality. One further advantage of the 'Resazurin–Rennet' test is that it shows up the presence of gas-forming bacteria by giving a 'stormy' fermentation. Mastitic milks can be demonstrated at farm level by the use of bromocresol purple papers on a 'strip cup', or by the use of a conductivity meter which indicates a high chloride content of the milk. The California Test for mastitic milks is also a simple, rapid test, in which an anionic detergent is mixed with the milk which becomes viscous if mastitis is present. In the laboratory, the somatic cell count of incoming individual milks is able to show up abnormal milks (mastitic milks), for such milks will have high cell counts. This approach makes use of the Breed Smear method for single samples, but electronic counters, such as the Coulter Counter or the Fossomatic Instrument (Andersen *et al.*, 1993), can be employed for counting cells in the milks from a large number of herds.

The testing of milk for specific micro-organisms requires more time than is available in the normal reception routines for milk. In addition, selective media are usually needed to identify the organisms, and this procedure requires more elaborate bacteriological facilities than for more routine 'blanket' tests; such facilities are often available to a group of cheese plants rather than to an individual unit. Although the cheesemaker will have already made the cheese before any test results are available and must, therefore, rely on the heat treatment of milk to reduce the effect of contaminants in the milk supply, the tests can provide valuable routine information for future action.

Although the routine testing of milk is essential, the regular bacteriological testing of equipment surfaces is equally important. Problems can arise from imperfectly cleaned equipment, and the implementation of a HACCP regime (Gavron and Luck, 1990), including the routine testing of equipment, is essential for success in cheesemaking.

6.5 Principles of HACCP

In theory, the only way of ensuring that every cheese produced in a given factory is safe, from a chemical or microbiological standpoint, is to test each cheese! Clearly, such a suggestion is totally ludicrous, so that, instead, a representative group of cheeses is withdrawn against an agreed sampling plan and examined as appropriate. However, whilst this approach will indicate trends in standards of hygiene or isolate a major outbreak of contamination, it cannot ensure that some unsound cheeses do not reach the market. Consequently, although routine laboratory examinations remain essential, the emphasis within quality assurance has now turned to the avoidance of potential problems, a concept that forms the basis of HACCP – Hazard Appraisal Critical Control Points. Many people replace the word Appraisal with Analysis but, either way, the HACCP system aims to identify specific hazards that may adversely affect the safety of a food, and allow the implementation of preventative measures for their control (Pierson and Corlett, 1992). In particular, the system identifies seven aspects of production that merit attention, and these aspects are enshrined in seven principles:

1. Identification of the potential hazards associated with food production, running from the growth/collection of raw materials, through processing/manufacture and distribution up to the point of consumption. Assessment of the risk that a hazard may arise, and determination of the preventative measures necessary to reduce the risk.
2. Determination of the precise points in the above sequence that can be controlled in order to eliminate a hazard or minimize the risk of occurrence. If failure to control a particular hazard is likely to lead to a risk to public health, then the step in the process is regarded as a Critical Control Point (CCP); if no major risk is involved, the step may be identified as a Control Point (CP). For example, the pasteurizer in a fluid milk line is a CCP, but the hand-washing facility in the same area could be treated as a CP because, even if control over hand-washing is temporarily lost, it is not likely to result in a consumer health problem. Similarly, it is important that a manufacturer has control over the composition of a processed food, the filled weight of a carton or bottle and the details on the label, but again such points need only be graded as CPs.
3. Establishment of targets and tolerances which must be achieved in order to maintain control of a CCP/CP.
4. Establishment of a monitoring system of scheduled observations and/or laboratory testing to ensure that control is in place.
5. Establishment of a procedure of corrective action to be taken if the monitoring procedure indicates that a CCP/CP is not under control.
6. Establishment of procedures for verification that the HACCP system is working efficiently, e.g. the introduction of additional tests or supplementary checks to ensure that the principal components of the system are operating to the required standard.
7. Establishment of a system of documentation that records accurately the details of all operations, including not only aspects likes times/temperatures and microbiological parameters, but also the responsibilities of the individual operators.

At first glance, this approach may appear daunting, but if each stage in a manufacturing process is identified and considered, at least initially, as a separate entity, then implementation can bring considerable benefits to a manufacturer. For example, vat failure from phage attack was, at one time, a major problem for the cheese industry, but a carefully controlled system of culture rotation should now ensure that the levels of phage compatible with a given strain of *Lactococcus* never become critical. Another advantage is that retailers have more confidence in a company that has proper control over its procedures, so that the introduction of a flexible and viable system of HACCP is fast becoming an essential component of the measures employed to monitor the microbiological aspects of cheesemaking.

6.6 Removal of bacteria from cheese milk

In spite of care during milking and storage, it is inevitable that milk will be contaminated with both vegetative bacteria and the endospores of *Bacillus* and *Clostridium*. As mentioned earlier, spores of the latter genus can cause problems during the production of cheeses, such as Edam, Emmental and Gouda, and, while the addition of nitrate or lysozyme to the cheese milk can be effective (Tamime, 1993), physical removal of the bacteria is possible. Two systems have been advocated, namely:

(i) Bactofugation: if whole milk is subjected to a process of centrifugation that removes 2–3% of the initial volume, then the fat fraction will contain most of the bacterial endospores. This fat-rich material can then be heat-treated at UHT temperatures (130–140°C for a few seconds), prior to mixing with the bulk of the cheese milk which will have been pasteurized under standard conditions. This combination of treatments means that the endospores are killed by the UHT process, but because the volume of milk concerned is low, no adverse effects with respect to flavour or coagulation performance are evident (Berg and Daamen 1989).
(ii) Microfiltration: the pore size of membranes employed for microfiltration ranges from 0.01 to 10 mm (Grandison and Finnigan, 1996), so that, while macromolecules like protein will pass through a membrane, endospores will be retained as will many fat globules and somatic cells. If the pore size is at the lower end of the range (0.8–1.4 mm), then vegetative bacterial cells will be removed as well, and this approach is the basis of the Bactocatch System developed by Alfa-Laval in Denmark. In this process, the cheesemilk is warmed to 40–50°C to facilitate separation of the cream, and the skim-milk portion is then micro-filtered to give milk with a low bacterial count, including spore-formers. The retentate from the filtration unit can, after UHT treatment, be blended with the cream and skim milk to provide the final process milk to be delivered to the cheese vat.

Trials with Swedish and Dutch types of cheese suggest that the process could offer an effective solution to the problems of late blowing, and the absence of any need for additives could find favour with consumer groups.

References

Andersen, T., Brems, N., Borglum, M.M., Kold-Christensen, S., Hansen, S., Jorgensen, J.H. and Nygaard, L. (1993) In: *Modern Dairy Technology*, Vol. 2 (ed. R.K. Robinson), Chapman & Hall, London, pp. 381–416.

Anonymous (1986) *The Oxoid Manual*, Unipath Ltd, Basingstoke, Hampshire (UK).

Berg, G. van den and Daamen, C.B.G. (1989) *North European Food and Dairy Journal*), **55**(3), 63–66.

Bergere, J.L. (1979) *Revue Latiere Française*, **378**, 19.

Bramley, A.J. and McKinnon, C.H. (1990) In *Dairy Microbiology*, Vol. 1 (ed. R.K. Robinson), Chapman & Hall, London, pp. 163–208.

Gavron, H. and Luck, H. (1990) In *Dairy Microbiology*, Vol. 2 (ed. R.K. Robinson), Chapman & Hall, London, pp. 345–392.

Grandison, A.S. and Finnigan, T.J.A (1996) In *Separation Processes in the Food and Biotechnology Industries* (eds A.S. Grandison and M.J. Lewis), Woodhead Publishing Limited, Cambridge, England, pp. 141–153.

Graves, D.C. and Schipper, I.A. (1966) *Applied Microbiology*, **14**(4), 535.

Griffiths, M.W., Phillips, J.D. and Muir, D.D. (1986) *Dairy Industries International*, **51**, 31.

Harrigan, W.E. and McCance, M.E. (1976) *Laboratory Methods in Food and Dairy Microbiology*, Academic Press, London.

International Dairy Federation (1980) *Factors influencing the Bacteriological Quality of Raw Milk*, Bull. Doc. 120, IDF, Brussels.

International Dairy Federation (1996a) *The Bacteriological Quality of Raw Milk*, Publ. S.I. No. 9601, IDF, Brussels.

International Dairy Federation (1996b) *Production and Utilization of Ewes and Goats Milk*, Publ. S.I. No. 9603, IDF, Brussels.

Kaplan, M.M., Abdussalam, M. and Biilenga, G. (1962) In: *Milk Hygiene*, WHO Monograph No. 42, Geneva, pp. 12–74.

Keceli, T. and Robinson, R.K. (1997) *Dairy Industries International*, **62**(4), 29.

Marshall, R.T. (1992) *Standard Methods for the Examination of Dairy Products*, American Public Health Association, Washington.

Matteuzzi, D., Trovatelli, L.D., Beavate, B. and Zani, G. (1977) *Journal of Applied Bacteriology*, **43**, 375.

Panes, J.J., Parry, D.R. and Leech, E.B. (1979) *Report of a Survey of Farm Milk*, Ministry of Agriculture, Fisheries and Food, London.

Pettifer, G.L. (1982) *Journal of the Society of Dairy Technology*, **35**(2), 59.

Pettifer, G.L. (1983) *Dairy Industries International*, **48**(5), 21.

Pettifer, G.L. (1993) In *Modern Dairy Technology*, Vol. 2 (ed. R.K. Robinson), Chapman & Hall, London, pp. 417–454.

Pierson, M.D. and Corlett, D.A. (1992) *HACCP – Principles and Applications*, Van Nostrand Reinhold, New York.

Prentice, G. and Neaves, P. (1993) In *Encyclopaedia of Food Science, Food Technology and Nutrition* (eds R. Macrae, R.K. Robinson and M. Sadler), Academic Press, London, pp. 887–891.

Tamime, A.Y. (1993) In: *Modern Dairy Technology*, Vol. 2 (ed. R.K. Robinson), Chapman & Hall, London, pp. 49–220.

Underwood, H.M., McKinnon, C.H., Davies, F.L. and Cousins, C.M. (1974) *XIX Int. Dairy Congress*, Vol. IE, 373.

Von Brockelmann, J. (1969) *Meddelanden Svenska Mejeriernas Riksforening*, Report No. 91, 15 pp., Lunds Teknista Hogskola, Alnarp, Sweden.

7 Tests for acidity and chemical analysis in process control

7.1 Acidity *versus* pH

The biological, biochemical and physical changes which take place from the time milk leaves the udder until the cheese is finally ripe are often controlled or aided by the pH (hydrogen ion concentration) of the liquid or curd at that time. The delicate physical equilibrium between the constituents of milk due to ionization of components, the binding of protons and the dispersion of salts and proteins between the phases, leads to a buffering capacity in respect of pH change. The acid–base equilibrium of the milk, whey or curd is a useful parameter for control of the processing of cheese.

The measurement of pH can be made during the processing from liquid to solid but there are some difficulties in the use of such measurements, which often need interpretation. The pH of milk has a greater temperature dependence than that of, say, phosphate buffers. Between 20 and 30°C, the pH of milk will fall by 0.10 units, whereas a phosphate buffer falls only by 0.03 units. For this reason, the comparison of pH values between processes should be between those values measured at the same temperature.

The buffering capacity of milk over the pH range 4.5–6.7 normally encountered in cheesemaking mainly involves proteins, phosphates, citrates and carbonates. The calcium and magnesium in the system are present both as free ions and as complexes with casein phosphates and citrates. Furthermore, both the calcium and magnesium influence titrations of milk, since they can precipitate as colloidal phosphates. These changes are time dependent so that titrations should be carried out fairly rapidly.

7.1.1 *Indicator dyes*

The pH can be measured using indicator dyes only if the colour change is easily distinguishable in the medium (milk or whey) and in a range that is relevant. The use of indicator dyes for pH measurement should be undertaken with caution, because of possible errors due to absorption of colour and/or to chemical or physical reactions.

The buffering power of milk is relatively low at the end-point of phenolphthalein (pH 8.3) which makes it very useful for titration of acidity in milk and whey. Many indicators have been used in the laboratory for adjustment of the pH of media, but bromocresol purple is useful for indicating acidity changes in tests for bacteriophage activity in whey samples.

7.1.2 Titratable acidity

A 'Soda' test had been used to assess acidity in milk until, in 1899, Lloyd advocated the use of an acidimeter to determine the amount of acid in dairy liquids, which made it possible to lay down more precise methods for the manufacture of cheese. Lloyd suggested the use of phenolphthalein as an indicator in the titration of milk because of its easily discernible end-point. Since then it has been traditional in the dairy industry to regard the neutral point of phenolphthalein (pH 8.3 under the conditions of the test) as the fixed point against which titrations are made. (It so happens that sodium lactate solution also has a pH of approximately 8.3.) Ninth molar (0.111 M) sodium hydroxide is used for the titration, since the number of millilitres of 0.111 M NaOH solution used to titrate 10 ml of the milk to the end-point of phenolphthalein (pink) can be used to express 'the equivalent percentage of lactic acid' in the milk (i.e. 1.0 ml of 0.111 M NaOH = 0.1% lactic acid in the sample). Older textbooks may use normality (N) instead of molarity (M); for NaOH the values are the same.

The expression 'equivalent percentage lactic acid' is an arbitrary one since all the 'acid' titrated in milk is not lactic acid. Proteins and fatty acids as well as phosphates, citrates, carbonates and sulphates of calcium and magnesium are also involved. Indeed, the dissolved carbon dioxide as carbonates accounts for 0.02% acidity since milk can contain almost 4% by volume of carbon dioxide in solution. The acidity value is in reality a measure of the buffering capacity of the sample between the sample pH and the pH at the end-point of the titration.

There are problems with the measurement of titratable acidity. The strength of the indicator affects the visibility of colour change (i.e. pH 7.85 colourless to pH 8.5 red). The indicator also tends to form deposits when left to stand. An amount of 80 mg of indicator in 100 ml of milk has been suggested as a standard amount, so aliquot additions are made to milk to give this concentration. Some indicator solutions use 0.5% by weight of phenolphthalein in 50% ethanol. However, ethanol itself may affect the equilibrium of the milk components and hence the pH (1 ml of alcohol in 10 ml of milk may change the pH by 0.10 units).

If 1 g of phenolphthalein in 60 ml of ethylene glycol monomethyl ether and 40 ml water is made up as the indicator solution, the solution does not cause a deposit and the colour change is quite intense.

Since the final pH of the solution titrated is dependent on the colour of the indicator, the use of a standard volume of indicator and a standard colour for the end-point of all titrations is desirable. Rosaniline hydrochloride, rosaniline acetate, fuchsin or safranine (0.01% w/v in 96% alcohol) have been used as a colour standard when added to another sample of the same milk. This colour standard takes into account the colour of the milk sample. Coloured stirring rods or tiles may also be used but must be checked against a standard as, over a period of time, the colour can fade.

The alkaline solutions of sodium hydroxide must be protected from the ingress of carbon dioxide from the air since this will decrease the strength of the solution over a period of time. Apart from the use of 0.111 M NaOH solutions, other strengths are used in the Dornic, Soxhlet and Thorner methods. Since these methods express the results in degrees, there is likely to be confusion when comparing rates of acid production by the use of alternative methods. These are illustrated in Table 7.1.

The use of alternative methods makes the comparison of results of different workers difficult. For example, one method uses a fuchsin colour standard, 0.5 ml of a 2% w/v phenolphthalein solution in 70% alcohol for the indicator, and 0.1 M NaOH for the titrant. The result is expressed in acidity units of tenths of a millilitre of 0.1 M NaOH used in titrating 10 ml of milk.

The Thorner degrees may not be directly equivalent to other degrees of acidity, since this method uses dilution of the milk with water, which is likely to upset the salt balance in the milk and gives readings lower than expected when compared with other methods of measurement.

Since the buffering power of milks and wheys varies from sample to sample, the increase in acidity between samples over a period of time can be regarded as a rate of acid development for that particular sample. These readings are not strictly comparable with readings taken on other samples of different milks.

Dilution of whey with water, either wash water or hot water used for scalding the curd, produces differences in osmotic pressure between whey and curd. While this assists removal of lactose (e.g. in Cottage cheese), both the buffering capacity and the actual level of lactic acid will be reduced, thus masking the rate of acid development measured.

The cheesemaker is most likely to use milks of similar buffering power on each daily occasion, but this is not necessarily so from season to season, or for milks from other areas and milks from animals of different breeds. Thus, a degree of interpretation is needed, which forms part of the expertise of the cheesemaker.

Measurements of titratable acidity have to be carried out, as a routine, in or near the processing area, and the development of plastic burettes and other equipment has minimized the risk of glass breakages; stringent precautions must be taken to keep reagents away from the cheesemaking operation.

7.1.3 Hot Iron Test

Difficulties will be encountered in measuring the titratable acidity of curds, as the acidity of whey exuding from the curd will be less than that in the

Table 7.1 Comparison of different methods of expressing acidity

	English traditional	*Common*	*Soxhlet Henkel*	*Dornic*	*Thorner*
Sodium hydroxide solution	0.111 M	0.10 M	0.25 M	0.111 M	0.1 M
Quantity of milk titrated	10 ml	10 ml	50 ml	10 ml	10 ml + 20 ml distilled water
Phenolphthalein solution	1 ml 0.5% in 50% ethanol	1 ml 1.6% in aqueous ethylene monomethyl ether	2 ml 2% in ethanol	2 drops 2% in ethanol	5 drops 5% in ethanol
Results expressed as	% lactic acid = 10× titre	titre = acidity	°SH = 2× titre (i.e. per 100 ml milk)	°D = 10× titre	°Th = 10× titre (i.e. per 100 ml milk)
Typical value for fresh milk	0.18% lactic acid	0.20 acidity	8°SH	18°D	20°Th

curd, and the quantity of whey expelled will be much reduced as dry curds are acidified, especially with high quality cheddar curd. The hot iron test was a very early measure of the effect of the pH on the functional properties of the curd proteins.

An iron bar was heated to black heat and a piece of dry curd gently pressed against it until it melted. The curd was then slowly removed from the bar and the length of the strands before breakage was estimated; the longer the strand, the lower the pH and the higher the acidity.

7.1.4 pH measurement

pH is a measure of the hydrogen ion concentration:

$$\mathrm{pH} = -\log_{10}[\mathrm{H}^+]$$

Neutral solutions have a pH of 7, with pH dropping as acidity increases.

The direct measurement of acidity development in terms of pH has gained slow acceptance. Measurement of pH is dependent on measuring a very small potential (voltage) difference, between the sensor and a reference electrode that maintains a standard potential. Early pH meters tended to be somewhat erratic and the probes were prone to fouling, both by the formation of fat films and protein adsorption onto the surface. Glass electrodes are also prone to mechanical damage, though the spear point types are much more resilient and may be used directly on moist curd with only a small penalty in time to reach equilibration.

Glass electrodes, while suitable for laboratory use, pose foreign body hazards if used in the cheesemaking plant. Non-glass systems have now been developed, for instance using field effect transistor technology to produce sensitive, reliable and portable systems. There still remains the need to ensure that the electrode is kept clean and to check the calibration against standard buffer solutions. Most modern systems use virtually automatic recalibration systems.

Unless otherwise stipulated by the manufacturer, pH probes may be easily cleaned of fatty deposits using a warm mild detergent solution, while the more tenacious protein films may be removed by soaking in a protease based detergent solution or in 0.1 M hydrochloric acid solution. Extended soaking in strong alkaline solutions may damage the probe.

The pH value is a measure of the hydrogen ions dissociated in a solution. Since the pH of a solution varies with temperature, it is necessary to standardize the electrodes within a narrow pH range and at a fixed temperature. A temperature sensor is often fitted alongside the pH probe to provide an automatic compensation, but for accurate measurement in milk systems, standardization of the measuring temperature is desirable.

After the heat treatment of milk or the addition of calcium chloride solutions, changes in the equilibria between the different components in the milk occur, such that serious errors may be introduced and time must elapse before satisfactory readings can be taken.

The major advantage of pH measurement, apart from the speed and sensitivity of the most recent systems, is that the parameter being measured is that which has a direct effect on the functional properties of the milk proteins and hence on the curd. It must be emphasized that there is no direct link between pH readings and those of 'titratable acidity'. Unlike titratable acidity readings, there will be no sudden change at cutting, the pH development during the cheesemaking process taking the form of a smooth curve. As with titratable acidity, measurement of the pH of the whey will give a reading remote from the pH within the curd particle. This common problem may be avoided, however, by either taking a sample of the curd and gently wrapping the curd around the probe, or by inserting the probe into the curd. Direct measurement of curd pH provides an alternative to the old 'hot iron' test.

7.1.5 Rate of acid or pH change

The rate of acid production is measured by the change in acidity over unit time, rather than by a single value. In addition, the final pH value of a pressed curd may provide an indication as to how the maturation stage may progress.

Typical pH values for different groups of cheese are as follows:

	Soft (Camembert)	Semi-hard (Edam/Gouda)	Hard (Cheddar)	Blue-veined (Stilton)
Green curds	4.85	5.2–5.25	5.15–5.2	4.9–5.0
Ripe curds	5.5–5.8	5.3–5.4	5.45	6.5–7.0

The rate of acid development (bacterial growth and metabolism) at each stage of the cheesemaking process influences the next stages. The rate of change of acidity (which for interpretation of results can be conveniently calculated as acid produced per hour) varies from stage to stage for each cheese variety, but the rates for each stage within a variety of cheese are very similar (see chapter 13).

7.1.6 Acidities during processing

Table 7.2 shows acidities attained during the manufacture of some varieties of UK cheese. Some variation of acidities at stages of cheese processing may be due to the following causes:

1. The quality of the milk as growth medium for bacteria
2. The quality and activity of the starter in the cheese milk
3. The milk ripening time (including coagulation time)

Table 7.2 Summary of typical acidities for UK varieties during cheesemaking

Cheese	Acidities (% as lactic acid)						Time[1]
	at rennetting	*after cutting*	*whey removal*	*during texturing*	*salting*	*ex press*	(h)
Caerphilly	0.21–0.23	0.12–0.15	0.17–0.19	0.19–0.23	0.30–0.39	–	3.5
Cheddar	0.20–0.23	0.14–0.15	0.20–0.24	0.22–0.60	0.65–0.85	1.0–1.15	5–5.5
Cheshire spring	0.20–0.21	0.13–0.14	0.19–0.20	0.22–0.50	0.55–0.65	0.95–1.0	3
autumn	0.22–0.24	0.135–0.155	0.21–0.23	0.25–0.60	0.75–0.85	1.15–1.3	4
winter	0.21–0.22	0.13–0.145	0.20–0.21	0.22–0.55	0.63–0.65	1.0–1.15	3.5
Derby	0.18–0.19	0.12	0.15–0.19	0.18–0.24	0.45–0.65	0.85–0.95	4
Dunlop	0.18–0.20	0.12–0.135	0.21–0.25	0.24–0.55	0.65–0.75	0.80–0.95	5–5.5
Double Gloucester	0.19–0.20	0.12–0.13	0.19–0.23	0.23–0.50	0.60–0.75	1.0–1.1	4–4.5
Lancashire	0.16–0.18	0.11–0.125	0.18–0.22	0.22 + 1.6[2]	0.90–0.99	1.05–1.5	(2 days)
Leicester	0.19	0.12	0.19–0.20	0.20–0.35	0.45–0.65	0.80–0.90	4.5
White Wensleydale	as for Cheshire						
Blue Wensleydale	0.18–0.185	0.11–0.13	0.20–0.22	0.25–0.30	0.50–0.58	0.30–0.45	3.5
Dorset Blue	0.25–0.35	0.19–0.25	0.35–0.38	0.40–0.60	0.85–0.90	1.15–1.35[3]	18–24
Stilton	0.17–0.185	0.12	0.13–0.17	0.20–0.45	0.55–0.65	1.01–1.25[3]	16–24

[1] Time from rennetting to moulding before pressing.
[2] Mixed curds, half from previous day.
[3] Whey drainings.

4. The temperature of the milk
5. The size of curds at cutting
6. The temperature regimes employed, i.e. rennet and scald temperatures
7. Elapsed time and drainage time
8. Quantity of salt and method of salting
9. Pressing times and temperatures.

7.2 Acidity in process control

To quote an acidity measurement out of context has little meaning in cheese processing. The cheese process is one of continuing activity of bacteria and/or enzymes so that measurement of this activity by acid production must have reference to time. Knowledge of the rate of change in acidity (titratable or pH) is a necessity for control to be exercised during processing. Usually the acidity at various stages is quoted in recipes.

The rate of change of acidity against time can be most usefully portrayed on a graph. The rate of change of acid production by starter bacteria growing in milk is shown in Figure 9.1. Similarly, the rate of change of acid production during the processing of milk into a green hard cheese is given in Figure 7.1. The scales of the axes of the graph can be adjusted to make the smaller increases in acidity more apparent. Such graphs are more illustrative of the process than tables of figures, especially for the cheese-maker. If such graphs are constructed from the cheese 'logs' compiled in the

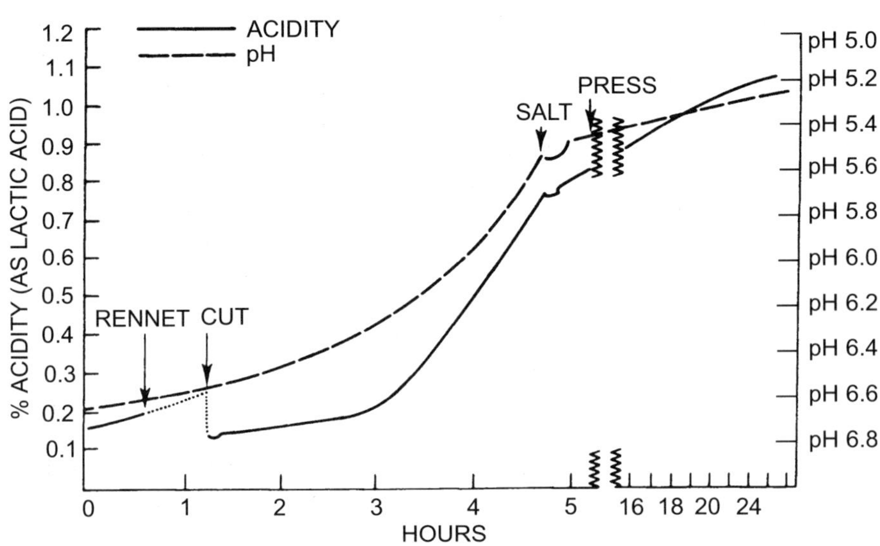

Figure 7.1 Graph showing the rate of change of acid development and pH change during the manufacture of Cheddar cheese in the first 24 h.

cheese room, they can prove of value for ascertaining trends in acid production in the cheese per unit of time. The most useful graphs will be obtained by plotting all the intermediate and logged points in the cheese make. 'Slow' or 'fast' makes do not produce the ideal raw curds for ripening, so the cheesemaker needs to ascertain the reasons for the divergence from the standard or recipe.

In the early stages of the process, the milk has a titratable value which is dependent on its composition (i.e. proteins, phosphates, calcium, salts, etc.), but also on any acid which may have been developed previously. Thus, unless a previous titratable value is known, it cannot necessarily be assumed that acid has been developed. Previous experience with milks from the same source will give the cheesemaker some information as to the likelihood of acid development. It is, therefore, an increase of titratability which is the real control information required. Similarly, it is a decrease in pH value between two measurements that indicates acid production. Thus the experienced cheesemaker uses the increase of acid production over a time span between two or more measurements (titration or pH) as the real information required for control purposes.

The pH of the milk and curd continues to decrease at a steady rate except when temperatures change or when salt is added, but the titratable acidity values do not rise evenly. There is a fall between the last acidity of milk and the first acidity of whey, primarily due to the removal of the buffering capacity of the casein. Figure 7.1 shows the pH values as well as titratable acidity values for a typical cheese, plotted against a time scale. Because of the differences between the systems being measured, there is no correlation between pH readings and titratable acidity values. The pH readings measure the hydrogen ion activity in the milk, the coagulum, the curds and finally the solid cheese. The titratable measurements are made on liquids, milk and whey, and only with difficulty can the titration be extended to more solid curds.

The titration of milks gives readings from which the increase in acidity per unit of time can be calculated, but the increase which takes place during coagulation needs to be extrapolated from previous readings. When the coagulum is cut, there is a separation into curds and whey, and it is only the acidity of the whey which is measured, since the curd is removed. From this stage until the curds are pressed, the titration values represent only the acid in the whey and, therefore, acid development in the curd is measured indirectly, via the whey. The difference between acid in the whey and acid in the curd is not a constant difference. This difference depends on: (a) the proportions of lactic acid-producing bacteria in the whey and curd, roughly 20% to 80%; (b) the temperature and size of curd particles which, in turn, control the rate of whey expulsion; and (c) the proportion of anti-microbial substances produced by bacteria growing independently in the separate media, whey and curds. In addition, the acidity in a curd may be produced normally and that in whey more slowly due to phage infection of the whey, although the reverse is more likely. For these reasons, the cheesemaker needs to interpret the results of titrations in the light of experience. Measurements of pH on the curd are, therefore, more satisfactory if a sensitive and accurate measuring system can be used.

The rate of acid increase in curd is usually greater than that in the whey and the more vigorous the acid production, the greater the difference. At the time of salting (in textured cheese) there is, for a period, a disturbance in titratable values due to expulsion of whey from the curd. However, after the first few hours, the acidity of whey from pressed curd is a good indication of the probable ripening activity of the green cheese. Low acidity indicates a sweet cheese and, therefore, one which is likely to be moist and develop off-flavours; whereas high acidity indicates an acid cheese which may mature slowly and is likely to become dry and hard. Furthermore, the titratable acidity of whey from the pressed cheese after the first flush of whey has left the cheese gives a good indication of the activity or otherwise of the starters used in the process.

The control of acidity development is obtained partially by the rate of scalding, but more by the temperatures employed. Figure 7.2 illustrates the effect of scald temperatures on the acid development in cheese made experimentally. Since most of the acid in the curd is produced by bacteria from the lactose in the curd, expulsion of whey from the curd removes the dissolved lactose and is one of the methods of controlling the final acidity. Apart from increases in temperature, which shrink the curd, increases of acidity within the curd itself will also promote shrinkage of the protein matrix and help to expel whey from the curd, for example in unscalded cheeses.

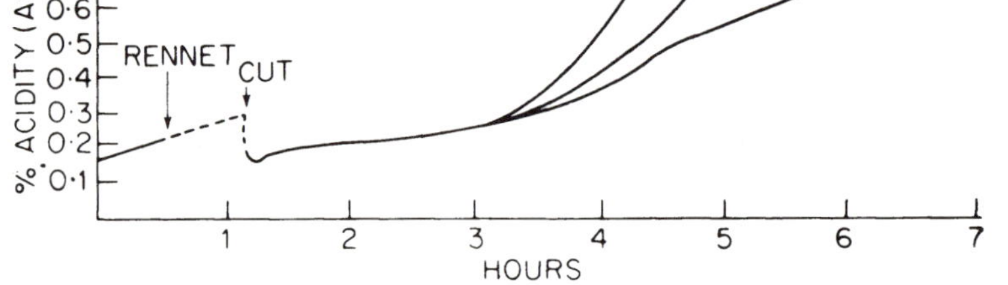

Figure 7.2 Graph showing acid development in cheese made experimentally to study the effect of scald temperatures on the cheese. Cheese A: low scald temperature (36°C), the cheese was acid and moist (44% water) and the keeping quality was poor. Cheese B: normal scald temperature (40°C), the cheese was of good quality when 9 months old. Cheese C: high temperature scald (43°C), the cheese produced was very slow maturing and the moisture content was 34%. Note: On cutting, the smaller the cubes the drier and slower maturing the cheese.

Success in cheese manufacture often depends on a cheesemaker's experience in the use of milk and starters and, where necessary, the adjustment of the process method.

7.3 Chemical analysis

The development of instrumental techniques has been of enormous benefit to the larger dairies, the classical methods being largely reduced to the essential core of analyses to provide reference samples for calibration. For smaller cheesemakers, the older titration-based methods may still be appropriate. Table 7.3 lists some of the methods used for fat, protein, lactose analysis and moisture testing of dairy products.

The economic returns from cheese manufacture depend on the quality as well as quantity of milk purchased, so that accurate chemical analysis of milk is a prerequisite, both for assessment of economic returns and to conform with legal requirements. Chemical analysis of the major components of milk, curd and whey are needed for processing purposes, while estimation of minor components is sometimes necessary in order to correct for faults (calcium content) or for process control (common salt). Some analytical methods require expertise not normally available amongst personnel in the cheese plant. In these instances, the use of an external or central analytical service is required.

7.3.1 Casein

The Walker formol casein test has been used in Australia and other cheesemaking countries for the estimation of casein in milk. The milk is titrated with 0.1 M NaOH to the end point of phenolphthalein ($\approx$ pH 8.3). To 17.6 ml of the neutralized milk, 4 ml of 40% formaldehyde solution is added. The milk turns white again and is titrated with 0.1 M NaOH to the end point of phenolphthalein. Since formalin itself is acidic, its acidity must be subtracted from the last reading taken from the milk to give a final figure due to the reaction between formalin and milk. This figure multiplied by 0.833 gives the quantity of protein (casein) in the milk. A similar 'aldehyde' test uses 0.05 M strontium hydroxide instead of the 0.1 M sodium hydroxide of the Walker test, and 100 ml of milk treated with 20 ml of 40% formalin. The acidity reading is multiplied by 0.17 to give the protein content. Disturbances due to calcium can be overcome by the addition of a solution of sodium oxalate to sequester the calcium before the titrations commence.

McGann *et al.* (1972) determined the casein in milk by precipitating the non-casein protein at pH 4 6; then the total casein less the non-casein protein gives the casein protein. Casein in milk can also be determined using the standard IDF No. 29 method (International Dairy Federation, 1964).

Table 7.3 List of methods used to estimate the major components of dairy products

Fat		
Reference	Röse Gottlieb	Ammonia hydrolysis, fat extraction, evaporation, drying and weighing. (IDF standards 1C:1987 for milk, 5B:1986 for cheese, 22B:1987 for skimmed milk and whey, 59A:1986 for whey cheese)
	Werner Schmidt	Similar to above but using hydrochloric acid to release the casein. As the Schmidt–Bondyinski–Ratzlaff method, it is described in provisional IDF standard 5B:1986.
Routine	Gerber method	Used for milk, whey and cheese according to BS 696 Parts I and II (1955) and IDF provisional standard 105:1981 for milk. Empirical test employing acid to dissolve the curd.
	Babcock method	Used for milk and whey, uses acid to dissolve the curds, but different apparatus and quantities to the Gerber method.
	Milko-tester®	Light scattering technique suitable for milk, whey and cream
	Infra-red	Absorbance method used for milk, whey, cream and possibly liquid preparations of other products – guidelines in IDF standard 141A:1990.
Protein		
Reference	Kjeldahl	Acid digestion, distillation then titration (IDF method 20B:1993)
Routine	Infra-red	Absorbance method used for milk, whey, cream and possibly liquid preparations of other products – guidelines in IDF standard 141A:1990.
	Dye absorption	Change in optical density of amido black (Pro-milk®) or similar pigment solution as a result of absorption by protein – IDF method 98A:1985.
	Titration	Formol titration

Lactose		
	Enzymatic	IDF method 79B:1991
	Titrimetric	Using chloramine T and potassium iodide (IDF method 28A:1974)
Routine	Infra-red	Absorbance method used for milk, whey, cream and possibly liquid preparations of other products – guidelines in IDF standard 141A:1990.
Salt		
	Titration	Potentiometric titration using silver nitrate with sample prepared in nitric acid (IDF method 88A:1988).
Total solids		
Reference	Gravimetric	Oven drying, used for milk (IDF 21B:1987), whey, cheese (IDF 4A:1982) BS 1741 (1951).
Routine	Gravimetric	Rapid moisture balances, microwave ovens, Carter Simon type oven.
Solids not fat		
Reference	Gravimetric	Total solids by oven drying then deduct fat value
Routine	Hydrometer	Calculated from the density of milk samples in conjunction with a Gerber test.
	Infra-red	Data from fat and protein estimation with factor for minerals automatically added.
Multicomponent analysis		
Routine	Infra-red spectrometry	Rapid analysis for fat, protein and carbohydrate in liquid samples.
	Near infra-red spectrometry	Rapid analysis for fat, protein and carbohydrate, including solid materials.

Fourier Transform Infra-red Spectroscopy now offers a rapid means of estimating casein levels in milk.

7.3.2 Salts

Routine estimations of sodium chloride in cheese and whey are necessary for control purposes, and also in the whey if it is to be used for pig food. The chloride/silver nitrate test uses a titration with standard silver nitrate and potassium chromate as indicator.

Alternatively, Lloyd (1963) has given details of an optical method of determining chlorides (NaCl). A rapid estimate of the salt content may be made using a sodium ion selective electrode and meter.

Where sodium nitrate is included in the manufacture of cheese, estimation of the residual nitrate and nitrite levels may be required. This may be accomplished by preparation of a fat and protein free serum from the cheese solution, reduction of a portion to nitrite and then interaction with sulphanilamide and *N*-1-naphthyl-ethylenediamine dihydrochloride to produce a red coloration that may be measured spectrophotometrically at 538 nm (IDF, 1984).

7.3.3 Moisture

Detection of added water in milk may be determined by the freezing point depression method (British Standards Institution, 1980: IDF, 1991).

The reference method for determining moisture in cheese by oven drying is too time-consuming for routine use. Faster oven methods using vacuum ovens and simple hot air ovens of the Carter Simon type have been used, but these have been largely replaced by simple bench-top moisture analysers which can be easily programmed for an appropriate time and temperature. Pieper *et al.* (1977) suggested the use of microwave oven drying for moisture in cheese. The rapid methods all need calibration against the reference method.

7.3.4 Miscellaneous substances

There is a growing tendency for official public bodies to require analysis for contaminants in dairy produce, which may include veterinary drugs, agrochemicals, metals, microbial by-products, disinfectants and environmental organics (Bluthgen *et al.*, 1984). These analyses will not be carried out in cheese plant laboratories, but in special service laboratories. These special analytical services may also provide facilities for non-urgent investigations, such as analyses for free fatty acid, amino acids, whey proteins, preservatives, colouring matters or the strength of detergent or sanitizing solutions.

Where admixture of dried skim milk in liquid milk is used for cheesemaking, the modified ferricyanide test of M. Yoshitake *et al.* personal communication to R. Scott) utilizing the protein bound sugar may be used for routine purposes. In some areas, mixtures of cow's, ewe's and/or goat's milk may be used for cheesemaking, or the cheese made from one type only. This has given rise to concern that the milk supplied or used may not be as specified. Immunological tests are now available to detect adulteration. Similar methods were suggested by Robertson and Cardwell (1979) and Hurrell *et al.* (1981). IDF (1995) reviews effects of heat treatment of milks, while chromatographic and enzymatic methods of measuring lactose are given in IDF (1994, 1995).

7.3.5 Instrumental and other analytical methods

The closing decades of the twentieth century have seen the increasing replacement of traditional tests based on 'wet chemistry' by rapid instrumental techniques. In cheesemaking, the dominant requirement has been for fat and protein analysis, and Baer *et al.* (1983) and, more recently, Andersen *et al.* (1993) reported on the principal instrumental methods of analysis now available.

The first instrumental technique for fat estimation in milk was by light scattering. The sample was prepared by dilution with a warm solution of disodium ethylenediamino tetra-acetic acid (EDTA) and a surface-active agent. The role of the EDTA was to chelate calcium from the casein micelles, leading to their dispersion, so that the only particles remaining in the sample were the fat globules. The prepared sample was then passed through a homogenizer to create a standard particle size distribution and thence through a flow cell where the degree of light scattering was measured and converted into a value for the fat content. The machines proved reliable and remained in use long after infra-red machines became widely available.

The dye-binding technique was the first to be applied to rapid protein estimation as it could be built into a relatively simple apparatus for routine use. The milk or cheese preparation was added to a standard volume of a dye, such as amido black. The mixture was filtered to remove the precipitated protein and the optical density of the dye solution measured in a flow cell. The drop in optical density was proportional to the protein content. Comments on the use of both dye-binding and spectroscopy for the determination of protein have been given by Grappin *et al.* (1980).

Dedicated infra-red spectrometers have effectively taken over the routine analysis of milk and whey in the larger dairies. There are three basic types:

(i) near infra-red reflectance spectrometers
(ii) multi-filter infra-red spectrometers
(iii) Fourier transform infra-red (FTIR) spectrometers.

The near infra-red technique has the advantage that it can be adapted for both solid and liquid samples, though great care is needed in sample preparation if accurate data are to be obtained (Egli and Meyhack, 1984; Weaver, 1984). The use of near infra-red reflectance has not become as widespread in the dairy as in some other sectors of the food industry.

Multi-filter infra-red spectrometers were developed using both single- and double-beam technology (Grappin and Jeunet, 1976; Biggs, 1979; Hofer and Szklenar, 1979; Finland, 1980; Sasano *et al.*, 1980). Initially they were used in larger plants and in central laboratories, but as capital costs were reduced and cost savings became apparent, the smaller machines have found wider use. Sample homogenization is required as an integral step to standardize particle sizes, and hence minimize interference through scattering (Robertson *et al.*, 1981). These instruments use filters selected to match the absorbance peaks of the components in the sample, e.g. fat 5.73 μm and/or 3.5 μm, protein 6.46 μm and lactose 9.55 μm, while 7.9 μm may be used for MSNF. The composition of the sample is calculated from the absorbance using multiple correlation techniques to give a direct readout.

The application FTIR spectrometry to milk and milk product testing offers greater flexibility and the potential for greater precision. Filters are replaced by a diffraction grating so that specific wavelengths may be programmed depending upon the properties of the components to be measured. Increased computing power is also used in both the setting up of the instrument and in calculation of values.

Spectrometer instruments need to be standardized by more traditional analysis as noted in Table 7.3, but abnormal amounts of components (or unusual ingredients) can upset the standardization of the instruments. For instance, carbohydrates other than lactose affect the observed value for lactose, though this is of minor importance for cheesemaking. Temperature affects the spectra from any sample, and it is usual to operate at a standard 40°C. Corrections for acid in the sample need to be applied.

Although there is a wide choice of instrumentation, it is essential that testing should be accurate whatever analytical methods are used. This applies especially with regard to legal requirements and the standardization of fat/protein ratios.

References

Andersen, T., Brems, N., Børglum, M.M., Kold-Christensen, S., Hansen, E., Jørgensen, J.H. and Nygaard, L. (1993) Modern Dairy Practice – 1: Chemical Analyses, in *Modern Dairy Technology*, Vol. 2, 2nd edn (ed. R.K. Robinson), Elsevier Applied Science, London, pp. 381–416.

Baer, R.S., Frank, J.F. and Loewenstein, M. (1983) *Journal of the Association of Official Analytical Chemists*, **66**, 858.

Biggs, D. A. (1979) *Journal of the Association of Official Analytical Chemists*, **62**(6), 1202, 1211.

Bluthgen, A., Heeschen, W. and Nijhiris, H. (1984) Special Publication No. 49, p. 206, Royal Society of Chemistry, London.
British Standards Institute (1980) BS 3095, Part L.
Egli, H.R. and Meyhack, U. (1984) Special Publication No. 49, p. 103, Royal Society of Chemistry, London.
Finland, (1980) Valtion Maitotalouden Tutkimuslaitos, Kone-ja Tarvikekoetuksia (Tests on the Multispec. M) No. 90.
Grappin, R. and Jeunet, R. (1976) *Revue Laitiere Francaise*, **347**, 663.
Grappin, R., Packard, V.S., Ginn, R.E. and Mellema, J. (1980) *Journal of Food Protection*, **43**(1), 52.
Hofer, H. and Szklenar, G. (1979) *Deutsche Molkerei Zeitung*, **100**(17), 634.
Hurrell, R. F., Guignard, G. and Finot, PA, (1981) *Experimentia*, **37**(6), 625.
IDF (1964) *International Standard for the Determination of Casein in Milk*, 29:1964, International Dairy Federation, Brussels.
IDF (1984) *International Standard for the Determination of Nitrate and Nitrite Contents*, 84A:1984. International Dairy Federation, Brussels.
IDF (1991) *International Standard for the Determination of Freezing Point*, 108B:1991. International Dairy Federation, Brussels.
IDF (1994) *Heat-treated Milk: Determination of Lactulose Content: High-performance liquid chromatography method*, International Dairy Federation, Brussels.
IDF (1995) *Milk: Determination of Lactulose Content: Enzymatic method*, International Dairy Federation, Brussels.
Lloyd, F.J. (1899) *Cheddar Cheesemaking*, Report of the Bath, West and Southern Counties Society to the Board of Agriculture, HMSO, London.
Lloyd, G.T. (1963) *Journal of the Society of Dairy Technology*, **16**, 232.
McGann, T.C.A., Mathiassen, A. and O'Connell, J.A. (1972) *Laboratory Practice*, **21**(9), 628–631, 650.
Pieper, H., Stuart, J.A. and Renwick, W.R. (1977) *Journal of the Association of Official Analytical Chemists*, **60**(6), 1392.
Robertson, N.A. and Cardwell, J. T. (1979) *Journal of Dairy Science*, **62**, Suppl. 1, 204.
Robertson, N.H., Dixon, A., Nowers, J.H. and Bruik, D.D.S. (1981) *South African Journal of Dairy Technology*, **13**(1), 3.
Sasano, M., Okada, M., Chonan, T. and Oura, Y. (1980) *Japanese Journal of Dairy and Food Technology*, **29**(3), 91.
Thomasow, J. and Paschke, M. (1981) *Milchwissenschaft*, **36**(2), 65.
Weaver, R.W.V. (1984) Special Publication No. 49, p. 91, Royal Society of Chemistry, London.

8 Additives used in cheese milks

Apart from starter bacteria, internal and external moulds, and the coagulants which will be considered separately, there are a number of additives used in the preparation of cheese milks and cheese curds.

The use of additives is controlled in some countries by law or by codes of practice; the following list includes those additives normally used in cheese processing.

(a) Salts to restore the calcium balance in milk.
(b) Salts which are inhibitory to undesirable organisms.
(c) Acids which are used alone or as supplements to lactic acid.
(d) Colours and bleaching agents.
(e) Added flavours in spices and herbs.
(f) Smokes.
(g) Addition of beverage flavours – beer, wine, liqueurs, etc.
(h) Addition of other foods, i.e. vegetables, ham, etc. into the cheese matrix.

8.1 Calcium balance

The calcium balance between the soluble, colloidal and complexed calcium is very delicate. Successful coagulation depends on this balance. Where there is a lack of balance or disturbance of the calcium due to heat treatment, cooling or disturbance in the milk itself, it has become common practice to add a calcium salt to the milk. This is especially necessary when some of the vegetable or microbial coagulants are being used.

A standardized solution of calcium chloride is the most common way of adding the salt to milk to allow easy distribution. Hostettler and Ruegger (1950) pointed to the effect of calcium chloride in reducing the rennet coagulation times. Accurate quantities must be used, because if too much calcium chloride is added the α_s casein–κ casein complex dissociates; the α_s casein no longer has protection from the κ casein and a precipitate forms. Slightly less calcium chloride will produce a harsh inflexible curd. Rarely is more than 0.02% of calcium chloride needed for satisfactory coagulation even when using highly heated milks. Dibasic calcium phosphate has been recommended for use with pepsin rennets; a 0.01% addition reduced the coagulation time from 30 min to 18 min, and 0.04% addition reduced it further to only 9 min. Some cheesemakers have used lime water (calcium hydroxide solution) in order to avoid the use of the stronger calcium chloride salt. Cheesemakers in the former Czechoslovakia used calcium lactate instead of calcium chloride for similar reasons. Retention of too much calcium chloride, apart from producing a hard unyielding curd, produces a cheese which is bitter in flavour and with a harsh body.

Phosphate in the form of sodium phosphate has been used to restore the salt balance (calcium/phosphate) where very disturbed conditions have prevailed. Milk from cows fed on potato silage was not coagulated by rennet until phosphate was added to the milk.

From 60 to 80% of the calcium in milk and from 50 to 60% of the phosphate is retained in hard cheese. In soft cheeses where lower pH conditions prevail, less calcium and phosphate are retained. When necessary, the addition of calcium salts is usually indicated by the recipe. However, if coagulants other than those specified are used, extra calcium may be necessary to form a firm coagulum in the specified time. It is essential to note that if the coagulation period is prolonged, the rate of acid production may be increased and firm curds may be crushed during cutting; this causes loss of curd dust and fat in the whey.

Dilution of milk with water, as used in the manufacture of some cheese, increases the disaggregation of the micellar casein. Since some of the casein is rendered more soluble by this practice, the yield of cheese may be reduced by the greater loss of the soluble components in the whey. The use of calcium salts as additives restores the micellar casein under these conditions and produces a curd which will drain more easily.

When milk has been cold-stored for long periods at 4–5°C, the amount of soluble casein may increase to 15% and, under these conditions, calcium additions increase the micellar size of the casein complexes. A sample of bovine milk could be expected to contain 0.123% of calcium, while the maximum clotting advantage occurs at 0.142%; therefore, the addition of 0.02% calcium chloride as permitted by many authorities should be a satisfactory aid to coagulation by rennet.

Foltmann (1966) indicated that the rate of the first phase of rennet reaction, in the absence of calcium, increases with temperature up to 42.5°C, but it is in the second, non-enzymic, phase that calcium plays an important role. The effect of calcium ions in milk is very complex. Thus Parker and Dalgleish (1981) have commented on the binding of calcium ions to casein, while Waugh (1971) has reported on the effect of calcium on the α_s–κ caseins. Waugh *et al.* (1971) noted that calcium binding is affected by sodium ions, i.e. an increase in sodium produces a decrease of calcium ions. Decreasing the pH also decreases calcium binding (Zittle *et al.*, 1958) but Fricker (1959) suggested that the decrease in rennet coagulation time by the addition of calcium was due to extra ions in the ultra-filtrate (serum). McGann and Pyne (1960) found that the size of micelles was related to the colloidal calcium phosphate complex.

The manufacture of cheese made from ultra-filtered milk has highlighted the question of cold storage of milk. Fricker (1958) found that storing milk at 4°C for 24 h increased the rennet coagulation time, and this was not restored by incubating at 50°C for 30 min. as happens with unconcentrated milk.

Schmutz and Puhan (1980) suggested that, in cold-stored milk, the casein became soluble, calcium and citrate migrated to the aqueous phase from the micelles, and the rennetting time increased by 30%. However, heating the milk to 60°C for 60 min. restored the casein but did not greatly affect the rennetting time. Ramet *et al.* (1981) in a review on cold-stored milks found that milk refrigerated at 4°C for 24 h can be restored to normal rennetting time by adding calcium chloride (10 g/100 kg of milk) after heating to 35°C for 30 min.

8.2 Inhibitory salts

It has been traditional practice in the production of some of the less acid types of cheese like Edam, Gouda and Svecia to use inhibitory salts in milk to prevent the growth of gas-producing organisms which cause 'blown' cheese (Figure 16.1). Sodium nitrate, an oxidizing agent, has been used to control the growth of the *Aerogenes* group of bacteria. Nitrate in combination with the salt in cheese helps to control the gas-forming butyric acid bacteria. The nitrate does not affect the growth of the lactic bacteria or of the gas-forming propionic organisms.

Sodium nitrate (saltpetre) is normally used at the rate of 20–30 g per 100 litres of milk. The reduction of the nitrate to nitrite may result in the production of colour defects in cheese, caused by the subsequent reaction with certain amino acids (i.e. tyrosine). This tended to limit the use of sodium nitrate in cheese milks, but the possible carcinogenicity due to the production of nitrosamines in the cheese has caused a further limitation on the use of nitrates. In spite of health risks, some recipes still carry a recommendation to use nitrate in cheese.

Goodhead *et al.* (1976) used 15 g $NaNO_3$ per 100 litres of milk for cheese which was stored at 13°C. After 4–5 weeks the nitrite level had fallen to 0.5 mg/kg of cheese. This treatment did not lead to nitrosamine formation. Birkkjaer *et al.* (1979) found that 10 g of KNO_3 per 100 litres of milk reduced the amount of butyric and isovaleric acids in Danbo cheese. This treatment produced better quality cheese and when ripe the level of nitrite was the same as in other cheese.

Nisin, an antibiotic produced by some strains of *Lactococcus lactis*, is used in processed cheese to inhibit the activity of gas-forming bacteria; however, in natural cheese the use of nisin is limited, due to the presence of bacteria which destroy or inhibit the activity of nisin. *Streptococcus thermophilus* produces nisinase, an enzyme which destroys nisin. Although nisin is active against species and strains from the genera *Staphylococcus*, *Bacillus*, *Clostridium* and *Corynebacterium*, certain bacteria found in milk, like *Enterococcus faecalis* and a strain of *Lactococcus lactis*, as well as *Propionibacteria* spp., also destroy the nisin activity.

Cheese ripening salts are sometimes added to the more acid types of curd like Quark. Sodium hydrogen carbonate and calcium carbonate mixtures may be added along with salt at the rate of 30 g/kg of curd. These salts are used to reduce acidity, i.e. increase the pH value from the low values ($\approx$ pH 4.6), brought about by the lactic acid bacteria. In some areas, monosodium dihydrogen phosphate has been used in place of the carbonate mixture.

8.3 Acidulants

The most usual acidulant in milk and cheese is lactic acid, which is produced *in situ* by the lactic bacteria growing in the milk and curd. The early practice in cheesemaking was to allow the milk to sour naturally (e.g. in Meira (Iraq), Kariesch (Egypt) or Queso Blanco cheese (Argentina)), and then to allow the whey to drain from the curds in cloth bags. Later (until the 1900s) whey was used as a source of bacteria in order to develop acid in the milk and curd. This produced off-flavours and defective curds due to contaminating organisms and was discontinued in favour of pure culture starters. However, the use of acids for chemically acidifying the milk is fairly widespread. Ricotta cheese is acidified by lactic or acetic acid. Lemon syrup and vinegar have also been used. Impastata cheese and Mozzarella cheese have also been made using vinegar (0.03%) before rennetting the milk. Queso Blanco cheese has been made using glacial acetic acid as acidulant (1.25 litres diluted to 10 litres and used for 450 litres of milk). Lime juice is used in India for Bandal cheese and vinegar from the water palm is used in the Philippines for making Kesong Puti curds.

D-Glucono-δ-lactone hydrolyses in aqueous solution to produce acid and it has been used for the production of acid in milk and curd. Unfortunately, since enzymes are not formed in this process the cheese is bland and lacks flavour, so the use of this acidulant has not found favour. Mabbit *et al.* (1955) reported on the use of glucono-δ-lactone in cheese. Metz (1980) took out a US Patent for the use of C2–C8 dione (glyoxal), using 0.05–0.50% together with 0.1–0.5% of H_2O_2 to oxidize the dione to acid. The coagulum formed can be used for Cottage cheese. No rennet is needed.

Occasionally, in extreme cases of poor coagulation of milk, phosphoric acid (food grade) has been added to the milk before rennet, colour or calcium chloride. The phosphoric acid (10% w/v) is diluted at the rate of

4 litres in 40 litres of water and added slowly to 1000 litres of milk while stirring vigorously. Excess acidity has been neutralized in some milks by the use of sodium bicarbonate or sodium carbonate, e.g. in the former Czechoslovakia for Olomouc cheese. Excess neutralizer gives rise to astringent, bitter flavours.

8.4 Cheese colours

The colour of milk and of cheese is an important factor in the consumer appeal of the product. The colour of milk is carried forward into the cheese with enhanced strength due to concentration in the curd. It is also common practice to add extra colour to pale-coloured milks.

Two colours are of importance in milk: that due to riboflavin, which is contained mainly in the serum portion (whey), and that due to carotene, a carotenoid pigment in milk fat.

Riboflavin is yellow in solution with a green–yellow fluorescence and tends to give curds a greenish tinge. Most of this colour is lost in the whey. Therefore, as a colour in cheese its effect is small. However, the contribution of the yellow–orange colour due to carotenoids, being retained in the curd, is more significant. There is a large difference between the colour of winter and summer milks since the latter have much higher levels of carotenoid carried over into the milk from the fresh grass. The cheesemaker, therefore, has tried to maintain an all-the-year-round colour of cheese and has used annatto or β-carotene colours in milk in the winter period. Saffron is also used as a yellow colouring for cheese like Box (Germany), Cacio-Fiore or Caciotta (Italy) and Luneberg (Austria).

Although the depth of colour is specific to each variety of cheese, it has no effect on the cheese flavour and the recipe usually gives the type of colour and quantity. There are some cheeses which are normally coloured a deep yellow–red colour (e.g. Leicester and whey cheese), while some are a lighter red–yellow colour (e.g. coloured Cheshire, Double Gloucester) dictated by some market requirements. Cheddar and Edam cheeses are also sometimes coloured for certain markets.

Annatto is a colour extracted by sodium hydroxide from the seeds of a South American shrub, *Bixa orellana*. The pigment in annatto is the acid bixin which in the alkaline extract becomes norbixin. The colour is composed of tints of yellow units and red units, and in cheese becomes a protein dye attached to the casein. Annatto is very susceptible to oxidation by hydrogen peroxide, released by some bacterial reactions, and by the air. Sulphydryl compounds found in ripening cheese also act as oxidative catalysts. Copper and iron accelerate the oxidation of the dye by hydrogen peroxide. Thus, bleaching of the red colour in patches in cheese is frequently found in poor quality moist or contaminated curds.

Annatto extracted by an oil, i.e. rape seed or other vegetable oil, is used as a fat dye mainly for butter, but has also been used for cheese of high fat content.

In order to make the blue–green veins stand out more prominently in blue-veined cheese, bleaching of the yellow–orange colour of carotene has been the practice over many years. The bleaching agent used is benzoyl peroxide. In the manufacture of some blue-veined cheese, the milk is separated, the cream homogenized and bleached and then added back to the skim milk for further treatment.

Proprietary agents are available from specialist dairy firms for the bleaching of cream or milk. Some of these agents are masking substances of complementary blue or green colours to mask the natural colour and so whiten the curd. Masking of the deeper yellow colour in some cheese is obtained by adding a green dye, chlorophyll, at the rate of 3–10 g/100 kg of cheese milk.

While bleaching treatments are applied to the liquid milk or cream used for some varieties of blue-veined cheese, in some cheeses the curd itself is coloured green. Chlorophyll extracted crudely from the ground-up fresh growing leaves of young wheat plants, kale leaves or curly green leaves has been used for colouring the curds in sage cheese, e.g. Sage Derby where chlorophylls and chlorophyllins (E 140) and their copper complexes (E 141) are permitted (UK Regulations, 1995). The sage-flavoured leaves used alone turn the curd an unappetizing brown colour. Leaves from some of the *Brassica* genus (cabbage) impart a flavour and must not be used.

Carotene has been tried for colouring Cheddar cheese, but was successful only at low levels. The carotene colour was too yellow for some of the more highly coloured cheese (Chapman *et al.*, 1980). Synthetic green colourings have been tried but the colours were unnatural and harsh and in some instances imparted a foreign, solvent-like flavour and aroma.

Both colour and flavour are added to Sapsago cheese from Switzerland or Germany. The powdered leaves of an aromatic clover are added to the ground-up ripened curd; this gives a mottled pale green colour and at the same time imparts a characteristic odour and flavour.

8.5 Legal usage of additives

Ingredients which may be used in cheesemaking are controlled by the regulations in force at any one time in the various countries. However, the FAO/WHO Committee has made a series of recommendations (which are revised from time to time) as follows:

Ingredients which may be used in any **accustomed quantity**:

Sodium chloride
Hydrogen peroxide and catalase
Lactic acid
Chlorophyll (including copper chlorophyll)
Citric acid
Riboflavin
Phosphoric acid
Pimaricin (antimycotic)
Propionic acid

The operative words are accustomed quantity and not any quantity. In the UK the term *quantum satis* is normally used.

Other ingredients have limits of usage. The maximum standards for some ingredients are as follows (see also Table 8.1):

Annatto and beta carotene alone or in combination, 600 mg/kg. However see Table 8.1 for levels of colours permitted in cheeses in the UK.
Calcium chloride, 0.02% w/w.
Sorbic acid or its sodium or calcium salts, 0.1% w/w calculated as sorbic acid (antimycotic uses).
Enzymes of animal or vegetable origin for flavour development, not more than 0.1% w/w.
Sodium or calcium nitrate, 0.02% w/w.
Sodium hydrogen carbonate or calcium carbonate, 3.0% w/w (used in Quark or sour milk or lactic curds).
Phosphate, maximum usage 0.002% w/w.
Nisin, not more than 0.25%w/w.
Whey proteins when added to cheese, not more than 20% in the fat-free dry matter of the cheese.

Table 8.1 Permitted colours in cheese (UK Regulations, 1995)

Cheese	Permitted colour		Maximum level
Sage Derby	E 140	Chlorophylls Chlorophyllins	*quantum satis*
	E 141	Copper complexes of chlorophylls and chlorophyllins	*quantum satis*
Ripened orange, yellow and broken-white cheese; unflavoured processed cheese.	E 160a	Carotenes	*quantum satis*
	E 160c	Paprica extract, Capsanthin, Capsorubin	*quantum satis*
	E 160b	Annatto, Bixin, Norbixin	15 mg/kg
Red Leicester cheese	E 160b	Annatto, Bixin, Norbixin	50 mg/kg
Mimolette	E 160b	Annatto, Bixin, Norbixin	35 mg/kg
Morbier cheese	E 153	Vegetable carbon	*quantum satis*
Red marbled cheese	E 120	Cochineal, Carminic acid, Carmines	125 mg/kg
	E 163	Anthocyanins	*quantum satis*

Note: These regulations implement European Parliament and Council Directive 94/36/EC and similar regulations will apply throughout the European Community.

Benzoic acid and the sorbic acid salts (sodium, potassium or calcium) are limited to coagulant preparations, i.e. 'rennets'.

Generally, the legislation in each country has due regard to the common or traditional usage of ingredients for cheese processing. The legislation on food additives, including colours, varies from country to country, and reference to the legislation of each country is necessary to ascertain the permitted usage of ingredients. Thus, some substances not accepted by the Food and Drug Administration in the USA are acceptable in other countries and vice versa.

The joint FAO/WHO Food Standards Committee, and the subsidiary body known as the Codex Alimentarius Commission, will in future hope to harmonize the legislation of all countries which belong to the Commission.

The aim of the Codex Alimentarius Commission is the production of food standards to protect consumer health and to ensure fair practices in the food trade. The provisions of the Codex apply to hygiene, additives, pesticides, and other agricultural residues (i.e. antibiotics), contaminants, labelling of foods, presentation packaging and standardization of methods of analysis. Reports on the work of the Commission appear regularly. Standards for cheese (A6) and whey cheese (A7) are now widely accepted standards.

References

Birkkjaer, H.E., Braun, H., Thomsen, D. *et al.* (1979) Beretning Statens Forsoegsmejeri, Report No. 236, Danish Research Station, Hilleroed, Denmark.

Chapman, H.R., Thompson, S.Y. and Slade, H.M. (1980) *Journal of the Society of Dairy Technology*, **33**(4), 162.

Foltmann, B. (1966) C. R. Trans. Lab., Carlsberg, **35**, 143.

Fricker, A. (1958) *Deutsche Molkerei Zeitung*, **79**(48), 1553.

Fricker, A. (1959) *XV International Dairy Congress*, Vol. 3, 1276.

Goodhead, K., Gough, T.A., Webb, K.S. *et al.* (1976) *Netherlands Milk and Dairy Journal*, **30**(3/4), 207.

Hostettler, H. and Ruegger, H.R. (1950) *Annuaire Agricole de la Suisse*, **51**, 669.

McGann, T.C.A. and Pyne, G.T. (1960) *Journal of Dairy Research*, **27**, 403.

Mabbit, L.A., Chapman, H.R. and Berridge, N.J. (1955) *Journal of Dairy Research*, **22**, 365.

Metz, F.L. (1980) US Patent 4 199 609.

Parker, T.G. and Dalgleish, D.G. (1981) *Journal of Dairy Research*, **48**(1), 71.

Ramet, J.P., El-Mayda, E. and Weber, F. (1981) *Milchwissenschaft*, **36**(3), 143.

Schmutz, M. and Puhan, Z. (1980) *Schweizerische Milchwirtschaftliche Forschung*, **9**(3), 39.

UK Regulations (1995) *The Colours in Food Regulations 1995*, SI 1995, No. 3124, HMSO, London.

Waugh, D.F. (1971) Formation and structure of casein micelles. In *Milk Proteins*, Vol. 2 (ed. H.A. McKenzie), Academic Press, London, pp. 3–86.

Waugh, D.F., Slattery, C.W. and Creamer, L.K. (1971) *Biochemistry*, **10**, 817.

Zittle, C.A., Della Monica, E.S., Rudd, R.K. and Custer, J.H. (1958), *Archives of Biochemistry and Biophysics*, **76**, 342.

9 Starter cultures

A starter culture in cheesemaking is a particular species of bacterium, or group of species from one or more genera, which, by growing and metabolizing in the milk and curd, assists in the production of a mature cheese. Up until 1880, most cheesemakers relied on the natural souring of milk for acid development, although some makers in Italy, for example, used sour whey. Today, starter cultures are obtained from a wide variety of sources. Sometimes individual species are purchased from culture banks held by research establishments or central dairy organizations, but more frequently, blends of species and/or strains are made available by commercial suppliers. These latter types of culture have now been developed for all the major types of cheese, and their reliable performance has ensured a dominant position within the cheese industry

9.1 Starter bacteria

Primarily, starter bacteria are required to convert the lactose into lactic acid, for this acid reduces the pH of the system and provides the conditions necessary for the multitude of reactions which occur in cheese processing. Secondly, and almost of equal importance, the enzymes from starter bacteria, living or dead, degrade some of the milk components and release precursors of compounds responsible for the flavour and aroma of the product (Law and Sharpe, 1977).

The three most important activities of starter bacteria are:

1. Glycolysis – conversion of lactose into lactic acid; acid production by purely chemical means (via glucono-δ-lactone, for example) cannot at present be satisfactorily used to replace starter activity, except for a bland product like Cottage cheese.
2. Proteolysis – degradation of protein chains into simpler substances, such as peptones, peptides and amino acids.
3. Lipolysis – hydrolysis of fatty acids from milk fat into keto acids, ketones and various esters, some of which are responsible for flavour and aroma.

Some micro-organisms used in starter cultures are shown in Table 9.1, and the starter – as added to the cheese vat – can be composed of a single strain of one species, or the culture might contain two, three or more separate species, including a number of strains of each species. The choice

Table 9.1 Organisms used in cheese as starters or for subsequent inoculation

Lactococcus lactis sub-sp. *lactis*	Many strains available, but nisin-producers should be avoided
Lactococcus lactis var. *hollandicus*	Edam, Gouda types of cheese
Lactococcus lactis sub-sp. *cremoris*	Flavour production
Lactococcus lactis biovar. *diacetylactis*[1]	Flavour production
Streptococcus thermophilus	Withstands higher temperatures
Enterococcus durans	Withstands higher temperatures
Enterococcus faecalis[2]	Sometimes used for its flavour production and growth at elevated temperatures
Leuconostoc mesenteroides sub-sp. *cremoris*	Flavour production
Leuconostoc mesenteroides sub-sp. *dextranicum*	Flavour production
Lactobacillus casei	Used in high-scald cheeses
Lactobacillus lactis	Used in high-scald cheeses
Lactobacillus delbrueckii sub-sp. *bulgaricus*	Used in high-scald cheeses
Lactobacillus helveticus	Used in high-scald cheeses
Propionibacterium freudenreichii sub-sp. *shermanii*	Used for gas and flavour production in some cheeses
Penicillium roqueforti	Blue mould growth (internal)
Penicillium glaucum	Blue mould growth (internal)
Geotrichum candidum (P. candidum)	White mould (surface growth)
Penicillium camemberti	White mould (surface growth)

[1] No longer recognized as distinct from *L. lactis* sub-sp. *lactis*.
[2] Some strains of this species, and the related *Enterococcus faecium*, are reported to be resistant to penicillin, and hence should not be used in human food in order to avoid the risk of transferring the 'resistance' to pathogenic organisms.

of culture depends on both the cheese variety for which it is to be used and, to some extent, local tradition. Thus for making Cheddar cheese, New Zealand and Australian cheesemakers have, in the past, preferred to use a precise blend of single strain cultures, while in Northern latitudes (USA and Europe), cheesemakers have preferred cultures containing a mixture of bacterial types. Tofte-Jespersen (1979), Cogan (1980), Speck (1981) and Tamime (1990) have reviewed the nature of starter cultures in some detail.

As mentioned earlier, geographical location plays a role in the choice of the starter organism, in that mesophilic bacteria tend to be associated with cheeses from Northern Europe, while thermophilic cultures are more widely used around the Mediterranean. However, this historical usage has been supported in recent times by more scientific considerations. Thus, mesophilic bacteria like *Lactococcus lactis* sub-sp. *lactis*, *Lactococcus lactis* sub-sp. *cremoris*, *Leuconostoc mesenteroides* sub-sp. *cremoris* and *Lactococcus lactis* sub-sp. *lactis* biovar. *diacetylactis* are often chosen for low-scald cheeses like Cheddar or Cheshire, but strain/species differences are important. For example, *Lactococcus lactis* sub-sp. *lactis* produces lactic acid quickly and so reduces manufacturing times, but some strains produce the bacteriocin nisin, which could affect the performance of other Gram-positive species present; other strains can give rise to bitter flavours in cheese. By contrast,

Lactococcus lactis sub-sp. *cremoris* grows more slowly in milk, but produces a well-flavoured cheese. Therefore, many starter cultures consist of suitably paired strains of *Lactococcus lactis* sub-sp. *lactis* and *Lactococcus lactis* sub-sp. *cremoris*. It can be important also that, according to Gordon and Shapton (1977), mixed mesophilic starters produce more acid than starters based on one individual species.

When a pronounced flavour is required in soft cheese varieties, *Lactococcus lactis* sub-sp. *lactis* biovar. *diacetylactis* is included in the starter culture. Both *Lactococcus lactis* sub-sp. *cremoris* and *Lactococcus lactis* sub-sp. *lactis* biovar. *diacetylactis* utilize citrate to produce diacetyl (a butter-like flavour), and CO_2 is also produced. *Leuconostoc mesenteroides* sub-sp. *cremoris* also produces diacetyl, but in addition it generates significant quantities of CO_2. Indeed so vigorous is the gas production that the heterofermentative species are sometimes used to produce a mass of small 'eyeholes' in a variety, such as Appenzeller, where the large 'eyes' associated with *Propionibacterium* spp. are not required. The co-production of CO_2 and diacetyl is useful also in the manufacture of more open-textured curds of the type associated with internally mould-ripened cheeses. However, leuconostocs are unlikely to find favour with the makers of Cottage cheese, since their presence may result in the production of 'gassy' floating curds, even though the inclusion of a small proportion of *Leuconostoc mesenteroides* sub-sp. *cremoris* in the starter culture will convert any acetaldehyde produced by *Lactococcus lactis* sub-sp. *lactis* (biovar. *diacetylactis*) into ethanol, and so remove the risk of adverse flavour development.

The advantage of a strongly growing lactic acid culture is that non-starter bacteria are largely inhibited (e.g. coliforms and *Staphylococcus aureus*), while some cultures of lactobacilli, for example *Lactobacillus delbrueckii* sub-sp. *bulgaricus*, accumulate H_2O_2 which inhibits the growth of pseudomonads and also *Proteus* spp. (Gilliland and Speck, 1975). These beneficial activities may be aided, in some cases, by a degree of synergism between components of the starter culture. The interaction between *Lactobacillus delbrueckii* sub-sp. *bulgaricus*, which displays marked proteinase activity, and *Streptococcus thermophilus*, which has only dipeptidase activity, is one of the best-known examples, for acid production is dramatically increased by their joint action. The thermophilic starters used for high-scald cheeses also produce metabolites which stimulate propionic acid bacteria essential to produce the characteristic 'eyeholes' in cheeses like Emmental (Langsrud and Reinbold, 1973). However, as *Propionibacterium freudenreichii* sub-sp. *shermanii* grows only slowly in milk, it has to be grown separately from the lactobacilli when used as a starter for Emmental and Gruyère-type cheeses.

The physical forms of culture available from commercial dairy laboratories are shown in Table 9.2, with types 6, 7 and 8 dominating the market. Full directions for use and for the rotation of cultures with respect to phage

Table 9.2 Types of starter cultures

1. Liquid cultures in skim or whole milk litmus milk or chalk milk. (The milk should be antibiotic-free).
2. Dried (lyophilized) cultures – normally liquid cultures vacuum-dried to 5% moisture (Tofte-Jespersen, 1974).
3. Liquid culture – frozen.
4. Spray-dried cultures (Jansen *et al.*, 1969).
5. Frozen cultures in special media frozen at −40°C (Stadhouders *et al.*, 1969).
6. Concentrated cultures frozen in sealed containers at −196°C (Tofte-Jespersen, 1977; Wigley, 1977; Johannsen and Hunger, 1979; Wigley, 1980).
7. Concentrated freeze-dried mixed cultures (Tofte-Jespersen, 1979).
8. Single strain freeze-dried cultures in foil sachets containing 5×10^{11} cfu/g in known units of activity. One unit of starter activity equals that required to produce 150 mmol of lactic acid in skim milk at 30°C in 4 h.

Types 1, 2 and 3 are used for inoculation of intermediate cultures, types 2, 4, 5, 6 and 7 for inoculation into bulk cultures, and types 4, 5, 6, 7 and 8 for direct-to-vat inoculation.

relationships, as well as details of special media to be used for bulk starter production, are available from the suppliers.

9.2 Coded starter cultures

The generally accepted system of classification, as used for instance in *Bergey's Manual of Determinative Bacteriology*, (Holt, 1984), does not really assist the practical cheesemaker to select starter bacteria that will be suitable for a given variety of cheese. In addition, the all too frequent changes in nomenclature can cause confusion when a literature search is selected to cover several years. For example, *Streptococcus citrovorus* was renamed *Leuconostoc citrovorum*, then became *Leuconostoc cremoris*, and is now cited as *Leuconostoc mesenteroides* sub-sp. *cremoris*, whilst the separation of biovar. *diacetylactis* from *Lactococcus lactis* sub-sp. *lactis* is no longer valid. Obviously the results of phylogenetic analysis of type strains of genera like *Leuconostoc* and *Lactococcus* support recent changes in the taxonomy of the constituent species, but problems arise for the cheesemaker because commercial sources of cultures still retain a system of nomenclature that is most helpful to the industry. Thus, in practice, some strains of *Lactococcus lactis* sub-sp. *lactis* will generate diacetyl from citrate and others will not, so that for the convenience of the customer, a culture catalogue may continue to distinguish them as *Lactococcus lactis* (citrate −) and *Lactococcus diacetylactis* (citrate +). In this way, the cheesemaker is aware that he ordering an acid-producer or a flavour-producer, and it will be many years before the industry drops the traditional names.

In order to avoid confusion, most strains of starter bacteria are now stored in one of the major Culture Collections, such as the National Collection of Industrial and Marine Bacteria held in Aberdeen. On deposition, purification and testing, each strain is given a code number that it retains, so that anyone purchasing, for example, strain No. 8662 of *Lactococcus lactis* sub-sp. *cremoris* can always be sure of receiving a culture with well-defined characteristics.

However, although mixed single-strain cheese starters are used, the majority of cheesemakers prefer to work with a culture supplier who can supply a multi-strain blend suitable for a specific cheese, or a range of blends with identified characteristics with respect to acid production or proteolytic activity. Such commercial cultures are often given special codes to aid selection by the potential user, and an early example of such a system was developed by Lawrence *et al.* (1971) for the description of single strains of bacteria that behaved in milk to give Cheddar cheese with fairly precise characteristics. They formed the strains into four groups according to the flavours which developed in the cheese (satisfactory, good, fruity or malty), as well as rate of acid production, i.e. S=slow acid, F=fast acid:

Group 1	Group 2	Group 3	Group 4
Satisfactory flavour	Good flavour	Fruity flavour	Malty flavour
TR(F)	H2(F)	ML8(F)	MLI(F)
WS3	P2(F)	Hl(F)	C13(F)
E8(S)	HP	SK3	
AM1(S)	Z8	WH 1	
AM2(S)	KH	BAI	

The advice might be, therefore, that a producer should select two single-strain cultures which would complement each other, e.g. AMI(S) and H2(F), are slow- and fast-acid producers, respectively. AM2(S) and P2(F) are a similar pair, and the use of fast and slow strains paired together reduces the possibility of producing bitter-flavoured cheeses. However, if two slow-acid strains like AM2(S) and E8(S) are used together, then the cheese milk will ripen slowly, and the interval may allow coliform bacteria to proliferate. Pairs from Groups 3 and 4 are likely to produce off-flavoured cheeses.

The selection of strains of bacteria also depends on their ability to withstand scald temperatures. Sullivan *et al.* (1974) studied the effect of heat on these strains, and while strains E8, KH, HP and Z8 were inhibited at 42°C, strains AM2, WS3 and AM1 were affected at 33°C and 35°C. Thus, some combinations were suitable only for low-scald cheeses (35°C). It was suggested also that pairs of strains, provided that they are phage-unrelated, should be used in rotation on a 4-day cycle (Lawrence and Pearce, 1972). This work also provided the basis for the development of commercial phage-insensitive, multi-strain starter cultures (Daniell and Sandine, 1981).

Mills and Thomas (1980) added a new aspect to the identification of cheese starters in rating some strains as proteolytic (identified as prt +) and others as non-proteolytic (prt –). They suggested that a method of reducing bitterness in cheeses was to use non-proteolytic strains of lactococci, even though some of their results suggested that manufacturing times might be extended.

This approach to the classification of starter cultures based upon rates of acid production, resistance to scald temperatures, phage type and proteolytic ability is now widespread, and most suppliers indicate salt tolerance as well. Consequently, most commercial cultures are blends of strains of one or more species, with the selection being geared to give a cheese with specific characteristics (Tamime, 1990). So long as the manufacturer can standardize the conditions within the factory, including the chemical composition of the raw materials, then the number of sub-standard batches of product should be reduced to a minimum.

One of the earlier attempts at commercial starter production was reported by Christensen (1969a,b), who incubated selected strains of starter organisms in a nutrient medium prior to concentrating the cells by centrifugation. The concentrate was frozen in containers kept in liquid nitrogen at −196°C. As a result of this concentration, the propagation of 1350 litres of bulk starter required only a few grams of the frozen concentrate. Ziemba (1970) also described a process for the production of deep-frozen cultures for storage in liquid nitrogen at −196°C. Alternative systems for producing concentrated cultures were also examined, and Osborne (1977) described a fermentation process in which high cell counts could be achieved by extracting the waste metabolites of growth (mainly lactates) using a diffusion technique. The medium was composed of enzyme-digested skim milk and yeast extract, and the cells were harvested periodically and formed into frozen granules by dropping them into liquid nitrogen. Osborne and Brown (1980) noted that lactate was the factor that limited cell growth.

The significance of these developments relates to the number of viable cells, and it has been reported that a normal culture dried under vacuum may only have from 2 to 8% of the original cells; such a culture would require sub-culturing six or eight times to bring it to normal activity. Similar liquid cultures (4×10^5 cfu/ml) concentrated by centrifugation could contain from 2×10^9 to 4×10^{11} cfu/ml viable cells, and these concentrated cultures can be used for the direct inoculation of bulk starter vessels or the cheese milk itself.

Stanley (1977) reported on the success of concentrated cultures, although Satji and Kroger (1979) noted that the activity of some commercial starter cultures varied. For example, a bulk starter produced from a Marstar culture might be more proteolytic and lipolytic than a starter produced from a Superstart culture, while the latter might produce more acid – even though both cultures were from the same supplier. However, such variations are not unexpected for, as mentioned earlier, cultures are blended to give a specific performance, and so long as that performance is standard, then differences between blends extend the choice available to the cheesemaker.

Potentially more significant were the findings of Chapman (1978) and Robinson (1981) that, because these starters have a different rate of activity from a liquid bulk starter, the use of direct-to-vat freeze-dried cultures for the manufacture of Cheddar cheese required an adjustment to the normally accepted process. Equally important, however, was the observation that the quality of the mature cheese was excellent. Early studies with frozen cultures were reported by Lloyd and Pont (1973), Hurst (1977) and Obermann *et al.* (1978), while Bouillane *et al.* (1979) found, perhaps unexpectedly, that freeze-dried cultures had four times the activity of frozen cultures. However, modern concentrated cultures designed for direct-to-vat usage are just as active as a typical liquid bulk starter, and the incorporation of growth stimulants and cryoprotective agents into the freezing media has long overcome the problem of extended lag phases (Tamime, 1990, 1993; Sandine, 1995).

9.3 Starter propagation

Starter cultures in general, and lactococci in particular, have very exacting nutritional requirements. Traditionally, milk from a supply similar to the normal cheese milk was employed for culturing bacteria for the production of a bulk starter, although reconstituted skim-milk powder or, in Italy, whey have been widely used. Attempts at improving the growth of bacteria in a bulk starter medium have usually involved a modification of the nutrient levels in the base milk, and it was found early on that peptone, for example, added to milk assisted the early growth of lactococci. Similarly, Law (1979) found that some streptococci used extracellular peptidases for the uptake of small peptides, and the requirement for organic nitrogen has been confirmed in numerous studies (Figure 9.1). Shehata *et al.* (1980) used cell-free filtrates from *Bacillus circulans* to stimulate starter growth, while Bergere and Hermier (1968) suggested a nitrogen-rich medium (Yeast extract–Tryptone–Lactose) maintained at pH 6.5. In the same vein, Hup and Stadhouders (1977) obtained a US patent for a starter medium based upon diluted whey with added soya or casein, and buffered to maintain pH.

Equally important was the observation that phages needed calcium to proliferate, and Hargrove (1959) advocated a medium of skim milk and 2% sodium phosphate; the phosphate sequesters any ionic calcium and renders it unavailable for phages. Richardson *et al.* (1977) advocated a whey-based medium with 1.0% mono-ammonium phosphate, 1.0% disodium phosphate and 0.32% yeast extract; the pH was maintained at 6.0 by injection of anhydrous ammonia. A number of commercial companies took this approach further with the marketing of so-called phage inhibitory media (PIM). The intention was that cheesemakers would reconstitute the dry

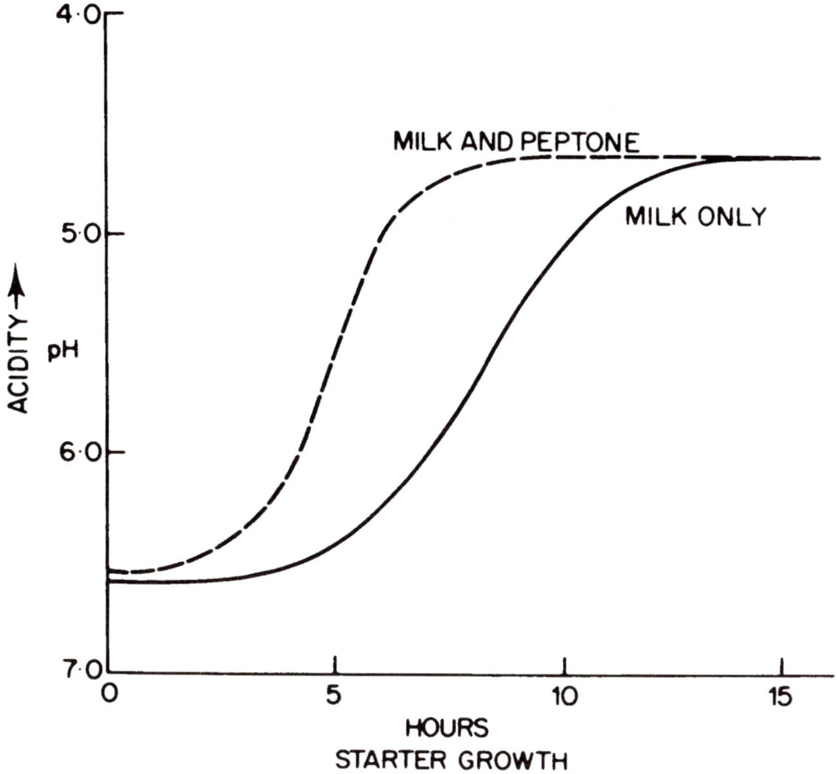

Figure 9.1 Graph showing the development of acidity in milk by a starter, with and without the addition of perptone.

medium and use the mix just for starter production, so reducing the risk of phage proliferating in the bulk tanks; additional growth factors are often added to such media to encourage active growth the lactic acid bacteria. Heap and Lawrence (1981) suggested that disodium glycerophosphate should be added to skim milk for the preparation of a base for starter cultures and, although there has been a tendency towards the use of more complicated and expensive commercial media, liquid milk or reconstituted skim milk remains the normal choice of many cheesemakers for the production of bulk starter cultures.

Since a starter culture needs to be easily dispersible in the cheese milk, the heat treatment of the bulk starter milk is important not only to destroy unwanted bacteria and phage, but also for the production of a smooth, dispersible culture. Thus, some ranges of heat treatment render milk-based media inhibitory to starter bacteria, whilst low temperature treatments tend to produce granular starters; agglutination of cultures can vary according to the heat treatment (Elliott and Emmons, 1979).

Heat treatments in the ranges of 65–72°C for 30 min or 90–100°C for 1–2 h are alleged to cause milk to be stimulatory to some bacteria, but ranges of 72–82°C for 40–90 min or over 120°C for more than 30 min can cause inhibition of some bacteria (Greene and Jezeski, 1957). In practice, temperatures of 105°C and above for 15 min are used for in-bottle or in-flask sterilization of starter milks, while the Lewis Method, using polythene bottles with Astel seals for culture preparation, is based on a temperature of 72–74°C for 45 min (Lewis, 1956; Cogan and Accolas, 1990). Overheating the milk media can lead to a long lag phase of growth, due, perhaps, to the expulsion of CO_2 (sterilized milk contains 0.3% CO_2 compared to the 2–3% needed by some species for optimum growth (Reiter and Oram, 1961)) or more likely to the build-up of Maillard products. It is relevant also that Wilkowski *et al.* (1954) found that the heat resistance of phage active against lactococci varied from 12 min at 75°C to 10 min at 62.5°C, while Tsaneva (1975) found that the heat resistance of phages of *Lactococcus lactis* sub-sp. *lactis* fell into three ranges, namely 72–74°C, 68–70°C and 62–64°C. Consequently, the heat treatment of cheese milks at temperatures lower than those used for pasteurization of liquid milk is not likely to destroy all the bacteriophage particles present. Under these conditions, it is vital that new starters do not introduce more phage (Cogan, 1995).

Inoculation of the starter cultures follows normal aseptic bacteriological techniques, and the incubation temperatures are adjusted according to the type of organisms: for instance, 22°C to 27°C for mesophilic cultures, and 30°C to 42°C for thermophilic cultures.

The build-up of starter volume from stock cultures to bulk starter for vat inoculation has been described in detail by Tamime (1990), and the practical schedule might proceed as follows:

Day 1 – Stock culture used to inoculate Mother culture; 1% inoculation rate and an increase in volume of 1 to 100.
Day 2 – Mother culture used to inoculate Intermediate culture; 1% inoculation rate and an increase in volume of 1 to 100.
Day 3 – Intermediate culture used to inoculate Bulk culture; 2% inoculation rate and an increase in volume of 1 to 100, with the final volume being calculated to reflect the anticipated volume of cheese milk.
Day 4 – Inoculation of cheese milk, 0.5% to 2.5% inoculation rate depending upon the time of year, variety of cheese and other factors.

The precise system can be subject to modification that suits the individual factory, but it has been suggested that batches of stock cultures should be produced weekly, and Mother cultures daily in groups of three; one as a reserve, one for activity and purity testing and one for inoculation of the Intermediate culture.

Each stage of starter growth will reach the end of the logarithmic phase at a time that depends on the incubation temperature. At this point, the viable count may be 2×10^5 to 2×10^7 cfu/ml and the culture, if not needed immediately, may be cooled to > 7°C and held for up to 18 h prior to use. This system of culture production has a number of advantages, namely:

- the bulk culture should be extremely active – so long as incubation is halted at the correct phase of growth;
- the bulk culture can be checked prior to inoculation of the cheese milk;
- the cheesemaker can select the strains of bacteria most appropriate for the variety and the local conditions.

The reverse side centres on the demands for laboratory and culture facilities, and hence it is not surprising that, where access to concentrated cultures for direct-to-vat inoculation is feasible, many cheesemakers are avoiding the tedium of producing cultures on-site. Nevertheless, in many parts of the world, the availability of commercial cultures is limited, and the build-up of bulk starters in the factory is the only option available. Careful control of starter production is, therefore, essential.

9.4 Starter control

The control of starters by the cheesemaker is usually limited to three tests: (a) activity test, (b) phage test and (c) test for tolerance to scald temperatures. Acid production by a starter is usually measured by the increase in titratable acidity over a 4-h period at a temperature of 22°C or 30°C, or 37°C for high-scald cheeses. The precise level of acidity produced will depend on the composition of the starter, but average figures might be 0.35, 0.45 and 0.60%, respectively. Regular records of these acidities provide a rapid indication of abnormal activities, as does the acidity of the whey leaving the press. Thus, such records may highlight contamination – the acidity of the whey may rise at first to an unusually high level, and then decline rapidly as contaminants become dominant.

Similarly, if a bulk starter is allowed to develop excessive acidity, the balance of species/strains within the culture may alter and reduce the capacity of the culture to produce acid. Starters can vary in activity in vat milks from different geological locations (Keogh, 1974), so that a culture 'borrowed' from a neighbouring factory could give disappointing results. Furthermore, the previous activities of non-starter bacteria can influence starter behaviour in the cheese milk, and the release of free amino acids in the milk by *Pseudomonas fluorescens* or *Bacillus cereus*, for example, could influence the growth of lactococci inoculated subsequently.

The simplest test for bacteriophage in whey consists of inoculating sterile milk with 1% of starter culture plus the addition of 0.1% of whey. Bromocresol purple can be used as an indicator of starter activity in the milk and, therefore, of the presence of active phage in the whey. Laboratory tests for phage consist of growing the starter on an appropriate agar medium, and then dropping some whey or milk into a 'well' cut into the colonized agar; the production of a clear zone around the 'well' indicates phage activity against the bacteria in the culture.

A quick test for bacteriological purity of a starter culture is the catalase test, in which 1 ml of hydrogen peroxide (3%) is added to about 5 ml of culture. As the lactic acid bacteria are catalase-negative, the release of oxygen bubbles indicates gross contamination. More specific tests of culture purity, tolerance to scald temperatures, or balance between strains in mixed or multi-strain starters are normally the responsibility of central laboratories.

The volume of the vessels used for bulk starter propagation depends on the quantity of starter required each day, and jacketed, stainless steel tanks are the vessels of choice. Thus, the circulation of water through the jacket allows for the starter milk to be heated to 90–95°C for 30 min to eliminate phage and unwanted bacteria, and then for the milk to be cooled to inoculation temperature; the water in the jacket can be employed also to maintain the incubation temperature. Equally important is the fact that the 'sterile' milk needs to be protected from an ingress of phage particles during the cooling and inoculation procedures. Protection of the milk during the cooling stage may involve applying a slight positive air pressure to the starter room using filtered air, and/or ensuring that air entering the culture vessel is filtered through a High Efficiency Particulate Filter to remove air-borne phage. The avoidance of contamination during inoculation may be achieved by the use of a ring of gas burners around the inoculation ports or having the ports submerged in a solution of sodium hypochlorite (100 ppm active chlorine), and Tamime (1990) has given a detailed account of the vessels used for starter propagation.

The location of the starter vessels in positive-pressure rooms can offer additional security, and some authorities advocate 'cleaning' the external atmosphere of a starter room with hypochlorite sprays (1 g available chlorine to 28 m^3 (1000 ft^3) (Kosikowski and Mocquot, 1958). However, abuse of this approach could have an adverse effect on workers and corrode metal surfaces.

9.5 Non-lactic starters

The spores of selected moulds are inoculated into milk, or rubbed onto curds along with salt, to effect proteolysis and lipolysis in mould-ripened cheeses. The facilities required to produce the spores are rather different from those of a normal starter laboratory, and hence most cheesemakers will not attempt to produce their own supply of spores; the low volumes needed are best purchased as freeze-dried cultures direct from a commercial supplier. Occasionally local custom dictates otherwise, and the spores of *Penicillium roqueforti*, for example, are still grown-up on bread prior to their use in the manufacture of Roquefort.

The following list is given to indicate the use of some important moulds:

Geotrichum candidum – white mould that grows on the surfaces of soft cheeses like Brie and Camembert, although it is never the dominant mould. The varied biochemical activities enable the species to contribute to the flavour of many smear-coated cheeses as well.
Penicillium camemberti – another white mould which grows on Camembert cheese for flavour and aroma. The species has an extensive system of proteases and peptidases, and their action on the caseins is essential to soften the curds of Brie or Camembert. The lipolytic activity of the species is important also, and it is these biochemical properties that give Camembert its typical flavour and aroma characteristics (Seth and Robinson, 1986).
Penicillium roqueforti and *Penicillium glaucum* – blue–green moulds growing in the curds of blue-veined cheeses like Stilton, Roquefort, and Danish Blue (or Danablu) (Bakalor, 1962; Robinson, 1995).
Mucor rasmusen – a mould which grows on Norwegian skim milk cheeses, e.g. Gammelost.

A number of bacteria and yeasts are also involved in the ripening of cheeses, and yeasts in particular are essential for coat development in the surface mould-ripened cheeses. For example, in Camembert, the total viable counts of yeasts and *Geotrichum candidum* may reach 5.0×10^8 cfu/g of cheese coat during the first few days of maturation, and there is no doubt that a population of this magnitude does influence the development of flavour. These non-starter organisms are resident in cheese stores, and are automatically inoculated onto the cheeses during handling and turning. Many aerobic bacteria like micrococci or corynebacteria are also major components of the surface floras of surface mould-ripened cheeses (Lenoir and Tourneur, 1993), but it is on the rinds of smear-coated cheeses like Limburger, Livarot and Romadur that yeasts and bacteria dominate. Yeast genera, such as *Kluyveromyces*, *Saccharomyces*, *Debaryomyces*, *Candida*, *Pichia* and *Rhodotorula*, are widely found on smear-coats, and their action in lowering the pH of the rind is essential to allow the acid-sensitive bacteria to develop; their contributions to flavour are less easily determined.

The bacterial flora consists, as mentioned before, of micrococci, e.g. *Micrococcus* and *Staphylococcus* spp., along with the genera *Arthrobacter*, *Corynebacterium* and *Caseobacter*. However, the most important organism is, perhaps, *Brevibacterium linens* which produces an orange–brown growth over the rind of many cheeses. Apart from the colour imparted to the coat, the sheer numbers of bacteria – 10×10^8 or 109/g of rind – has a major impact on the flavour and aroma characteristics of the variety. Thus, both the proteolytic and lipolytic activities of these bacteria contribute to the complex flavour profile of the mature cheeses, e.g. Port Salut or Munster, as does the breakdown of amino acids like methionine and serine (Lenoir and Tourneur, 1993).

It could be argued, of course, that the yeasts and bacteria mentioned above are 'casual contaminants' rather than starter cultures *per se*, but, given that the surface floras are essential for maturation and are acquired as a matter of routine from the cheese room, their inclusion in this section may not be inappropriate.

9.6 Bacteriophage

Just as it is important to appreciate the activities of a bacterial cell in cheesemaking, so it is also important to understand some of the problems associated with bacteriophage (virus) proliferation which can cause lysis of starter bacteria.

These viruses can be separated from bacteria by filtration – 0.45 μm pore size (Neve, 1995), and phage of lactococci usually have the form of an isometric or prolate head with a narrow tail that ends in a base plate from which protrude a number of 'spikes'; both the head and the tail are composed of protein. Inside the head region is a core of deoxyribonucleic acid (DNA).

Phage attack can affect bacterial metabolism according to the type of phage and state of the bacterial cell, and the reactions of cells of the lactic bacteria may be grouped as follows:

(a) Insensitive, i.e. cells not affected by phage.
(b) Phage sensitive – cells will undergo complete lysis; the virus responsible is known as a virulent phage.
(c) Phage carriers – such cells may produce reduced amounts of lactic acid, but often the presence of the phage can only be detected in the laboratory through the action of UV light or mutagenic agents. In this situation, the phage is described as temperate.

If the cell is phage sensitive, the protein tail will 'recognize' specific receptor sites on the bacterial cell wall and, if the 'match' between the tail and the receptor site is exact, the virus becomes attached to the host wall. The nucleic acid is then injected into the bacterial cell (Erickson, 1980). If the components in the cell wall are not receptive towards the phage, the bacterial cell is said to be 'restrictive' towards that phage (West, 1983). However, assuming that the phage DNA enters the host cell, the normal metabolism of the cell ceases and is replaced by the synthesis of phage DNA and proteins. The formation of entire phage particles from these components completes the building phase, and a lysin (enzyme) encoded on the genetic material of the phage causes the cell wall of the host to disintegrate. The new virus particles are then released into the medium and can continue the infection of the host cells. In addition, the lysin present in the medium can also cause lysis of host cells even in the absence of infection, so that a susceptible bacterial culture can rapidly be inactivated and be incapable of producing the acidity necessary for cheese production.

The growth of a virulent phage depends on the host–phage relationship, and this interaction is characterized by a so-called burst size and latent period, i.e. how many new phage are released as a result of infection by one original phage, and how long does the build-up of this number take. For *Lactococcus* spp., the burst size can vary from < 10 to > 300 particles, and the latent period from 10 min to more than 2 h (Klaenhammer and Fitzgerald, 1993). These values are, of course, of great importance for the cheesemaker, because a heavy infection by a phage that releases 300 new particles every 10 min can so devastate a starter culture that the cheese milk may fail to coagulate. Conversely, a light infection by a phage with long latent period could passed unnoticed, as the process will be well-advanced by the time the phage begins to proliferate.

If the phage is a temperate one, then the phage DNA becomes integrated with the chromosomal DNA of the host, and as the host DNA replicates, so does the phage material. No lysis of the bacterial cell takes place, and this lysogenic behaviour – in contrast to the lytic activity described above – renders the cell immune from infection by other closely related phages. In practice, this lysogenic state may be common, and Davidson *et al.* (1990) suggest that most strains of the sub-species of *Lactococcus lactis* are carriers of one or more temperate phages; indeed, multilysogeny could be an important source of phage-resistant strains of the lactococci.

It is important also that phage are resistant to normal environmental factors, and Whitehead (1958) reported that phage could survive for 22 years in the laboratory; Keogh (1983) noted the survival of a phage of *Lactococcus lactis* sub-sp. *cremoris* for 6 years under frozen (−18°C) conditions. Thus, it must be assumed that phage can exist in crevices, walls and floors of a dairy for many years, and remain capable of contaminating cheese milk in the vat. This point could be of especial relevance to a cheesemaker planning to establish a plant to manufacture a traditional cheese from raw milk. Thus, Hull and Brooke (1982) found that phage grew better in raw milk than in heated milk and, as phage is active at 30°C, the inoculation of phage-contaminated raw milk could run into difficulties. Indeed, even though the starter cultures are carefully selected and prepared phage-free, the vat milk may be contaminated by phage from the atmosphere, cheese vat, pipes and other equipment.

References

Bakalor, S. (1962) *Dairy Science Abstracts*, Rev. Art. 107, Parts 1 and 2, 24, pp. 529, 583.

Bergere, J.M.A. and Hermier, J. (1968) *Lait*, **48**, 1.
Bouillane, C., Landon, M., Auclair, J. and Mocquot, G. (1979) In *Food Microbiology and Technology* (eds B. Jarvis, J.H.B. Christian and H.D. Michener), Lab de Tech., Institut National de la Recherche Agronomique, Jouy-en-Josas, France, pp. 89–94.
Chapman, H.R. (1978) *Journal of the Society of Dairy Technology*, **31**(2), 99.
Christensen, V.W. (1969a) US Patent 3,483.087.
Christensen, V.W. (1969b) *Food Engineering*, **41**(5), 104.
Cogan, T.M. (1980) *Lait*, **60**(597), 397.
Cogan, T.M. (1995) in *Dairy Starter Cultures* (eds T.M. Cogan and J.P. Accolas), VCH Publishers, New York
Cogan, T.M. and Accolas, J.P. (1990) Starter cultures: types, metabolism and bacteriophage, in *Dairy Microbiology*, Vol. 1 (ed. R.K. Robinson), Chapman & Hall, London, pp. 77–114.
Daniell, S.D. and Sandine, W.E. (1981) *Journal of Dairy Science*, **64**, 407.
Davidson, B.E., Powell, I.B. and Hillier, A.J. (1990) *FEMS Microbiology Reviews*, **87**, 79.
Elliott, J.A. and Emmons, D.B. (1979) Paper at the Biennial Marschall Int. Cheese Conference, Madison, USA.
Erickson, R.J. (1980) *Dairy Industries International*, **45**(3), 37.
Gilliland, S.E. and Speck, M.L. (1975) *Journal of Milk and Food Technology*, **37**, 107.
Gordon, J.F. and Shapton, N. (1977) *Journal of the Society of Dairy Technology*, **30**, 15.
Greene, V.W. and Jezeski, J.J. (1957) *Journal of Dairy Science*, **40**(9), 1053.
Hargrove, R.E. (1959) *Journal of Dairy Science*, **42**(3), 906.
Heap, H.A. and Lawrence, R.C (1981) *New Zealand Journal of Dairy Science and Technology*, **16**(1), 91.
Holt, J.G. (ed.) (1984) *Bergey's Manual of Determinative Bacteriology,* 8th edn, Williams and Wilkins Co., Baltimore.
Hull, R.R. and Brooke, A.R. (1982) *Australian Journal of Dairy Technology*, **37**(4), 143.
Hup, G. and Stadhouders, J.J. (1977) US Patent 4 053 642.
Hurst, A. (1977) *Canadian Journal of Microbiology*, **23**, 936.
Jansen, L.A., Stadhouders, J. and Hup, G. (1969) *Voedingsmiddelen Technologie*, **3**(18), 407.
Johannsen, G.B. and Hunger, W. (1979) *Deutsche Milchwirtschaft*, **30**(6), 953.
Keogh, B.P. (1974) *Australian Journal of Dairy Technology*, **29**(3), 105.
Keogh, B.P. (1983) *Australian Journal of Dairy Technology*, **38**(1), 32.
Klaenhammer, T.R. and Fitzgerald, G.F. (1993) in *Genetics and Biotechnology of Lactic Acid Bacteria* (eds M. Gasson and W. De Vos), Chapman & Hall, London, pp. 106–168.
Kosikowski, F.V. and Mocquot, G. (1958) *Advances in Cheese Technology*, FAO Study No. 38, Rome.
Langsrud, T. and Reinbold, G.W. (1973) *Journal of Milk and Food Technology*, **36**, 531.
Law, B.A. (1979) *Journal of Applied Bacteriology*, **46**, 455.
Law, B.A. and Sharpe, M.E. (1977) *Dairy Industries International*, **42**, 10.
Lawrence, R.C. and Pearce, L.E. (1972) *Dairy Industries International*, **37**(2), 73.
Lawrence, R.C., Pearce, L.E. and Wills, I. (1971) *New Zealand Journal of Dairy Science and Technology*, **6**(3), 129
Lenoir, J. and Tourneur, C. (1993) in *Encyclopaedia of Food Science, Food Technology and Nutrition* (eds R. Macrae, R.K. Robinson and M. Sadler), Academic Press, London, pp. 3146–3153.
Lewis, J.E. (1956) *Journal of the Society of Dairy Technology*, **9**, 123.
Lloyd, G.T. and Pont, E.C. (1973) *Australian Journal of Dairy Technology*, **28**(3), 104.
Mills, O.E. and Thomas, T.D. (1980) *New Zealand Journal of Dairy Science and Technology*, **15**(2), 131.
Neve, H. (1995) in *Dairy Starter Cultures* (eds T.M. Cogan and J.P. Accolas), VCH Publishers, New York, pp. 157–190.
Obermann, H., Piatkiewicz, A. and Libudzisz, Z. (1978) *XX International Dairy Congress*, Vol. E, 577.
Osborne, R.J.W. (1977) *Journal of the Society of Dairy Technology*, **30**, 40.
Osborne, R.J.W. and Brown, J.V. (1980) *Journal of Dairy Research*, **47**(1), 141.
Reiter, B. and Oram, J.D. (1961) *Journal of Dairy Research*, **28**(2), 175.
Richardson, G.H., Cheng, C.T. and Young, R. (1977) *Journal of Dairy Science*, **60**, 378.
Robinson, R.K. (1981) *Dairy Industries International*, **46**(12), 11.

Robinson, R.K. (1995) *Colour Guide to Cheese and Fermented Milks*, Chapman & Hall, London.
Sandine, W.E. (1995) in *Dairy Starter Cultures* (eds T.M. Cogan and J.P. Accolas), VCH Publishers, New York, pp. 190–206.
Satji, J.P. and Kroger, M. (1979) Biennial Marschall Int. Cheese Conference, Madison, USA.
Seth, R. and Robinson, R.K. (1986) *Developments in Food Microbiology – 4*, Chapman & Hall, London.
Shehata, A.E., Magduob, M.N., Fayed, E.O. and Hofi, A.A. (1980) *Milchwissenschaft*, **35**(1), 28.
Speck, M.L. (1981) *Food Technology*, **35**(1), 71.
Stadhouders, J., Jansen, L.A. and Hup, G. (1969) *Koeltechniek*, **62**(1), 29.
Stanley, G. (1977) *Journal of the Society of Dairy Technology*, **30**(1), 36.
Sullivan, J.J., Schlipalius, L.E., Mou, R. and Jago, G.R. (1974) *Australian Journal of Dairy Technology*, **29**(2), 74.
Tamime, A.Y. (1990) In *Dairy Microbiology*, Vol. 2 (ed. R.K. Robinson), Chapman & Hall, London, pp. 113–156.
Tamime, A.Y. (1993) In *Modern Dairy Technology*, Vol. 2, (ed. R.K. Robinson), Chapman & Hall, London, pp. 49–220.
Tofte-Jespersen, N.J.T. (1974) *International Cheese Production*, Chrs. Hansen Lab. Symp., pp. 69–86, Hansen, Milwaukee.
Tofte-Jespersen, N.J.T. (1977) *Journal of the Society of Dairy Technology*, **31**(1), 47.
Tofte-Jespersen, N.J.T. (1979) *Journal of the Society of Dairy Technology*, **32**(4), 190.
Tsaneva, K.P. (1975) *Comptes Rendus de l'Academie Bulgare des Sciences*, **28**(4), 529.
West, R.A. (1983) *Dairy Industries International*, **48**(6), 21.
Whitehead, H.R. (1958) *Journal of Dairy Research*, **25**(2), 191.
Wigley, R.C. (1977) *Journal of the Society of Dairy Technology*, **30**(1), 45.
Wigley, R.C. (1980) *Journal of the Society of Dairy Technology*, **33**(1), 24.
Wilkowski, H.H., Nelson, F.E. and Parmless, C.E. (1954) *Applied Microbiology*, **2**(5), 250.
Ziemba, J.V. (1970) *Food Engineering*, **42**(1), 68.

10 Preparation of cheese milks

10.1 Milk storage

Milk storage is an important feature of cheesemaking. It is a necessary part of the cheesemaking routine to ensure that tanks and ancillary equipment be used effectively and kept clean. Except for the smallest quantities of milk or those milks used for cheese while still warm, most milks are subject to storage for longer or shorter times; even during transportation, milk is accepted as being in storage. With the advent of larger cheese factories, and especially those concerned with collecting milk ex-farm in bulk cold tanks, the storage of milk has become a most important stage in the process. The advent of a 5-day working week for creamery workers makes the storage of milk even more important.

The bacteriology of raw milks has been discussed in Chapter 6 and Table 6.2 and Figure 6.2 show the possible growth of bacteria in cold-stored milks. Later, the bacteriology of heat-treated milks will be dealt with (see Figure 10.6); these milks often find temporary storage in tanks. Cheese milk ripening in storage tanks by starter, before transfer to cheese vats, is also used in some systems. Apart from these uses, the storage of milk provides time for bacteriological and chemical testing, mixing of milks from other collecting areas, standardization of the milks and any other treatments as necessary.

It is essential that the tanks be constructed of such materials and in such a conformation that they are easily cleaned. Furthermore, the tanks, pipelines and pumps must not add to the bacterial population of the milk. The tanks must be properly insulated against heat loss or gain and, in some instances, they may be required to have a refrigeration system built-in for cooling the contents (e.g. starters). It should be possible to mix the milk in the tanks, at whatever level of filling, to avoid cream separation.

Cleaning-in-place (CIP) systems are essential for cleaning milk silos, which may be from 70 to 200 000 litres capacity. The tanks should be cleaned every time they are emptied, and not refilled with more milk, since contamination of the second filling of milk can occur from the first milk drainage. The tanks should be rinsed with water to remove as much soiling as possible and then washed with suitable detergent solutions, rinsed and disinfected. Detergent solutions are compounded for different duties from a mixture of chemicals, in many instances alkaline. CIP systems are covered by Romney (1990). The detergents used should be those normally and easily available in the area of the cheese plant and, therefore, they will vary in composition from country to country.

The use of heat in the large tanks must be limited to that required for efficiency of cleaning, since the body of a tank can take up excess heat which has to be removed before cold milk can be stored in the tank again. Disinfection of these tanks by steam is only carried out in extreme cases, as stresses set up in the metal of a tank may cause deformation of the tank; therefore, chemical disinfection is commonly employed. Those tanks which have a refrigeration system attached need to have the refrigerant removed before hot cleaning is practised.

The heat retained by tanks after normal cleansing does not cause an appreciable increase in the temperature of well-cooled milk, 4°C, when the tanks are filled, but can cause an unacceptable increase in partially filled tanks. A temperature of 5°C seems to be the critical temperature for short-term storage of milk. This is illustrated in Figure 10.1; the optimum temperature for storage would be 4°C.

The usual method of mixing in the large tanks has been by mechanical rotary blade type mixers. The mixer must be capable of moving the mass of milk from top to bottom to prevent cream separation; the position and direction of the mixer is important. To prevent damage to the milk by over-mixing in overnight storage, the mixers should be set to operate intermittently for short periods rather than continuously. Mixing by means of

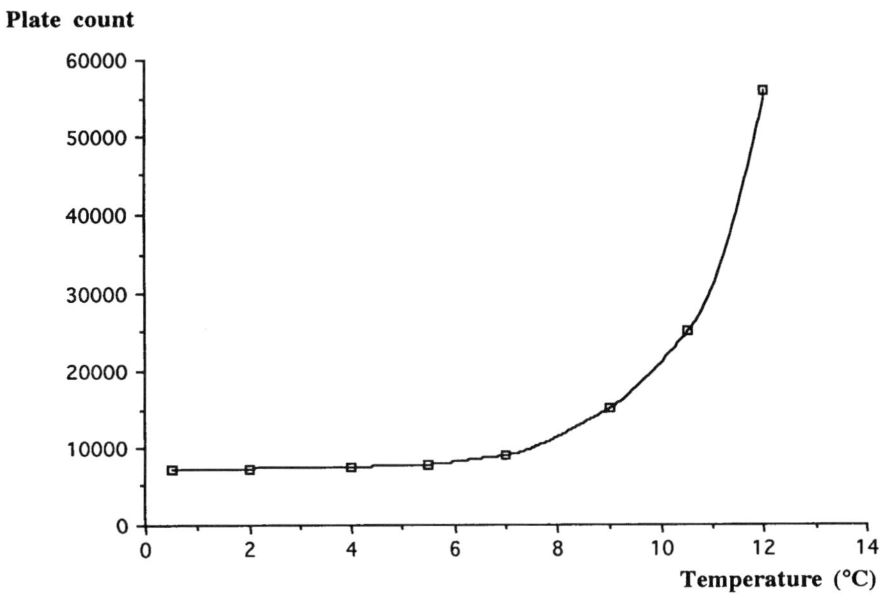

Figure 10.1 Numbers of bacteria in a clean farm milk supply when stored for 24 h at various temperatures (plate count on yeastrel milk agar incubated at 37°C).

cold air bubbles has been used in some creameries. This method can cause some oxidation of fats, and a rise in milk temperature if the air is warm. Milk in the silo has also been mixed by jets of milk recycled from the silo tank.

Proper mixing of cheese milks in storage tanks is important because when the storage tank milk is distributed over several cheese vats, some may otherwise receive more fat than others. This is particularly so when tanks are used for ripening the milk by starter, after which the ripened milk is filled into more than one rennetting vat.

10.2 Standardization of cheese milks

The standardization of milk is aimed at two separate objectives: (1) to achieve a cheese which has a composition which complies with agreed standards (e.g. see Table 3.2), (2) to achieve the most economic use of milk components consistent with consumer acceptance. This last objective really combines the objective of economic cheese yield from the milk with the objective of consumer acceptance of the cheese (in respect of fat, water and body of the cheese (protein), apart from flavour and aroma). There may be a conflict between these objectives.

10.2.1 Formulae for standardization and cheese yields

Much has been written concerning yields of cheese from a given volume of milk. Thus 'cheese sold minus milk bought minus x equals profit' is a deceptive economic argument, since x, representing overhead expenses, will vary. The selling price of cheese depends on its quality or marketable value to consumers. The statement that 'milk of a specific quality will produce a certain quantity of cheese' lacks the definition of 'quality' of cheese.

The following formulae have been reported, from time to time, for the calculation of the yields of hard or semi-hard cheese from milk. Other formulae have been devised for soft cheese.

$$\text{Yield of 'green' or raw cheese} = 1.63 \times (\text{fat\%} + \text{casein\%})$$

$$\text{Yield of 'green' or raw cheese} = 1.22 \times (\text{fat\%} + \text{casein\%}) + 2.32$$

$$\text{Yield of 'green' or raw cheese} = 1.07 \times \text{fat\%} + 2.35 + \text{casein\%}$$

$$\text{Yield of 'green' or raw cheese} = 2.7 \times \text{fat\%}$$

A more complicated formula is:

$$\text{Yield of raw cheese} = 0.91 \times \text{fat\%} + 0.77 \times \text{protein\%} + 0.48 + \text{water factor}$$

where:

$$\text{Water factor} = \text{water\%} \times (0.77 \times \text{protein\%} + 0.48)/(100 - \text{water\%})$$

The percentage of casein has been based on (fat% in milk × 0 4) + 1.

Dolby and Harkness (1955) have given various formulae for the adjustment of the fat/casein ratio in milk for cheesemaking. The accuracy of these formulae depends on the analytical and sampling methods. Some formulae do not take account of the milk constituents lost in the whey and, therefore, the methods of manufacture, the type of cheese and the ripening of the cheese may cause the formulae to be misleading. The results differ from one factory to another, and so any formula worked out for one situation will not necessarily apply to another. Indeed the computation of cheese yields from an analysis of milk has been little more than a mathematical exercise, and it was only with the advent of infra-red analytical techniques that relevant on-site milk assays could be carried out. The ratio of cheese sold against milk used is a more realistic day-to-day exercise.

Lelievre (1982a) suggested that cheese yield is related to the starter and the composition of the milk and cheese in respect of % fat and % casein in both. However, Lelievre (1982b) also suggested that there is a question of balancing cheese yield against profit from the cheese. Chapman (1974) found that a ratio of Solids-Not-Fat (SNF) to fat of 0.33–0.35 is the lowest for cheese with 48% fat in dry matter, but that a ratio of 0.46–0.48 is the best for economic production if quality grading is at 6 months.

The lists of cheese varieties being produced by the FAO/WHO Committees give the following details in respect of cheese: country of origin, name of cheese, shape, size, weight, minimum fat percentage, dry matter, type of milk (i.e. cow, goat, sheep), but, as indicated in Table 10.1 and, for the UK, in Table 10.2, there are differences in the legal standards in respect of Minimum Fat-in-Dry-Matter (FDM) and Minimum Dry Matter for various cheeses in different countries. It is therefore important that the ratios of milk constituents (i.e. fat and protein (SNF)) in cheese milks are such as to meet the relevant standards and to maintain the 'quality' of cheese from the consumer's point of view.

Table 10.1 Standards in respect of fat and dry matter in some cheese varieties

	Fat-in-Dry-Matter (FDM %)	Dry matter (%)
Parmesan type cheese	30–35	68–70
Emmental type cheese	43–50	55–64
Gruyère type cheese	40–45	58–61
Cheddar type cheese	45–50	58–63
Gouda type cheese[1]	30–50	54–60
Edam type cheese[1]	20–40	52–56

[1] Since there is manufacture of cheese of different fat contents for Gouda and Edam cheese, the variation in minimum fat percentage is wider than in other varieties.

Table 10.2 UK food standards for cheese. Descriptions are covered by The Food Labelling Regulations 1996 (SI 1996 No. 1499)

'Cheese' means the fresh or matured product intended for sale for human consumption, which is obtained as follows:

(a) in the case of any cheese other than whey cheese, by combining, by coagulation or by any other technique involving coagulation, of any of the following substances, namely milk, skimmed milk, partly skimmed milk, concentrated skimmed milk, reconstituted dried milk, butter milk, materials obtained from milk, other ingredients necessary for the manufacture of cheese provided that those are not used for replacing, in whole or in part, any milk constituent, with or without partially draining the whey resulting from coagulation;

(b) in the case of whey cheese:
   (i) by concentrating whey with or without the addition of milk and milk fat, and moulding such concentrated whey, or
   (ii) by coagulating whey with or without the addition of milk and milk fat.

Schedule 8 Part II

	Water (max %)
Cheddar	39
Blue Stilton	42
Derby	42
Leicester	42
Cheshire	44
Dunlop	44
Gloucester	44
Double Gloucester	44
Caerphilly	46
Wensleydale	46
White Stilton	46
Lancashire	48

Lelievre (1983a) has suggested that, in Cheddar cheese, the Moisture-in-Non-Fatty Solids (MNFS) can be kept constant by changing the manufacturing conditions when the ratio of casein/fat in milk varies. Ratios of 0.55:1 to 0.75:1 are quoted as giving cheese with 57–55% FDM and a fixed 54% MNFS. The fat in such cheese might vary from 37.9 to 33.2% and the moisture from 33.5 to 36.1%.

Whitehead's work (Whitehead, 1948) suggested that, when the fat and dry matter varied by 2%, the MNFS varies by 1%. Practical cheesemakers often assume that more moisture tends to be retained in curd when there is more fat present. It would appear that there is bound to be a trade-off between cheese yield, grade of cheese, cost of equipment for standardization and the profit motive for the practical cheesemaker (McDowall, 1936; Whitehead and Harkness, 1954; Lelievre, 1983b).

In recent years, the term Fat-in-Moisture-Free-Substance (FMFS) has been used instead of Fat-in-Dry Matter (FDM). The two percentage values FMFS and WFFS (Water-in-Fat-Free-Substance) have been used to classify cheese into different groups, but they are also useful to the cheesemaker to indicate cheese with a variation in composition due to fat content or high moisture. Optimum values for all cheese can be worked out. Thus, the

Table 10.3 Effect of fat in milk on production of a soft cheese

Milk fat	Approximate composition of cheese		
(%)	FDM (%)	DM (%)	WFFS (%)
1.4–1.6	30	38	69.9
2.4–2.8	40	42	69.7
2.8–3.0	45	44	69.8
3.5–3.7	50	47	69.2
4.8–5.2	60	54	66

WFFS value for Cheddar cheese is 54%, and for other cheese the WFFS values are as follows:

Very hard cheese, e.g.	Parmesan	42%
	Emmental	52%
Hard cheese	Edam	60%
	Gouda	58–60%
Semi-soft cheese	Romadur	68%
	Munster	68%
Soft cheese	Camembert	70%
	Cottage cheese	85%
	Blue-veined cheese	58–62%

The amount of fat in the various cheese qualities, i.e. 30, 40, 50% FDM, does not greatly affect the WFFS value.

The qualities of milk required for the production of soft cheeses (e.g. Camembert) of different fat content are illustrated in Table 10.3. The WFFS is almost 70% for optimum consistency except with the higher fat range of cheese (60% FDM).

10.2.2 Methods employed for the standardization of milks

It has been a long-standing practice in many countries to standardize for fat content. This was an economic advantage in some cases, since the milk fat obtained during standardization could be used for butter production. The sale of the butter provided a financial gain for the processor. Cheeses of various fat levels have become commonplace, e.g. full fat, standard three-quarters fat, half fat and quarter fat (more properly expressed as cheeses with 50, 45, 30, 20% FDM). Some major cheesemaking countries (e.g. The Netherlands and Australia) standardize the cheese milk, whereas some other countries' creameries standardize only the milk for cheese to which fat standards apply (i.e. Edam, Gouda).

In many countries, cheesemaking has been a major industry in its own right, but where liquid milk has been the main outlet for milk, cheese has been subsidiary to the liquid milk trade. In the UK, for instance, there has

Pearson's Square

Let:

The desired fat percentage in the cheese milk be C %,

The fat percentage in the raw milk be A %,

The fat percentage in the skim milk (or cream) to be added be B %.

Then:

X is the proportion of raw milk of A % fat need to be added to

the proportion Y kg of skim milk of B % fat to give

Z of cheese milk of C % fat, as follows:

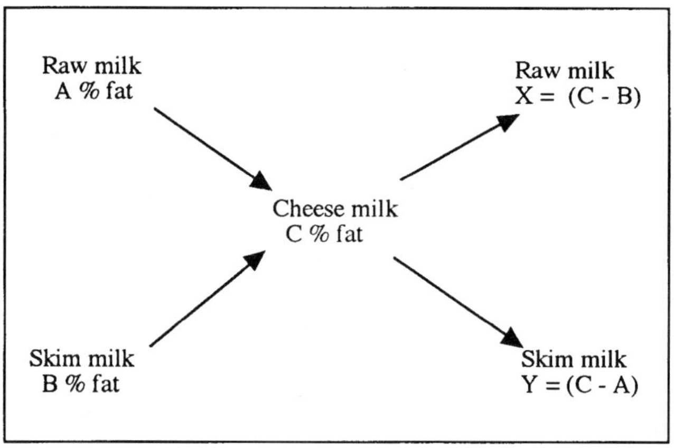

Note: It is the differences between A, B and C that are taken, irrespective of whether C is the larger or smaller.

The unknown can be found from the equations:

$$X / (B - C) = Y / (A - C) = X / ((B - C) + (A - C))$$

Figure 10.2 Illustration of the use of the Pearson Square for standardization.

been great reluctance to introduce standardization of cheese milk, even though other dairy products (like condensed milk) have legal standards based on the use of standardized milk.

As already discussed, milk varies in composition from season to season, in addition to other factors, so that the cheese produced at some times of the year suffers from imbalance in the fat to protein ratio. Since these two components constitute the main body of cheese, it seems reasonable to standardize the milk at the ratio of fat to protein (casein) which gives satisfactory quality cheese and an even quality of production during all seasons.

The aim of standardization of milk was originally to produce a constant fat level in the milk in the belief that the protein was constant. The methods employed were as follows:

1. Milk with a too high fat percentage may be adjusted by: (a) adding skim milk, (b) adding low fat milks, (c) skimming part of the milk and then adding the skim milk to the remainder of the high fat milk, (d) the use of a standardizing centrifuge to take sufficient cream off the milk so that the remaining cream and milk together are of the correct fat standard.
2. Milk deficient in fat may be adjusted by (a) adding cream, (b) adding milk richer in fat, (c) skimming part of the milk and adding the cream so produced to the remainder of the low fat milk.

Few of these methods, which are irksome and time consuming on a small scale, take account of the protein content as well as the fat content. Small quantities of milk can be standardized, at least in terms of fat, by the use of Pearson's Square method as shown in Figure 10.2.

Milk for standardizing the fat to protein ratio for cheese manufacture can be calculated if the casein content of the milk is known; a normal casein to fat ratio for Cheddar cheese is 0.70:1. Thus, if the casein is 2.7% in the milk, the fat percentage required is $2.7/0.7 = 3.85\%$ fat, or with 2.1% casein in the milk $2.1/0.7 = 3.0\%$ fat. Standardization equations may be solved algebraically, particularly using spreadsheet programmes, to standardize both protein and fat levels so that more or less constant fat/casein ratios may be maintained.

The use of large capacity silo tanks has made the standardization of milk easier. It is also less complicated practically when the milk is standardized to a lower fat percentage through the use of a standardizing centrifuge, which skims off the surplus fat as cream through a control valve which can be preset to extract a known quantity of cream. Modern milk standardizing systems produce skim and cream so there is no problem in standardizing to higher fat levels if needed; the surplus becomes skim milk rather than cream (Bird, 1993). Examples of in-line methods of standardizing milks are shown in Figures 10.3 and 10.4. Control may now be by in-line cream density measurement and proportional control of the cream and skim flows.

In addition to complying with legal regulations or Codes of Practice for milk composition, the cheesemaker must have due regard to the quality of the cheese. One of the main characteristics of cheese is its 'body', which is that feature of a cheese which is detected by feel rather than vision. Each variety of cheese has a distinct body or feel or 'chewability'. To a large extent the body is dependent on the ability of casein to enclose and contain the other constituents – fat, water, salts and whey proteins. The ratio of casein to fat is, therefore, one of the most important ratios of components in cheese milks as illustrated in Fig 10.5.

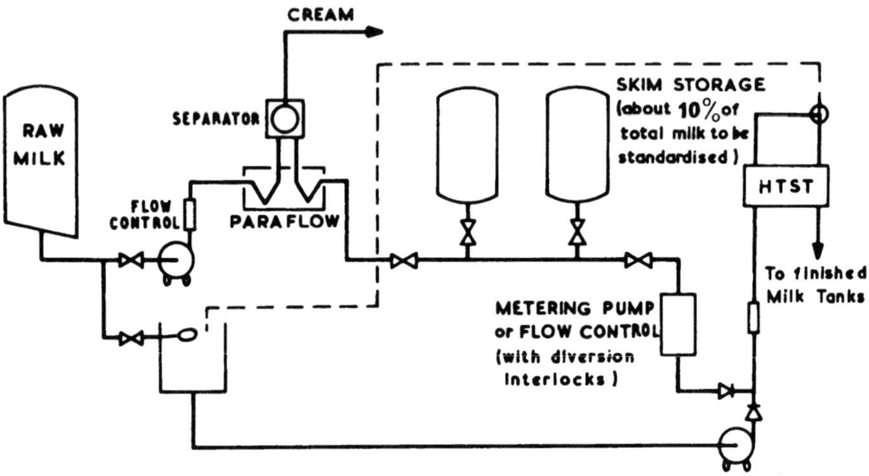

Figure 10.3 Simple milk standardizing system, using a small capacity separator.

Various workers have suggested SNF to fat ratios from 1:0.36 to 1:0.47. Since these ratios include lactose, as well as whey proteins which are lost in the whey, the fat to casein ratio of 1:0.69 is probably more meaningful. Due to higher losses in mechanical systems a ratio of 1:0.70 is recommended for Cheddar cheese milks. (Sometimes reports show the reciprocal of these ratios, either fat to casein of 1:0.70, or casein to fat of 1:1.42.) Since standardization may affect the salt balance in the milk, the addition of calcium salts may be necessary.

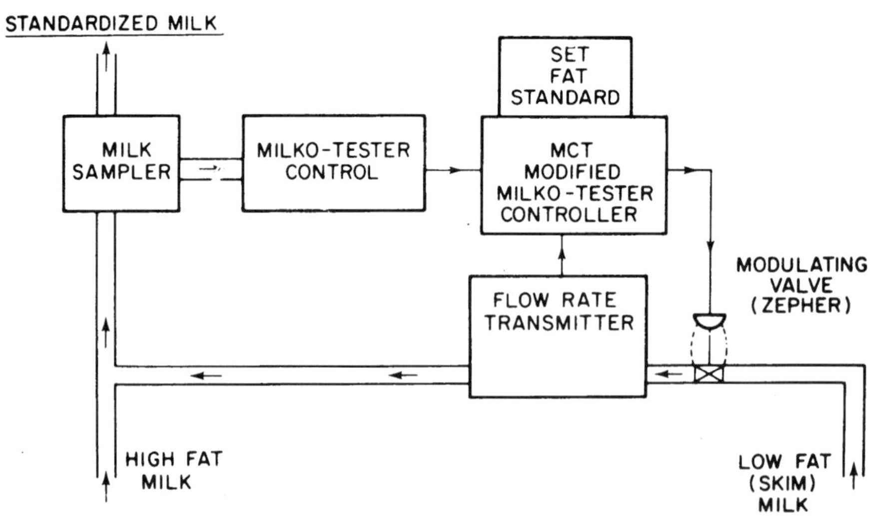

Figure 10.4 Automatically controlled system for milk standardization.

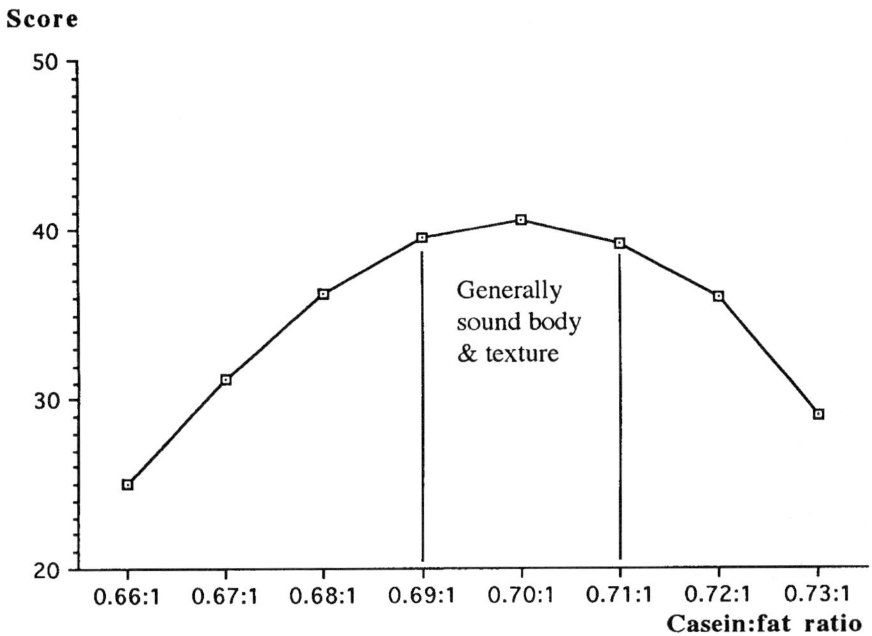

Figure 10.5 Graph showing probable relationship between the fat/casein ratio and grading points for body and texture in cheese.

Standardization of milk for cheese requires that other products are produced from the by-products (e.g. butter or milk powder) and that adequate laboratory test facilities are available.

10.3 Homogenization

While homogenization of the skim-milk portion of milk disturbs the calcium–casein–phosphate complex in milk and the salt balance, its primary function is to break the fat globules into smaller sizes. The original fat globule membranes needed to cover these smaller globules are inadequate and need to be supplemented from the serum by other protein and lipid components. Heating milk prior to homogenization is responsible for some of the disturbance to the salt/protein balance. Furthermore, the time of travel between the homogenizer and the heater extends the holding time at a higher temperature, and thus extends the total heat treatment. Since the fat globule membranes are disturbed, a more rapid penetration by lipolytic enzymes to the fatty core is accomplished. Some cheese with a flavour resulting from fatty acid breakdown, i.e. the blue-veined cheese, will ripen more quickly if the fat is homogenized.

Table 10.4 Homogenization pressures expressed in different units

MPa	Atmospheres	kg/cm^2	psi ($lb/inch^2$)
3.4	34	35	500
5.5	54	56	800
6.9	68	70	1000
10.3	102	105	1500
13.8	136	140	2000
17.2	170	175	2500
20	200	206	2940

Normally the milk is separated into skim and cream (e.g. 25% fat) and only the cream is homogenized. The skim and cream are combined before a final heat treatment. Homogenization pressures used for this treatment vary according to the previous experience with the milk supply. Pressures from 5 to 17 MPa have been used, but pressures from 7 to 10 MPa are usually sufficient. Since homogenization pressures are quoted in a number of units, Table 10.4 gives a comparison between units.

Two-stage homogenizers are often used, the second stage operating at 3.5–5.5 MPa. The optimum temperature of homogenization has been cited at 69°C, but lower temperatures, e.g. 57°C, may be used for cheese milk, coinciding with that used for separation.

10.4 Centrifugal clarification

Centrifugal clarification has been used for many years to remove from milk the debris which ordinary cloth filters will not remove. Any form of rapid mixing, such as that which occurs when milk enters a rapidly rotating centrifugal separator or clarifier bowl, breaks the clumps of bacteria in the milk into smaller aggregates. While this appears to give a larger numbers of bacteria by plate count than in the original milk, there is no actual increase in real numbers of viable cells. However, the smaller groups have better access to nutrients and grow more rapidly. This tended to make milk sour more quickly, and the use of clarifiers was discontinued in cheese processing though the centrifugal principle was developed into a bactofuge machine.

Bactofugation of cheese milk was intended to remove by centrifugation the bacteria and spores from milk. Trials were successful in this respect, as shown by Table 10.5 indicating the percentage reduction of the bacteria and spores. Unfortunately, the volume of bactofugate removed from the milk in the process was approximately 3% of the total milk volume and contained 6.8% of milk protein; this constitutes a high loss of protein from cheese milk. The temperatures used for bactofugation are normally in the range 56–57°C, and sterilization of this bactofugate, before adding it back into the milk, did not affect the quality of the cheese. Bactofugation may be used along with

Table 10.5 Reduction of spore count and total bacterial counts in milks with and without bactofugation

	% Reduction in numbers	
	Pasteurized	*Pasteurized after bactofugation*
Total count of bacteria	99.2	99.8
Aerobic spore count	13	77
Anaerobic spore count	0	79

heat treatment where spore counts are very high in raw milk (see Table 10.7 for a list of bacteria producing spores). Experiments using bactofugation of milk for cheese have often used the traditional inhibitory salt (sodium nitrate) at the rate of 30 g per 100 litres of milk, but with a reduced bacterial population, it ought not to be necessary to use inhibitory salts.

Microfiltration now provides an alternative method for reducing the bacterial load in the aqueous portion of the milk (see Chapter 6).

10.5 Hydrogen peroxide/catalase method

The hydrogen peroxide/catalase method has been used as an alternative approach to removing unwanted bacteria from cheese milk. The use of hydrogen peroxide, which is both a bacteriocidal and bacteriostatic agent, has not been widely accepted. However, in those instances where small quantities of milk are involved, and in rural areas where full heat treatment is difficult or impossible, the method has been used for cheese milks.

The use of a catalase enzyme to remove the residual hydrogen peroxide (H_2O_2) ensures that there are no toxic effects due to the H_2O_2. Testing of the treated milk to ensure the H_2O_2 has been removed is imperative.

Unfortunately, the effects of treating milks with hydrogen peroxide are variable, depending on the amount of H_2O_2 used, its distribution, the length of time of the application before adding catalase, the temperature, the normal milk catalase content, the bacterial population and the cell count (leukocyte) of the milk. In those areas where the method is used, both the bacterial population and leukocyte content are likely to be high.

Oxidation of proteins, aldehydes, ketones and the vitamins A and C occurs in milk so treated. The most serious effect in cheese milk is the splitting of minor components as α_{1-5} from the casein complex. This results in a longer phase of rennet activity during coagulation, a longer setting time and softer curd; excessive dosages with hydrogen peroxide should be avoided. The use of 0.25% H_2O_2 of 130 volume strength for 24 h at 28°C in cheese milks produced soft pasty cheese; whereas 0.1% of H_2O_2 (130 volume strength) for a similar time and temperature allowed the production of cheese with negligible defects. One recommended treatment for milk has been 0.08% v/v H_2O_2 at 49–51°C for a period of 30 min. instead of pasteurization of the milk. This treatment should be given only to tuberculosis-free milk, since the bacterium (*Mycobacterium tuberculosis*) can survive the treatment for longer than 24 h.

Stadhouders and Hassing (1974) found that heating milk above 77–80°C destroyed lactoperoxidase (which reacts with H_2O_2), thus inhibiting acid production. They found it necessary to add peroxidase to the milk in order to restore acid production. This observation may affect starter production.

Since the peroxide treatment is aimed primarily at preserving the cheese milk, a normal usage would be 0.07–0.1% of H_2O_2 for not more than 40 min. at a temperature of 50–54°C. Treatment with peroxide for longer than 1 h tends to produce milks giving gummy sticky curds which are not digestible. For maximum effect, the milk should be treated at body heat immediately after milking. Less H_2O_2 (0.042%) is necessary for destruction of coliform bacteria in milk at temperatures of 30–32°C than at 15°C (0.13%) or at 4°C (0.175%).

The admixture of peroxide-treated, overnight-stored milk and fresh milk, with its natural catalase, in equal quantities has been used to destroy the residual H_2O_2 in the treated and stored milk.

Although less catalase is required to decompose the H_2O_2 at higher temperatures, it is usual to treat the cooled milk with catalase for a longer period of time. It is essential that sufficient catalase be added to the milk and that enough time be allowed for the catalase to destroy the H_2O_2 before use. Before further processing, the treated milk should be tested for residual H_2O_2. The usual rapid test method uses potassium iodide to test the milk or whey after precipitation of protein with sulphuric acid. When peroxide is present a yellow colour develops. Alternatively, potassium iodide/starch paper can be used. A blue colour indicates a positive test. A more sensitive test uses the red/pink colour which develops in the presence of peroxide when a solution of vanadium pentoxide (V_2O_5) in sulphuric acid is added to the milk (20 drops of 1% vanadium pentoxide solution in 6% sulphuric acid).

It is essential that only food grade hydrogen peroxide is used and that it be diluted with water just before dosing the milk (since it loses strength in diluted form). One of the early difficulties arose through the use of impure hydrogen peroxide which contained trace amounts or iron, copper or even arsenic.

Various micro-organisms exhibit different resistances to dilute hydrogen peroxide, and it is particularly important to apply the correct dosage. Pathogenic bacteria do not die out in unripened cheese, and thus their destruction in the milk is important. Table 10.6 shows some of the times and H_2O_2 quantities for inactivation of micro-organisms normally found in cheese milks.

Table 10.6 Times required for inactivation of bacteria normally found in cheese milks by various H_2O_2 treatments

Bacteria	H_2O_2 used in the treatment and time (min) required for inactivation		
	0.005%	0.03%	0.05%
Lactococcus lactis spp.	360	240	150
Lactobacillus delbrueckii sub-sp. *bulgaricus*	360	270	150
Bacillus subtilis	2160	1440	1080
Bacillus cereus	1440	840	420
Escherichia coli	345	45	45
Aerobacter aerogenes	360	60	30
Alcaligenes viscosus	120	60	40

Data from Luck (1962).

The catalase activity of certain groups of bacteria, mainly the aerobic cells, is responsible for the variable destruction rates due to H_2O_2 in cheese milks. The following genera produce catalase when growing in milk:

Micrococcus spp.
Staphylococcus spp.
Bacillus spp.
Corynebacterium spp.
Propionibacterium spp.

Lactobacillus and *Lactococcus* spp. rarely produce catalase in detectable amounts. These are the organisms that are, of course, required in cheese milks, and it is of note that they are quickly and easily destroyed by the peroxide treatment; *Bacillus* spp. are more resistant.

10.6 Heat treatment of cheese milks

Heat treatment of cheese milks is aimed at destroying the micro-organisms or enzyme systems which would be harmful to the cheese process, or in the case of pathogenic bacteria, to the consumer of the cheese. It is not a method of standardizing the bacteriological quality of the milk even though it assists in that direction. A number of the bacteria and enzymes which would be useful in the cheese process are unfortunately destroyed by the heat treatment.

Some cheesemakers prefer, therefore, to use lower than pasteurization temperatures in order to preserve some of the enzymes already in the raw milk (mainly the lipases). On the other hand, where heat treatment apparatus exists for other purposes, cheese milk has been subjected to temperatures higher than Pasteurization, e.g. 75–80°C. Milk 'Pasteurization' is a term normally employed to describe a statutory process for the treatment of milk, while 'pasteurization' (lower case 'p') is used for heat treatments within the concept laid down by Louis Pasteur in his research. EC regulations for heat treatment of milk specify the range of temperature and time where alkaline phosphatase is inactivated but lactoperoxidase activity remains (European Community, 1992). This range is essentially 71.7–78°C for 15 s or equivalent. Since there are so many different treatments used, it is preferred to use the term 'heat treatment' rather than 'pasteurization', and to specify the time and temperatures employed.

Heat treatment of milk is composed of two separate parts. One is the temperature to which the milk is heated, and the second is the time during which milk is held at that temperature. This latter phase is normally called the holding time. This holding time is an essential part of efficient heat treatment, since not all the unwanted organisms or enzymes are destroyed instantly at the treatment temperatures. Some of the bacteria may simply be heat shocked and they may recover later. Some cheesemakers prefer to neglect the holding time and rely on temperature only. However, where the system is an enclosed mechanical system, the passage of the heated milk along to the cooler constitutes a holding period, even though it may be less than 1 s.

Organizations or bodies, whether statutory or not, have been set up to control industrial practices in most countries. Such bodies have been concerned with the temperatures employed in treating cheese and other milks. Sometimes the treatment of cheese milk is the same as for liquid milk, but sometimes a lower temperature is allowed to be used. Thus, in the UK, for Pasteurization of liquid milk a temperature of 71.7°C and a holding time of 15 s (HTST) or any equivalent combination is required; the milk must have a negative phosphatase test (UK Regulations, 1995). Some cheese milks heat treated to higher temperatures have had a negative phosphatase test at 4°C, but when held for ripening at 31°C, the test became positive due to reactivation of the enzyme.

A second temperature treatment was also used in the UK for liquid milk: 62.8–65.6°C with 30 min. holding time. This treatment has been used for small quantities of cheese milk.

Following problems with *Staphylococcus aureus* toxins in cheese, a Code of Practice was drawn up in the UK requiring cheese milks to be heat treated to at least 69°C with 15 s holding time. In spite of this minimum treatment, many factories use a normal HTST treatment since some of the milk may be diverted for other purposes. Note that the Code of Practice temperature is near to the Thermization treatments now being advocated for cold-stored milks.

In other countries of the world, the heat treatment temperatures and holding times for both liquid milk (consumer milk) and cheese milk vary considerably. The practice even varies from town to town in some countries, according to the status of local authorities.

Three systems are in use in countries concerned with cheese making:

(1) Flash heating (i.e. no holding time) to temperatures from 75 to 95°C.
(2) High-temperature, short-holding systems using temperatures from 71 to 75°C and holding times from 15 to 40 s.
(3) Low-temperature, long-holding systems using temperatures from 61 to 65°C and holding times from 20 to 40 min.

The equipment used for heat treatment of milks in most countries needs to be approved for use, to have certain specified recording thermometers and, in the case of HTST plants, a flow diversion valve to prevent inadequately heated milk passing forward to the cooler section.

In the case of cheese milks, the regulations may be relaxed and, in some areas, no regulations for cheese milks exist. However, it is generally accepted that the minimum standard of heat treatment for cheese milk should be the same as for liquid milk, so that pathogenic organisms are eradicated from all milks.

This wide range of heat treatments has evolved because of the advantages or disadvantages found by using the various kinds of treatment. Thus, some cheesemakers, desirous of maintaining the natural milk lipase activity, have preferred to use low-temperature heat treatments, about 65.5°C, with or without holding times; while these treatments preserve the lipases, they allow harmful bacteria (pathogens and some coliforms) to survive with cheese milk. Long-keeping, long-ripening cheeses allow time for pathogenic bacteria to die out; but the shorter-life cheeses or soft cheeses may still contain viable pathogens – if pathogens were present in the cheese milk. This is a contentious issue.

The number of bacteria in milk to be heat treated depends on the efficiency of the hygiene routines during milk production, as well as on the increase in numbers during transportation and subsequent storage. The increase in numbers during storage, as already noted, is dependent to a large extent on temperature. Any increase in individual types of bacteria depends on their optimum growth temperature.

After efficient heat treatment, the number of organisms should have been reduced by 92–99%. If the original numbers are high, then the numbers of survivors after heat treatment will be high. A total kill of 99% can only be expected under ideal conditions, but the percentage will depend on the heat resistance of the microflora present; a population of *Pseudomonas* will be far more easily inactivated than one of lactobacilli. The survivors will be capable of proliferation if incubated near to optimum growth temperature; therefore, cheese milks should not be held warm for appreciable periods before further use. Figure 10.6 shows the bacterial population of milk heat-treated at different temperatures, showing that it is not so much the numbers of bacteria present as the types of bacteria which are so important in the manufacture and ripening of cheese curds. Table 10.7 shows the heat resis-

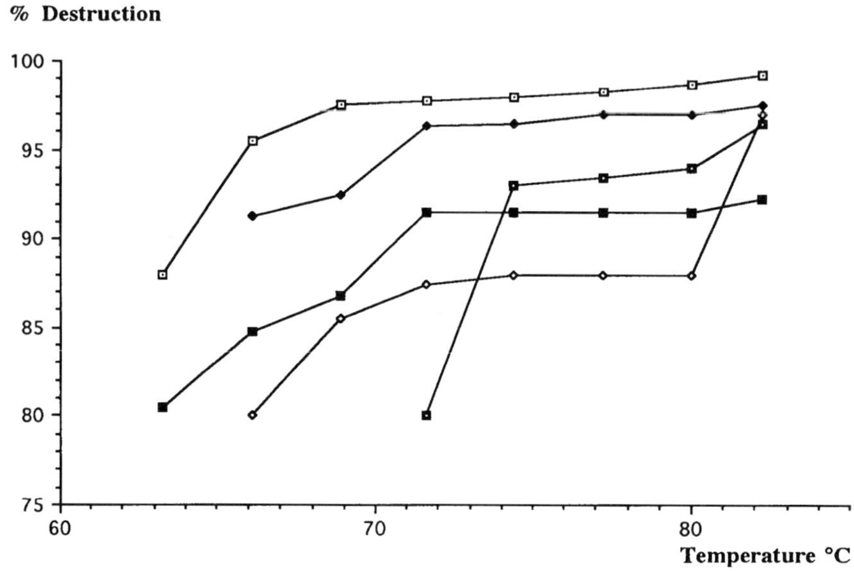

Figure 10.6 Graph showing the effect of heat treatment on the numbers of bacteria surviving in five different samples of milk.

tance, under laboratory conditions, of some bacteria isolated from milks. The table shows that the spore-forming bacteria, through their spores, show a high resistance to heat although their vegetative forms are less resistant.

Heat treatments of more than 65°C seem to activate at least some of the spores into active growth, and when milk is kept warm after heat treatment, as in a period of ripening by starter bacteria, these activated vegetative forms grow rapidly. Heat-shocked, but not destroyed, bacteria exhibit a long lag phase of growth which may extend for some hours beyond the time for curd formation. Such bacteria growing in ripening curd may give rise to discoloration in patches, or to small areas of bitter tasting curd.

Brucella melitensis, which has not been included in Table 10.7, is of interest in goat's milk rather than other milks where low-temperature, long-hold heat treatment might be used. The bacteria are destroyed at 62.8°C in 5 min, within the normal heat treatment range.

The bacteria tested for thermal death points were growing in liquid media (i.e. non-coagulated). Tests on milks containing these organisms heat treated through plate heat exchanger equipment showed the need for temperatures slightly higher (0.8°C) than the laboratory tests indicated. However, the phase of growth also affected the thermal death point. Bacteria in the lag phase of growth, or in the resting phase (48 h culture), showed slightly more heat resistance than actively growing cultures.

The thermal death point of some staphylococci at 75°C is 2 s; this makes the use of flash heat treatment reasonable for destruction of these bacteria,

Table 10.7 A list of bacteria isolated from milks showing their thermal destruction temperature when held at that temperature for 15 s before cooling takes place (laboratory tests)

Micro-organism	Thermal destruction temperature	Comments
Bacillus cereus	Vegetative cells 69°C	Spores survive 80°C
Bacillus subtilis	Vegetative cells 70°C	Spores survive 80°C
Bacillus licheniformis	Vegetative cells 70°C	Spores survive 80°C
Bacillus mycoides	Vegetative cells 69°C	Spores survive 80°C
Micrococcus spp.	Variable	Many survive 80°C
Micrococcus luteus spp.	65°C	
Entococcus faecalis	Survived 73.9°C	
Lactococcus lactis spp.	63.3–67.8°C	(various strains)
Lactococcus lactis sub-sp. *cremoris*	63.3°C	
Lactobacillus casei	70.6°C	
Lactobacillus plantarium	67.2°C	
Lactobacillus brevis	67.8°C	
Corynebacterium pyogenes	70.6°C	
Staphylococcus aureus	64.4–68.9°C	(various strains)
Staphylococcus citreus	68.3°C	
Proteus vulgaris	63.3°C	
Pseudomonas fluorescens	64.4°C	
Achromobacter lacticum	67.8°C	
Serratea marcescens	64.4°C	
Escherichia coli	65.6°C	(heat-resistant strain 68.9°C)
Klebsiella cloacea	68.3°C	
Klebsiella aerogenes	62.8°C	
Pediococcus cerivesae	62.2°C	
Salmonella spp.	70°C	
Coxiella burnetii	70.6°C	
Mycobacterium tuberculosis	70.6°C	
Brucella abortus	71.7°C	

as well as for others of similar low heat resistance. Although the thermal death point of the staphylococci was 75°C in 2 s when growing in liquid media, the heat resistance reached 80°C when the culture was in coagulated milk. Thus the heat penetration into coagulated milks is slower than in liquid media. The heat resistance of different strains of growing organisms varied by 2–6°C, apart from the phase of growth. Some coliforms (*Escherichia coli*) seemed to acquire heat resistance during growth, and while some were destroyed at temperatures as low as 62.2°C, a more resistant strain required 68.9°C for 15 s for complete destruction.

Normal heat treatment at 71.7°C for 15 s reduced the numbers of organisms responsible for hydrolysis of casein in milk, but the survivors grew rapidly in milk held at 30°C when longer periods of ripening were employed due to low inoculation rates of starter bacteria (i.e. less than 0.25% ripened for 2 h). While some hydrolysis of casein is desirable in ripening cheese curds, hydrolysis or proteolysis of casein prior to coagulation by rennet is likely to cause unsatisfactory curd structure after coagulation, i.e. soft, water retaining curds, or curds so soft that the fat and curd dust are released in the whey.

Both proteolysis and lipolysis by the Gram-negative pseudomonads which grow at low temperatures, 2–7°C in cold-stored milks, can cause severe difficulties in cheese milks at the coagulation stage and, subsequently, in the maturing cheese. Thus, cheesemakers have been advised to heat treat milks with large numbers of such organisms before cold storage (i.e. Thermization). This practice, which is contrary to the older theory that cheese milk should only be heat treated just prior to use in the cheese vat, is probably sound in view of the different flora now found in these cold-stored milks.

One of the major difficulties with regard to the growth of starter bacteria, apart from antibiotics or similar, inhibitory substances, has been the incidence in the milk of bacteriophage which destroys the starter organisms (see Chapter 9). As far as heat treatment is concerned, bacteriophage particles will withstand temperatures as high as 70–75°C for 30 min, and therefore heat treatment of cheese milk is not effective. However, some of the viruses responsible for animal and human diseases are destroyed by normal heat treatment at 71.6°C for 15 s.

If whey cream is added to milk, either to increase the fat content or for fat standardization, it is essential to heat the cream to 93°C for at least 30 min to destroy phage particles in the cream. Normally 50% whey cream is used for these purposes to reduce the amount of adsorbed cheesy flavours in the cream from the previous cheese.

The flora of milk fluctuates from day to day and so it is inevitable that the flora of heat-treated milks also varies. Figure 10.6 illustrates the variability of the flora in a sample of milk after heat treatment at different temperatures. While it is not possible to determine the bacterial flora of milks made into cheese in the time available for testing, the cheesemaker should be aware of these variations which may or may not produce variations in the cheese. It is equally important to regard the resident flora in the milk as a source of enzymes, which break down the constituents in the milk, and which provide those substances which give rise to flavour, aroma and variations in the texture and body of the cheese. Heat treatment destroys bacteria in the raw milk but it does not destroy all the enzymes which have been secreted by those bacteria into the milk. These enzymes continue to be active in the milk and curds when the conditions for activity arise (e.g. the correct pH). The enzymes from *Pseudomonas* spp. are more heat resistant than the bacterial cells responsible for their synthesis.

If the heat resistance of the bacteria shown in Table 10.7 is compared with the proteolysis, lipolysis and other reactions of those bacteria listed on page 75, it will be seen that the heat-resistant bacteria (i.e. bacilli and micrococci) which survive heat treatment at 71.6°C for 15 s are in the group which are highly proteolytic and lipolytic, capable of attacking both proteins and fats.

Cold storage of the raw milks may have caused the vegetative forms of bacilli to turn into spores which lie dormant. However, heat treatment of the milk awakens some of the spores, and, as vegetative forms, the bacteria multiply and become active in the milk. *Bacillus cereus* is responsible for the bitty cream floating on the surface of milk. *Alcaligenes faecalis*, although non-sporing, survives heat treatment and is responsible for the persistent cooked or burnt flavour in cheese milks. It ought to be the aim of cheesemakers to select raw milks with a minimum of spores, and other heat-resistant organisms. Farm advisory visits from the creamery can be a help with these problems although the visits are time consuming.

Removal of such spores by bactofugation is not feasible for the occasional spate of spores in milks which occur in dry weather periods, or when dusty conditions prevail. However, some creameries have found the use of bactofuge helpful in maintaining cheese quality, when the bactofugate which is removed is sterilized and remixed with the cheese milk.

Biochemical reactions in heat treated milks are difficult to forecast, especially those assisted or catalysed by enzymes. The following lists are given to indicate, in summary, the probable effects of heat treatment on the various components in milk.

1 **Enzymes**

Alkaline phosphatase	Destroyed at 71.7°C in 15 s.
Acid phosphatases	Not destroyed.
Milk lipases	Destroyed at 66°C in 15 s.
Bacterial lipases	Many survive heat treatment of cheese milks.
	Bacteria surviving heat treatment add more lipase to the milk.
Xanthine oxidase	Survives normal heat treatment.
Peroxidase	Survives normal heat treatment.
β-amylase	Survives normal heat treatment.
Proteases	Survive normal heat treatment.
Enzymes entering milk from the bloodstream – aldolase, salolase, rhodonase, etc.	Do not normally survive.

2 **Taints in milk**

Taints arising from animal foods, atmosphere or micro-organisms, etc.	Taints are not removed by heat treatment, but some volatile substances can be removed by atmospheric or vacuum cooling after heat treatment.

3 **Inhibitory substances**

Antibiotics	Not inactivated by heating.
Lysozyme	Not inactivated by heating.
Lactenins	Lactenin$^{(1)}$ is destroyed but Lactenin$^{(2)}$ survives.

4 **Salt balances in milk**	
Calcium	Heat-sensitive milks have high soluble calcium phosphate or calcium citrate contents. Increasing temperature drives the calcium phosphate into the colloidal state. Availability of calcium is governed by combination with phosphate and protein and is temperature dependent. Equilibrium after heating can take 48 h to establish (hence use of $CaCl_2$).
5 **Proteins**	
Micelle size	Temperature dependent, but also dependent on the equilibrium between calcium, magnesium and the phosphate and citrate ions.
Casein	Not itself affected by normal heat treatment of cheese milks but interaction with whey proteins is possible.
Whey proteins	Interaction with casein and milk salts is temperature and pH dependent. Interaction between β-lactoglobulin and the casein complex does not take place in normal heat treatment, but in disturbed milks some combination can take place. Denaturation of the whey proteins may also occur. The whey proteins begin to denature at 70°C and progressively denature at higher temperatures. Denatured whey proteins are enclosed in the cheese curd.
β-lactoglobulin	The most heat sensitive of the whey proteins. The heat sensitivity of milk is partly dependent on the amount present. Disturbed milks, i.e. from diseased udders or colostrum, have large amounts of β-lactoglobulin. When substantial amounts are present, heat treatment of cheese milks in the range 74–80°C causes harsh curd formation.
6 **Curd formation by coagulants in rennets**	
Free fatty acids	Liberation of the liquid free fatty acids and their absorption by proteins at sites required by rennets cause soft curds. Heat

	treatment can destroy natural milk lipases but not bacterial lipases.
Proteins	Denatured β-lactoglobulin and κ-casein interact during heat treatment. Since κ-casein guards the casein complex against coagulation, curd formation can be disturbed by this pathway. Colostrum decreases the stability of milk to heat due to the serum protein content, and may make drying out of cheese curds from heated milks difficult.
Calcium	Milks which need to be coagulated immediately after heat treatment often need calcium additions for satisfactory coagulation.
Proteolysis	Heat treatment cannot restore proteins which have been partially broken down by proteolytic enzymes which themselves are heat resistant. Heat treatment early in the life of cheese milks, by destruction of bacteria producing proteolytic reactions, can reduce the amount of protein breakdown and thus enable satisfactory coagulation of all the casein in milk.

Where lipases have attacked milk fat and released fatty acids, the observation by Anders and Jago (1970) that certain concentrations of oleic acid can cause a reduction in starter activity may be relevant. Also with regard to starters, Elliott and Emmons (1979) found that agglutination of cultures in skim milk occurred in 13 out of 20 tests and, as a result, the culture produced acid more slowly than normal. Heat treatment of the milk above 77°C for 30 min prevented agglutination.

10.6.1 Equipment for heat treatment of cheese milks

The equipment used for heat treatment of cheese milks is very variable. The small farmhouse cheesemaking dairy may be equipped with only simple vats or tanks, which may or may not be jacketed to carry hot water or steam. Sometimes the vat or pan may be placed over a slow fire to heat the contents, and such simple equipment has served well for many years, especially in isolated districts or in mountain huts.

Between the smallest volume equipment and the large scale factory equipment there are numerous other systems in use. Many small farmhouse

Figure 10.7 Plate heat exchanger used for heat treatment of cheese milk. (Courtesy of Wincanton Engineering Ltd)

dairies do not use heat treatment of the cheese milk and are, therefore, equipped with little more than vats, presses and a small hot water or steam boiler to provide heat for washing and other cheesemaking operations.

The larger scale factory operations require heat exchangers for most forms of heat transfer into or out of liquid milk or whey, since they provide the most economical usage of heat. The heat exchangers may be of the tubular type or more often the multi-plate type. An example of a heat exchanger for treating cheese milk is shown in Figure 10.7.

Once the throughput of a factory is large enough for standardization of the milk, then a composite plant built-up from stainless steel units is required. The units, including heat exchangers, pumps, separators and tanks, are so connected that air is excluded from the milk being processed. In such a system the cleaning and disinfection should preferably be automated. The range of equipment needed for large-throughput plant is much more extensive (e.g. as in Figure 14.5) and requires more highly trained operators than the simpler vat and draining table used for farm scale cheesemaking.

References

Anders, R.F. & Jago, G.R. (1970) *Australian Journal of Dairy Technology*, **25**(2), 68.
Bird, J. (1993) *Journal of the Society of Dairy Technology*, **46**(2) 35–37.
Chapman, H.R. (1974) *Dairy Industries International*, **39**, 329.
Dolby, R.M. & Harkness, W.L. (1955) *New Zealand Journal of Science and Technology*, **37**A,(1), 68.
Elliott, J.A. & Emmons, D.B. (1979) Proceedings of the Biennial Marschall International Cheese Conference, Madison, USA.
European Community (1992) *Council Directive 92/46/EEC Laying down the health rules for the production and placing on the market of raw milk, heat-treated milk and milk-based products.*
Lelievre, J. (1982a) *XXI International Dairy Congress, Moscow*, Vol. 1, Bk 1, 505.
Lelievre, J. (1982b) *XXI International Dairy Congress, Moscow*, Vol. 1, Bk 1, 504.
Lelievre, J. (1983a) *Journal of the Society of Dairy Technology*, **36**(4), 119.
Lelievre, J. (1983b) *New Zealand Journal of Dairy Science and Technology*, **18**, 63.
Luck, H. (1962) in *Milk Hygiene*, WHO, pp. 423–447.
McDowall, F. (1936) *New Zealand Journal of Dairy Science and Technology*, **18**, 237.
Romney, A.J.D. (1990) *CIP: Cleaning in Place*. Society of Dairy Technology, Huntingdon.
Stadhouders, J.H. and Hassing, F. (1974) *XIX International Dairy Congress*, Vol. IE, 369.
UK Regulations (1995) *The Dairy Products (Hygiene) Regulations 1995*. Statutory Instrument 1995 No. 1086, HMSO, London.
Whitehead, H.R. (1948) *Journal of Dairy Research*, **15**, 387.
Whitehead, H.R. & Harkness, W.L. (1954) *Australian Journal of Dairy Technology*, **9**, 103.

11 Coagulants and precipitants

As already mentioned in Chapter 1, the early tribes of the Middle East used natural fermentations to produce fermented milks (yoghurt, koumis, kefir). It was, therefore, a short step to the use of naturally soured milks or sour whey as a ferment to produce the acid necessary to give a precipitation of the casein which enclosed the other constituents of the milk. Separation of liquid serum (whey) from this coagulated milk leaves a curd (i.e. a soft lactic cheese).

The addition of vegetable latex or extracts from the stomach of mammals (a mixture of enzymes referred to as rennet) to milk produced a coagulum or 'junket' which, when drained, produced a cheese curd quite different in character to the sour, acid precipitated curds. Thus there are three types of curd as follows:

1. Precipitated curds, which may be sub-divided into:
   (i) salt-precipitated curds;
   (ii) acid-precipitated curds.
2. Enzyme-coagulated curds.

11.1 Salt-precipitated curds

The principle of 'salting out' proteins, also employed for many years in laboratory experimental work, has been used for the production of 'Baker's curd' from skim milk. To the skim milk, heated to 100°C for 1 h, is added a 'cream' made up of wheat flour and sodium sulphate (sometimes magnesium sulphate). The proteins are coagulated by the gelatinous flour 'cream' having been salted down by the sulphate. The curds are pressed to remove surplus whey and are then dried-off with more flour before being used for baking purposes (e.g. fillings). The curds tend to be open textured but suitable for mixing with other ingredients. Excess calcium chloride will also precipitate a curd, but the flavour is bitter and not usable.

11.2 Acid-precipitated curds

The precipitation of curds from milk by the use of acid has been common for many years. Vinegar, acetic acid, citric acid, lemon juice, lactic acid and even hydrochloric acid have been used to produce curds in the manufacture of some cheese. Ricotta cheese is made by the use of white vinegar or citric acid to precipitate curds from a mixture of whey and skim milk at a temperature of 85°C. However, Ricotta cheese, made from whole milk, used lactic starters to acidify the milk to 0.3–0.32% lactic acid before heating the acid milk to 80°C for flocculation of the curds. Because acid-produced curds are open in texture, stabilizers such as guar or carob gum may be used to assist curd structure.

Mozzarella cheese is also made with acetic or lactic acid as the acidulant, although lactic starters are also used. White vinegar is used for a second flocculation of the remaining proteins in Impastata cheese whey, after the first acidification with lactic starter (Kosikowski, 1977).

Queso Blanco in Latin America is often made from low-fat milks using lemon juice as the acidulant, although glacial acetic, citric or phosphoric acid has also been used.

Casein coagulation by acid begins at pH 5.3 and is complete at its mean isoelectric point, ≈pH 4.6. The coagulation is dependent on temperature and salt balance. Thus, spontaneous coagulation of skim milk of 0.25% acidity occurs at about 82°C, while at 0.35% acidity the coagulation occurs at 65°C. It is possible to add acid to skim milk at 2°C until it is near the isoelectric point without coagulation.

The calcium and phosphate is mainly in soluble form at pH 5.2 and the size of the casein micelle aggregates is increased by higher levels of available calcium; hence the use of calcium chloride additions to increase the aggregation of the micelles. Thus, 0.07 M calcium chloride at 5°C causes the micelles to aggregate into a form removable by a centrifuge. High heat treatment causes serum proteins to denature and to coat the casein micelles. The serum proteins are almost completely denatured in 5 min at 90°C, a condition often required for starter milks but usually avoided for food products since there is sufficient activation energy to initiate interactions between the reducing group of the lactose and free amino groups of the proteins. The products of this Maillard browning reaction include cooked flavours and brown pigments. Little colour is produced at 80°C but over 90°C the colour becomes more intense.

Lactic acid produced *in situ* in the milk by bacteria is a preferred method for acidification, whether for skimmed milk or whole milk cheese. Lactic starters using *Lactobacillus bulgaricus*, *Lactococcus delbrueckii* sub-sp. *thermophilus* and *Lactococcus lactis* spp. are commonly used for many of these cheeses. Sometimes acid whey from ultrafiltration of cheese whey has been used in preference to more concentrated acids, e.g. for the production of Queso Blanco cheese (Kosikowski, 1977).

When acid is used for curd production, a gradual reduction of the pH of milk from 6.8 towards 5.3 results in aggregation of the micelles to form weak clots, while continued acidification results in more complete gel formation. This method of curd formation is used in Cottage cheese, lactic cream and similar cheese. Glaser *et al.* (1980) found a steady aggregation of micelles during Cottage cheese production by acidification.

Emmons *et al.* (1982) reported that high heat treatment of milk for Cottage cheese tended to cause brittle curds and loss of 'fines' in the whey, while Kosikowski (1982) found that the high heat treatment of milk caused soft, weak-bodied Cottage cheese with a high moisture content. These results are consistent with the formation of complexes of casein micelles and denatured β-lactoglobulin leading to greater water retention.

Proteolysis by enzymes in the milk tends to cause soft curds which do not drain well and give poor yield. Similarly, curds from unstable milks (colostrum) also produce soft sticky curds on account of the increased whey proteins.

Reconstituted milk powders can be used to make curds using either organic acids or lactic starters as acidulants. The solubility of the milk powder (reflecting the heat treatment the powder has received during drying) affects the texture and yield of the curds. Low heat treated powders with high undenatured serum protein contents are normally preferred for curd production.

11.3 Enzyme coagulants

Mankind has used coagulants from vegetable and animal sources for centuries, but in recent years, coagulants from bacteria and fungi have also been used. However, the use of a natural coagulant, rennet, extracted from the stomach of a mammal to produce cheese curds from milk, was a natural choice for the early cheesemakers.

It is important to note that milk from the goat, ewe, buffalo, reindeer, yak, and llama, as well as the cow, is used for cheese curds. Even the milks from non-ruminant mammals (i.e. the camel, mare and ass) have also been used for curd making. Indeed, these milks are nutritionally and economically important for some peoples. Although the more likely usage of milk from the camel, mare and ass is in non-fermented milk products, the milk from the camel will coagulate with rennet; the milk from the mare and ass will not coagulate with rennet, and needs precipitation with acid.

The type of coagulum formed is dependent on the enzyme used, salt balance in the milk, pH, proportion of other milk ingredients (fat, whey protein), and the composition of the casein itself. Casein which has been previously attacked by enzymes cannot form so complete a curd as the native casein. A curd produced at low pH (4.6–5.0) tends to be granular, particulate and inelastic, due to precipitation of the casein at or near its isoelectric point where it is insoluble. The coagulum produced by an enzyme at higher pH (5.8–6.6) is softer, smoother, shrinkable and more elastic than the acid curds.

Although the individual components of milk interact with one another in a very complex system, the casein is the substrate on which the coagulant enzyme acts to produce, in the presence of calcium ions, an insoluble coagulum which encloses the remainder of the components of the milk (i.e. water, fat, salts, proteins). The coagulum may be consumed as 'junket', often sweetened and flavoured, but needs to lose moisture to become a cheese curd.

11.3.1 Animal rennets

Calf rennet is still regarded as the most important rennet for cheesemaking, and its performance is regarded in most quarters as the standard against which other rennets are assessed.

The production of rennet on a small scale, as practised by smallholder farmer cheesemakers, consists first of washing the fourth stomach (abomasum or vell) of the newly slaughtered milk-fed calf or other young ruminant (kid or lamb), and then slicing the stomach into strips which are extracted in sodium chloride solutions (12–20% salt) to give a solution of the rennet enzymes in salt solution. This is filtered, and purified by 'salting down'. The solution has been preserved by boric acid and salt in the past, but boric acid is not permitted in the EC. This crude extract from the fourth stomach of the calf is called rennet. Various extracts can be made and the coagulating power of each extract varies.

The enzyme present at this stage is pro-rennin which, on acidification, becomes rennin with an increased clotting activity. The term chymosin is used in place of the older name rennin to avoid confusion with another enzyme, renin, which can be extracted from the kidney. The rennet extract from the young milk-fed calf contains 88–94% rennin and from 6 to 12% pepsin, while in extracts from the older fodder-eating bovine, almost the reverse is true: 90–94% pepsin and only 10% rennin. There is, therefore, a variable amount of pepsin in rennets depending on the age and food of the calf from which the extract is obtained. Although most information available is about the calf rennets, extracts from the stomachs of other animals have been used, i.e. the kid, lamb, buffalo and pig.

Since rennets of different strength produce different cheese curds, it is important to be able to measure the strength of coagulating activity of the extract. The commerical manufacturers of rennets have used skim milk or a substrate made up from skim-milk powder – 12 g dissolved in 100 ml of a solution of 0.01 M calcium chloride. Alternatively, the unit of rennet activity (RU) is defined as the number of millilitres of milk which can be clotted in 40 min at 35°C.

The British Standards Institution method uses a freeze-dried purified rennet as a standard (BS 3624; 1963, British Standards Institution, London) for estimating the clotting power of rennets. Foltmann (1970) proposed a unit based on chromatographically pure chymosin B with a spectral absorbancy of 1.00 at 278 nm that contains 100 rennin units per ml. More recently, the IDF standard for determination of chymosin and bovine pepsin included a method for determining the clotting time (IDF, 1987). Molecular aspects of rennets have been reviewed by Foltmann (1993).

These methods give the clotting power of rennet/chymosin extracts but they do not measure the proteolytic activity of the rennet; there has been interest in the proteolytic activity of the rennets since the use of microbial rennet and pepsins became widespread. The clotting to proteolytic activity ratio (CPR) is a meaningful expression of the usefulness of an extract.

Buffalo rennet is used for Kesong Puti cheese in the Philippines, but this type of rennet has not been widely used in other types of cheese. The rennets from lamb and kid are more widely used for cheesemaking. Lamb rennet gives a sharper flavoured cheese than when calf rennet is used (e.g. Kefalotyri cheese). Some cheese markets (e.g. Greece and Italy) prefer a sharper flavoured cheese, but if an even higher lipolytic activity is required from the coagulant, then a paste made from the ground-up mucosa of the lamb or kid is used, along with the rennet. The lipases in the pastes produce a characteristic flavour in the cheese, but if rennet extract is used alone, then some lipase needs to be added before coagulation of the milk.

Rennets from the pig contain pepsin (i.e. porcine pepsin) which is used along with calf rennet for the coagulation of milk (i.e. 50:50 rennet). Calf rennin is most active at pH 6.2–6.4, whereas pepsin is most active in the high acid region (pH 1.7–2.3), so that the two activities are complementary, rather than additive. Coagulants which were mixtures of enzymes began to be used when milk production expanded and calf rennets became scarce (e.g. in wartime). The following coagulants have been used:

Calf rennet
Bovine rennet
Bovine pepsin
Porcine pepsin
Calf rennet/bovine pepsin, i.e. 50:50 rennet
Calf rennet/porcine pepsin, i.e. 50:50 rennet

For the latter two coagulants proportions other than 50:50 ratios have been used, i.e. 25:75.

When the microbial coagulants came into use, the following types of rennet became available:

Fungal extracts
Bacterial extracts
Calf rennet/fungal extract
Calf rennet/bacterial extract
Fungal extract/pepsin
Bacterial extract/pepsin

The proportion of the coagulant retained in the cheese curd after pressing is important with respect to the future maturation of the cheese. Of the rennet added to milk, approximately 30% of the calf rennet is retained in the curd before pressing, but only 5–8% is retained after pressing; 3–5% of microbial rennet can be expected in pressed cheese and 3–8% of pepsin. The whey from cheese, therefore, contains a high percentage of the coagulants, which can be a disadvantage if the whey is to be used for other products.

The amount of rennet added to cheese milk is governed by the recipe, which also specifies the temperature and acidity of the milk, the aim of the recipe being to obtain a coagulum ready to cut or break within a specified period of time. Temperature is an important factor in the time a curd takes to form. At 21–27°C, the curds tend to be soft and jelly-like; at 30°C the curds are firmer and do not break into dust on cutting, while at 33–36°C the curds are tough and rubbery and release whey slowly. Below 20°C and above 50°C, rennet activity is very low (Figures 11.1, 11.2 and 11.3). Similar data to that shown in Figure 11.2 was obtained by Ramet and Weber (1980).

The rate of clotting is also important since, during the coagulation period, the lactic bacteria are still actively growing. A long coagulation time may, therefore, result in acid production being higher than normal in the subsequent operations.

The acidity, or really the pH of the milk at the time of the addition of rennet, is also relevant to the formation of the curd structure (i.e. the

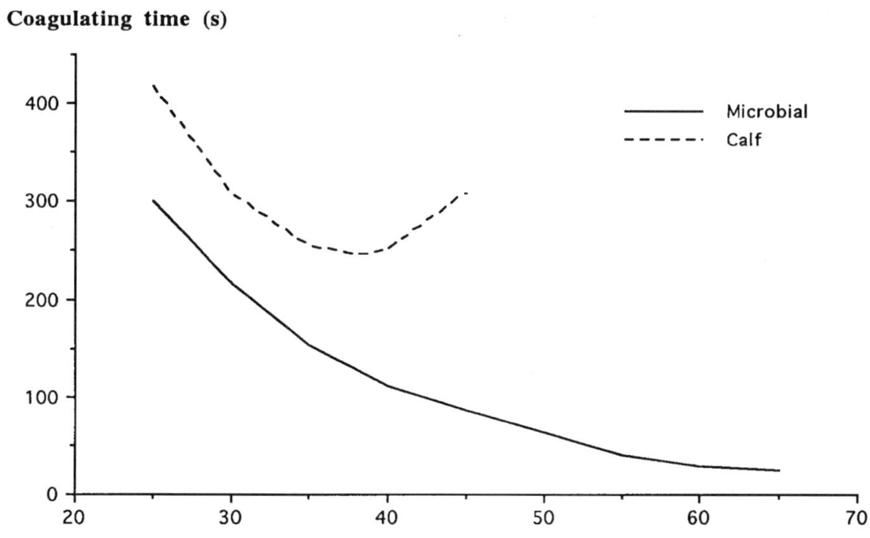

Figure 11.1 Graph showing the time taken to coagulate milk by calf and microbial rennets in relation to the temperature of coagulation.

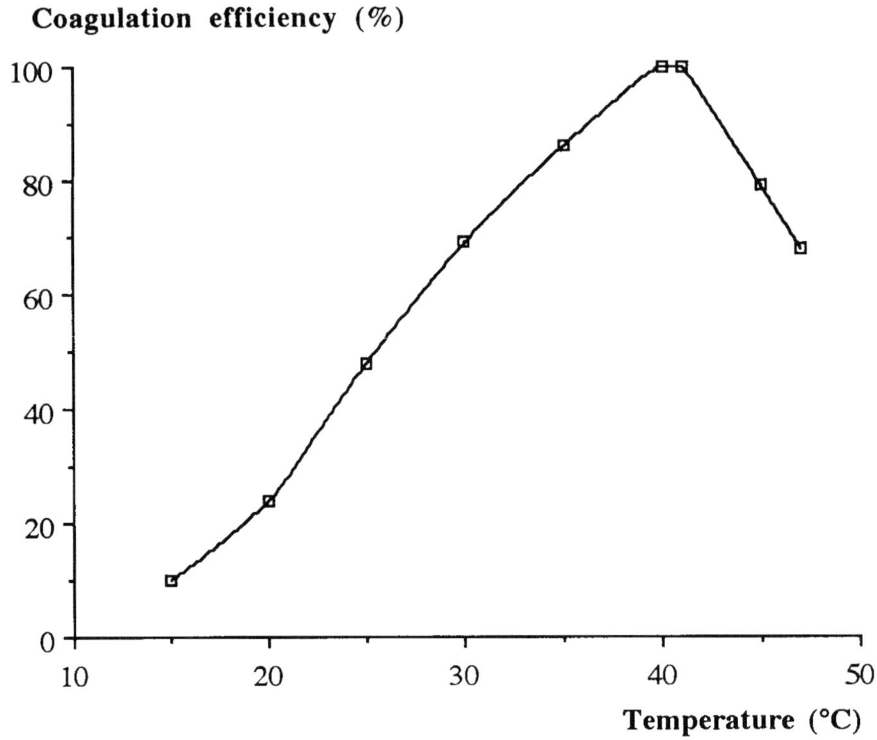

Figure 11.2 Graph showing the efficiency of rennet as a coagulant at different temperatures.

firmness, Figure 11.4) and, therefore, controls the time at which the coagulum is ready to cut or break. The quantity of rennet used varies from 10 to 45 ml per 100 litres; but in some cheese made by the acid precipitation of curd (e.g. Cottage cheese) a smaller amount of rennet is added (2–5 ml per 100 litres) in order to toughen up the precipitate. In all cases the rennet must be diluted with clean water (ideally ≈10 times) to aid even dispersion of the rennet in the milk.

Rennet loses strength rapidly after dilution. Light will also reduce the activity of rennet. Commercial manufacturers of rennets have individual methods of preparation, standardization and preservation of rennets. One method of purification of rennin is by ‘salting out’ the enzyme. If the ‘salting out’ process is carried on long enough, a dry powder can be produced. From this dry powder, tablets are made to a standard activity (e.g. one tablet per 45 litres of milk). Rennet tablets are particularly appropriate for small-scale production and for use in hot climates where there may be problems in keeping liquid rennets refrigerated.

In cases where the whey is used for human food preparation, preservatives would be an embarrassment, glycerine has been used in place of boric acid.

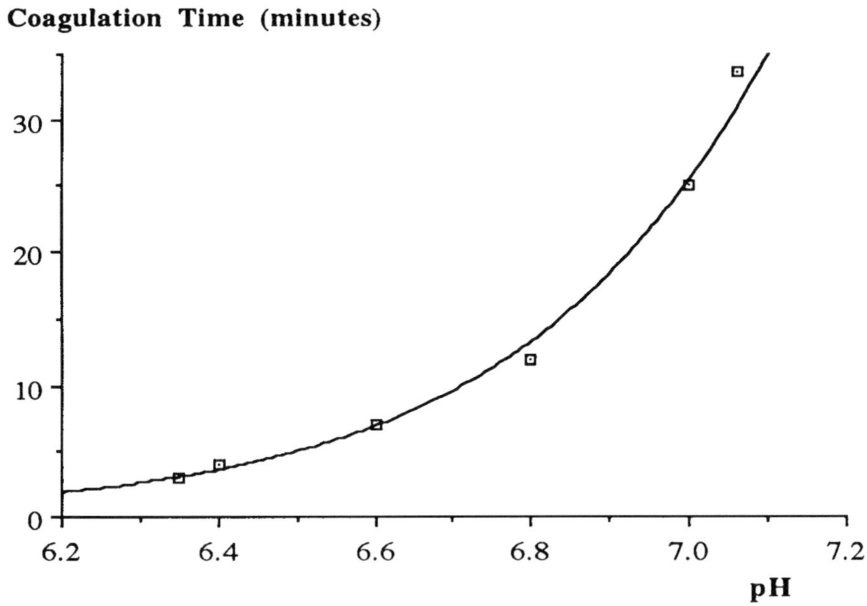

Figure 11.3 Graph showing the effect of pH on the rate of clotting of milk by rennet.

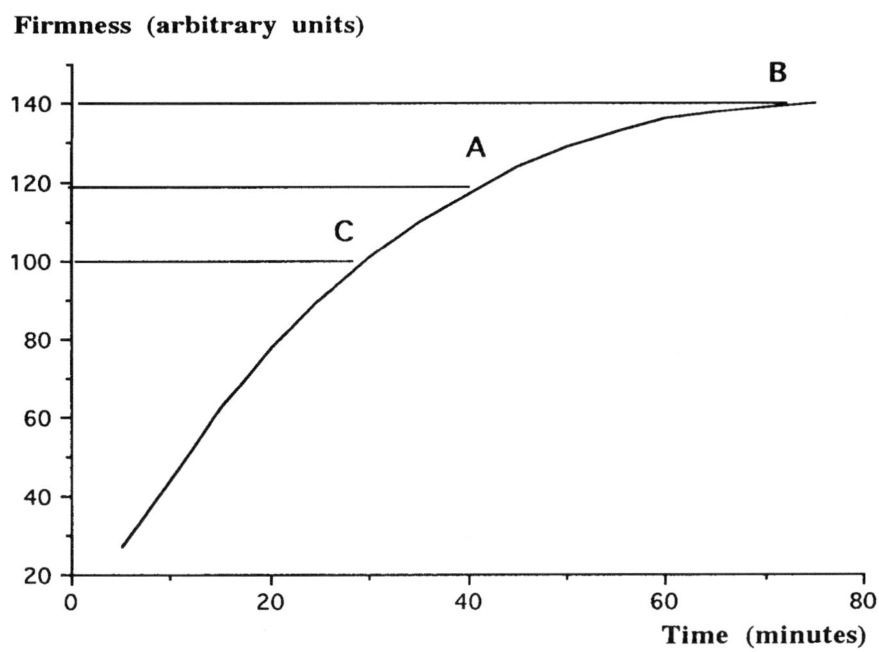

Figure 11.4 Graph showing the firmness of a coagulum in respect of the setting time. A, normal cutting time; B, longer time and hence a hard curd; C, too short a time and hence too soft a curd.

Alcohol, glycerol and iodine have been used to produce sterile rennets. Generally an extract with a 17–24% salt content has been sufficiently bacteriostatic to prevent deterioration of the rennet enzyme by the proteolytic bacteria present. Admixture of other enzyme systems to calf rennet may introduce proteolytic enzymes which degrade the rennet, which itself is a protein. Calf rennets will lose 1–1.5% of coagulating activity per month, but in the first few weeks after preparation the loss is much greater; therefore, rennets are normally stored for a period until reasonably stable.

11.3.2 Pepsin rennets

Pepsin is a natural proteolytic enzyme present in many calf rennets. This enzyme is present in the gastric juices of older bovines, where it replaces the chymosin of the younger animal. The ratio of the clotting activity to proteolytic activity of pepsins from bovine, porcine or ovine species differs from that of the calf chymosin, and recommendations to use bovine or ovine pepsins for cheesemaking have been made in respect of the CPR for their pepsins. Porcine pepsin is reported to have little or no clotting activity or pH greater than 6.68, and loses activity above 44°C (see scald temperatures, Chapter 13). Hence pepsins are not so active in 'high scald' cheese curds. Pepsins prepared from chicken entrails have been used in Israel for cheesemaking.

Cheesemakers used to traditional rennets may find that the clotting period before cutting the curd is longer with pepsin rennets than with calf rennets, although the curd firms up later. Pepsin in the whey is destroyed by moderate heat treatment and the whey can be used for other products afterwards.

11.3.3 Microbial rennets

Shortage of animal rennets, especially during the stress of World War 2, and the rapid subsequent increase of milk production, stimulated a renewed search for other types of coagulants for milk. Many hundreds of bacterial and fungal cultures have been investigated for milk clotting enzymes and their proteolytic abilities. Few have reached commercial production, although the search goes on and new sources are reported from time to time. Examples of those micro-organisms which have reached the commercial stage are listed in Table 11.1.

The rates of addition of the preparations vary slightly from 18.75 ml per 100 litres milk, up to 25 ml per 100 litres, but each commercial preparation has directions for use in the form of printed instructions. Some preparations are granular and need to be dissolved in water before use (e.g. Noury rennet, 1.2 g in 10 litres water per 100 litres milk). New sources of coagulants are developed from time to time.

Table 11.1 Some of the available microbial rennets

Name	Source of enzyme
Hannilase	*Mucor meihei*
Rennilase	*Mucor meihei*
Fromase	*Mucor meihei*
Miki	*Mucor meihei*
Marzyme	*Mucor meihei*
Novadel	*Mucor pusillus*
Noury	*Mucor meihei* (Lindt)
Meito	*Mucor meihei* (Lindt)
Emporase	*Mucor meihei* (Lindt)
Suparen	*Endothia parasitica*
Sure curd	*Endothia parasitica*
Mikrozyme	*Bacillus subtilis*
Chymogen[1]	*Aspergillus niger*
Chy-Max[1]	*Escherichia coli*
Maxiren[1]	*Klyveromyces lactis*

[1] Recombinant chymosins.

The following micro-organisms have been reported as sources of coagulants: *Bacillus polymyxa* (strain 13-13), *Bacillus mesentericus* (strain PB) and the mushroom *Russula discolorus* (strain 2). Various strains of these organisms behave differently; therefore, the cheesemaker should use only well tried and tested rennets for the types of cheese contemplated. Some microbial strains produce rennets for soft cheese, but produce bitter flavours when used to make riper cheese. Therefore, it is important to recognize that the activity of the various microbial rennets varies according to pH and the enzyme system.

The protease of the *Mucor miehei* coagulants splits the peptide bonds at the aromatic amino acid residues thus: at phenylalanine/valine, leucine/tyrosine, phenylalanine/phenylalanine, or at phenylalanine/tyrosine bonds. The enzyme degrades casein fairly rapidly in the pH range 5.5– 7.0; however, in spite of this, the incidence of bitter cheese using this enzyme is low, and this extract is used successfully for many types of cheese. Addition of sodium chloride to milk, as in some recipes, slows down the coagulation with the enzyme but the addition of calcium chloride reduces the clotting time considerably. The effect of the calcium ion concentration on clotting of milk by *M. miehei* rennet is given in Figure 11.5. However, the enzyme is very sensitive to temperatures in the region 37–45°C (optimum 42°C) and is destroyed at 70°C. Thus, the enzyme can be destroyed in whey heated above 70°C. Unfortunately, some whey proteins are denatured in this process, which may be a disadvantage in further usage of the whey. The enzyme is active in the pH range 5 5–7 5, and attacks both α and β caseins in this range, but incidence of bitter flavours in cheese curds is low. Although *M. miehei* rennets are often used alone, some cheesemakers have used a mixture of *M. miehei* and pepsin coagulants satisfactorily.

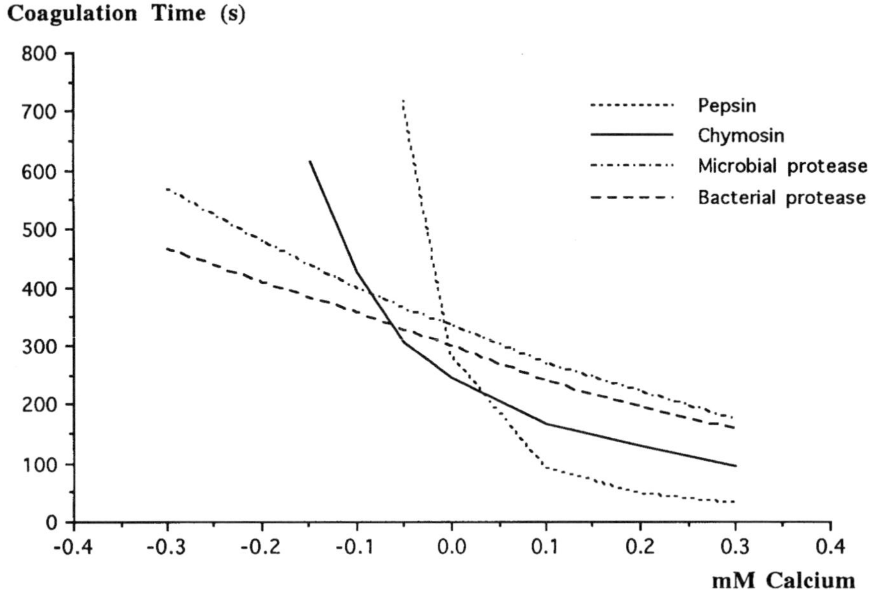

Figure 11.5 Graph showing the effect of calcium on the time taken to coagulate milk by various enzymes.

Mucor pusillus (Lindt) extract was one of the earlier microbial rennets to be used but the extract is more highly proteolytic than either calf rennet or the *M. miehei* extract. The clotting ability of the extract is three times that of calf rennet although the commercial coagulant is standardized before distribution. An increase in the calcium ion concentration in milk decreases the clotting time and added calcium chloride is almost essential. However, the activity is not so pH dependent as for other coagulants. The enzyme from *M. pusillus* splits the peptide bonds at the aromatic amino residues, but is not so specific as *M. miehei* extract. The enzyme produces large peptides, whereas calf rennet produces a number of smaller peptide chains and, for this reason, *M. pusillus* extract and calf rennet are sometimes used together.

The enzyme from *M. pusillus* tends to give hard curds because of its higher proteolytic activity and the curds tend to lose fat into the whey. Thus, 0.4–0.5% fat in the cheese whey is common and the cheese yields are lower than with other coagulants. This coagulent may be blended with pepsin (porcine) for some cheese varieties. *M. pusillus* enzyme alone tends to produce bitter flavours in ripe cheese and has thus been used for short cure cheese. Newer methods of purification of all microbial extracts tend to produce more satisfactory milk clotting extracts with less proteolytic activity. *M. pusillus* extract has optimal activity at pH 5.5, which is very acid for normal coagulation of milk, and its activity may persist in the whey, which is a disadvantage if the whey is to be used subsequently.

The enzyme extracted from *Endothia parasitica* is an acid protease but in some extracts a number of proteases and lipases may be present if the extract is not well purified. The *E. parasitica* coagulant is more caseolytic than *Mucor* or calf rennets and tends to produce more bitter flavours in high moisture cheese. The recipe used for cheesemaking with *E. parasitica* extract needs to be adapted, and probably the extract should be used with calf rennet to improve the curd yields. *E. parasitica* extract is destroyed at 60°C in 5 min and whey is usable if heated above this temperature.

Improvements have been made to the newer microbial coagulants by purification procedures to remove unwanted material. Purification procedures have included the use of silicates and bentonite as absorbents, or zeolites in the range between pH 3 and pH 9, to remove those substances and enzyme systems not required in the rennet.

11.3.4 Recombinant chymosin

Much of the development work on microbial rennets has been overtaken by the development of techniques for transferring the gene for the expression of prochymosin from calf abomasum tissue to a range of host microorganisms, for example *Klyveromyces lactis* and *Aspergillus niger*. The gene has also been synthesized and inserted into *Eschericia coli* for commercial chymosin production (Teuber, 1991). As with natural rennet, the prochymosin may be converted to chymosin by treatment with acid.

The structures of the recombinant chymosin are virtually identical to that of the original calf chymosin, and similar results to the natural rennet have been obtained in cheesemaking (Fox and McSweeney, 1997). These rennets are sold as products of fermentation, and examples have been included in Table 11.1.

11.3.5 Vegetable rennets

Amongst the early vegetable coagulants reported is the latex of the fig tree (*Ficus carica*). This extract has been used from the earliest times in those areas where the fig grows. Many extracts from vegetable sources will coagulate milk, but some are too proteolytic (e.g. papain from the paw paw, *Carica papaya*, bromelin from the pineapple, *Ananas sativa* and ricin from castor oil seeds, *Ricinus communis*).

Extracts from plants traditionally thought to coagulate milk enzymatically do not do so by this mechanism; these include marjoram, mint, rosemary, plantain, sorrel and ground elder. Table 11.2 gives a few of the plants reported to give extracts suitable for milk coagulation. The plants usually have an abundance of cell sap.

Some plants give extracts which are poisonous, e.g. hemlock (*Conium maculatum*) and castor oil seeds (*Ricinus communis*). A number of the plant

Table 11.2 Plants giving extracts for milk coagulation

Cardoon	*Cynaria cardunculus*
Burdock	*Articum minus*
Bitter sweet	*Solanum dalcamara*
Mallow	*Malva sylvestris*
Thistle	*Cirsium* and *Carlina* spp.
Fig tree	*Ficus carica*
Hogweed	*Herculeum spondylum*
Knapweed	*Centurea* spp.
Lady's bedstraw	*Galum verum*
Nettle	*Urtica diocia*
Ragwort	*Senecio jacobea*
Spearworts	*Ranunculus* spp.
Spurge	*Euphorbia lathyrus*
Teasle	*Dipsacus sylvestris*
Yarrow	*Achillea millefolium*
Withiana berry	*Withiana coagulans*

extracts are used in isolated places where other coagulants are not easily available. Some extracts, produced under crude conditions, appear to produce a combined acid and enzyme coagulation, which is used mainly for the softer curd cheese. The Serra da Estrela cheese of Portugal is an example of the use of a vegetable extract, made by coagulating ewes' milk with an aqueous extract of cardoon flowers.

Because of the multiplicity of coagulants, it is important for the cheesemaker to realize that each coagulant has its own characteristics, and that conditions for use of each respective coagulant must be observed when changing from one coagulant to another or when mixing coagulants together.

11.4 Observations on the use of some rennets

Calf rennet is acknowledged in many references as the best coagulant; indeed calf rennet has been regarded as a standard against which other, perhaps less expensive, coagulants have been compared, particularly in respect of

(a) cheese quality (body texture, aroma and flavour),
(b) cheese yield,
(c) loss of 'fines' in the whey, and
(d) the quality of the whey for further processing.

The retention of coagulant properties in the whey after heat treatment, is unsatisfactory for further usage. Hyslop *et al.* (1979) have given thermal inactivation temperatures for some rennets, thus:

M. miehei extract, 77°C at pH 6.0 for 15 s,
M. pusillus extract, 71°C at pH 6.0 for 15 s,

while calf rennet is inactivated at 67°C at pH 6.0 for 15 s.

In the 1970s, there were many references to a variety of enzymes used for making numbers of different types of cheese. De Koning (1980) reviewed rennet and its substitutes and Scott (1979) considered the use of rennets in cheese and indicated a variety of different cheeses made with numerous rennets. The general trends regarding coagulants were:

(a) the proteolytic activity of rennets increases in the following order: calf rennet < *M. miehei* extract < *M. pusillus* extract < bovine pepsin < porcine pepsin < *E. parasitica* extract, and
(b) the sensory quality of the cheese appears to decrease in the following order: calf rennet > *M. miehei* extract > *M. pusillus* extract > bovine pepsin > porcine pepsin.

However, many of these rennets have been reported to be satisfactory when used in mixtures with other enzymes. Some microbial rennets, however, have not found favour; thus de Koning *et al.* (1981) advised against the use of Noury rennet made from *M. pusillus* (Lindt); Dolizalek and Konradova (1980) found that Milkozyme produced bitter flavours, while Lactochyme gave alkaline peptides in soft cheese. El-Safty *et al.* (1980) used *M. pusillus* rennet for Domiati cheese, but at higher coagulation temperatures (48–50°C). Denkov and Vasileva (1979) found that Milkozyme (*Bacillus polymyxa*) was unsuitable for white pickled cheese. Gutfield and Rosenfeld (1975) reported on the use of chicken pepsin for cheese in Israel, but Stanley *et al.* (1980) found chick pepsin unsuitable for Cheddar cheese in Canada.

Since the appreciation of flavour quality in cheese varies from one community to another, it is evident that a coagulant may find favour in one area and not in another. During the 1980s attention was focussed on refining enzymes to make them more suitable for cheesemaking. Thomsen and Birkkjaer (1980) reported Rennilase SP215 and SP216 as having lower heat stability than normal Rennilase L14. Lynggaard-Jensen and Petersen (1982) reported on the clotting activity of destabilized *M. miehei* rennet. Although Tofte-Jespersen and Dinesen (1979) reported that microbial enzymes give lower cheese yields, it would seem that further refinement of microbial enzymes is required to make them more suitable as coagulants, even if they are only suitable for specific classes of cheesemaking.

11.5 Factors influencing coagulation

From the cheesemaker's point of view, the formation of a coagulum or gel from liquid milk is but a first stage in curd formation. Initially this coagulum encloses all the ingredients of the milk, i.e. fat, protein, lactose, salts, microorganisms and water. This coagulation occurs in two stages, only the first being enzyme mediated. In this first stage, the chymosin (or alternative protease) attacks the hydrophilic κ-casein peptide chain that protrudes from the casein micelle into the surrounding milk serum. This attack is at the 105–106 peptide bond (phenylalanine–methionine), specified by the amino acid sequence on each side of these residues. The hydrophilic macropeptide (residues 106–169) then diffuses off into the milk serum, raising the soluble nitrogen to $\approx$ 4% of total nitrogen and leaving behind a hydrophobic para-casein residue. This stage occurs rapidly, Walstra and Jenness (1984) suggest a turnover rate of 100/s. The reaction has a Q_{10} of $\approx$ 3, and will progress at low temperature.

Continuing hydrolysis of the κ-casein is accompanied by an increasing hydrophobicity of the casein in the milk. Under normal conditions, when $\approx$ 90% of the κ-casein has been hydrolysed, the paracasein/casein is sufficiently hydrophobic for flocculation (gelation) to take place. This step is temperature and calcium ion dependant. The Q_{10} $\approx$30, but is not a good guide as the energy of activation increases with drop in temperature, probably caused by temperature related changes in the configuration of the para-casein micelles. Gelation does not occur readily at low temperatures. Calcium ion activity is important; levels should normally be above 1.5 mM, but higher levels may be needed if the micelles are low in calcium phosphate.

Flocculation is a second-order reaction, so the rate will increase with casein concentration. A consequence of this is more rapid coagulation with concentrated milks, e.g. in milks concentrated by ultrafiltration. During flocculation the micelles give rise firstly to chains, and then to strands which form a three-dimensional net-like matrix in which the fat globules are trapped, the paracasein and fat taking up less than 7% of the volume of the gel in the case of cow's milk. The bridges contract as the gel becomes firmer, with formation of micellar clusters and finally amorphous masses, although fusion does not mask the identity of the micelles. Munya and Larsson-Raznikiewicz (1980) showed that the addition of calcium ions to native milk caused the micelles to form larger particles, and that it was the micelles which aggregated rather than the monomeric caseins. Quist (1979) showed that the addition of from 4 to 20 mg of $CaCl_2.2H_2O$ per 100 ml of milk reduces the rennet coagulation time (RCT). The secondary phase of rennet coagulation has been reviewed by Green and Grandison (1993).

Contraction of the coagulum leads to expulsion of serum from the gel. The rate of exudation may be increased by increasing the surface area of the gel. Van Leeuwen (1982) produced results indicating that the cutting of larger cubes of firmer curds produced a higher moisture in Cheddar cheese curds than if the curd were cut into smaller cubes. Hansson (1979) suggested that the process of coagulation should proceed to a specific hardness before cutting the curd. The elasticity of the curd is affected by:

(a) cutting time,
(b) quantity of rennet,

(c) pH,
(d) temperature, and
(e) the addition of soluble calcium.

A number of instruments have been devised to measure the rate of coagulation (Vanderheiden, 1976; Olsen and Kowalchyk, 1978; Marshall 1982). Zannoni and Annibaldi (1981) have described the Foss Electric Formograph for assessing the rennetability of milk.

Reuter *et al.* (1981) reported that the time for the first phase of rennet activity decreased with the concentration of the milk, and that increasing the amount of rennet decreased the time of the first phase of rennet action. Increase in the fat content of milk decreases the rate of curd formation (Marshall, 1982) and also the rate of syneresis of the curd (Storry *et al.*, 1983). Green *et al.* (1981) found that sensory firmness, crumbliness, granularity and dryness of curd increased with the concentration factor of the milk. They suggested that the basic structure of curd is laid down during the curd firming. The protein network increased in coarseness with the concentration.

Disturbances of the milk constituents also affect the coagulum and curd therefrom. Humbert *et al.* (1980) found that homogenization reduced the clotting time of milk and that high homogenization pressures (40–60 MPa) could result in disagreeable flavour production, whereas a pressure of 20 MPa was satisfactory. Storry and Ford (1982) noted that homogenization reduced syneresis of the curd by binding fat and altering the curd structure. Green *et al.* (1981, 1983) suggested that homogenization of milk tended to slow down curd formation and decrease the rate of syneresis of the curd. Curds from homogenized milk did not fuse so readily as those from normal milk.

Processing methods affect the characteristics of the curds according to Emmons *et al.* (1980). The protein networks made from milk in which only the cream was homogenized were more open but smoother, softer and more elastic than normal. In low fat cheeses, the curds tended to be firm and crumbly.

Apart from processing, the structure of curds may be affected by the coagulants used. Eino *et al.* (1979) found that curd structure was more open when made with pepsin than with calf rennet. Enzymes which remove peptides from casein, temperature regimes which alter the solubility of casein, and salt balances in the milk (calcium and phosphate), as well as the ionic strength and pH of the milk, all affect the ability of micelles to form a network – the basis of satisfactory cheese curds.

11.6 Direct acidification

In some cheeses, the coagulation is primarily the result of acid precipitation of the casein. Direct acidification of milk to produce unripened cheese has been practised as an alternative to the growth of lactic acid bacteria for a long time (Fox, 1978). Cottage cheese, Ricotta type cheeses Impastata, Queso Blanco, and Mozzarella (Pizza) cheese are some of the cheeses made by direct acidification methods. Citric, acetic, lactic, phosphoric, tartaric and hydrochloric acids have also been employed to produce curds for these soft unripened cheeses, the flavour of which is acid but mild and lacking the distinctive flavours associated with ripened cheese.

Ricotta-type cheeses are made from partly skimmed milk, skim milk, or skim/whey mixtures heated to approximately 88°C and acidified to pH 5.0 with citric, lactic or acetic acid. The resultant curds are scooped off the whey and shaped or moulded. The cheese may be stored until dry and then used as a grating cheese (Kosikowski, 1977).

Lactic acid cheese starters (bacterial) have, however, been preferred as precipitants for whole and skim-milk cheeses on account of the rich nutty flavour, good cheese yield and clean flocculation of the curds.

The use of direct acidification in cheese manufacture arose out of a desire to eradicate starters. Unfortunately, Cheddar cheese is difficult to manufacture even by biological acid production, since body and texture are almost as important as flavour. Attempts to make Cheddar cheese by direct acidification with hydrochloric acid to pH 5.8, to assist rennet coagulation, and then to use glucono-δ-lactone (an acidogen which hydrolyses to gluconic acid in aqueous solution), to reduce the pH to 5 2, gave better results than the use of gluconic-D-lactone only. The body and texture of the cheese was reported as a normal Cheddar curd but flavour failed to develop (Mabbit *et al.*, 1955). Cheese curds made by the use of glucono-δ-lactone do not produce a typical flavour of cheese. When texture and body were of secondary importance, a Cheddar-type flavoured cheese for processed cheese manufacture was suggested by Breene *et al.* (1964). Milk was acidified to pH 5.4–5.6 with lactic acid and then, prior to rennetting, 2% lactic starter was added to reduce the pH of the curd to 5.2 after the scald at 39–42°C. The growth of starter bacteria to produce acid gave time for flavour development in the curds.

Direct acidification for Cottage cheese follows a fairly rigid pattern, in that cold milk at 4–5°C is acidified with lactic or phosphoric acid to pH 4.5 followed by heating, cutting the curd and scalding before cooling the washed curds. Acidification of milk to pH 5.1 with food grade lactic acid and subsequent acidification to pH 4.5 with glucono-δ-lactone has also been tried for Cottage cheese. Since directly acidified cheese has less consumer appeal than starter acidified cheese, Born and Muck (1969) suggested culturing milk with starter to pH 5.5 before directly acidifying to pH 4.6 at 4°C with hydrochloric acid.

With regard to the manufature of Pizza cheese by direct acidification, Shahata *et al.* (1967) suggested that lactic acid gave curds with higher moisture content; hydrochloric acid could be used, but phosphoric acid gave better retention of calcium in the curds and a better cheese yield. Citric acid curds retained less calcium than curds precipitated with phosphoric acid.

A USA patent method for Cottage cheese was reported by Metz (1980), in which the direct acidification of milk by the addition of an aliphatic dione (C_2–C_8 glyoxal) at 0.05–0.50% and with 0.1–0.5% H_2O_2. The hydrogen peroxide oxidizes the dione to acid. The coagulum is formed and no rennet or proteolytic enzyme is required. With regard to Cottage cheese, Mehta *et al.* (1980) found direct set curd was better than starter set using standardized protein milks. Direct set gave 5% higher cheese yields.

References

Born, B, and Muck, G. (1969) *Cultured Dairy Products Journal*, **4**(3), 11.

Breene, W., Price, W.V. and Ernstrom, C.A. (1964). *Journal of Dairy Science*, **47**(8), 840.

De Koning, P.J. (1980) *International Dairy Federation Bulletin* No. 126, 11.

De Koning, P.J., van den Berg, G. and de Vries, E. (1981) *Zuivelzicht*, **73**(21), 469.

Denkov, T. and Vasileva, S. (1979) *Nauchi Trudoue Institute po Mlechna Promyshlenost*, **9**, 158.

Dolizalek, J. and Konradova, B. (1980) *Prumysl Potrauin*, **31**(4), 214.

Eino, M.F., Biggs, D.A., Irvine, D.M. and Stanley, D.W. (1979) *Canadian Institute of Food Science and Technology Journal*, **12**(3), 149.

El-Saty, M.S., Basily, S. and Abd-El-Latif, F. (1980) *Egyptian Journal Dairy Science*, **8**(1),81.

Emmons, D.B., Kaleb, M., Larmond, E. and Lowrie, R.J. (1980) *Journal of Texture Studies*, **11**, 15.

Emmons, D.B., Becicet, D.C. and Modler, H.W. (1982) *Proceedings XXI International Dairy Congress, Moscow*, Vol. 1, Book 1, p. 407.

Foltmann, B. (1970) In *Methods in Enzymology* (eds S. Colwich and N. Kaplan), Vol. 19, Academic Press, New York.

Foltmann, B. (1993) In *Cheese: Chemistry, Physics and Microbiology*, Vol 1, 2nd edition (ed P.F. Fox), Chapman & Hall, London.

Fox, P.F. (1978) *Dairy Science Abstracts*, Review Article, 177, **40**(12), 727.

Fox, P.F. and McSweeney, P.L.H. (1997) In *Microbiology and Biochemistry of Cheese and Fermented Milk*, 2nd edition Ed. B.A. Law, Chapman & Hall, London.

Glaser, J., Carroad, P.A. and Dunkley, W.L. (1980) *Journal of Dairy Science*, **63**, 27.

Green, M.L. and Grandison, A.S. (1993) In *Cheese: Chemistry, Physics and Microbiology*, Vol. 1, 2nd edition (ed. P.F. Fox), Chapman & Hall, London, 101–140

Green, M.L., Turvey, A. and Hobbs, D.G. (1981) *Journal of Dairy Research*, **48**(2), 343.

Green, M.L., Marshall, R.J. and Glover, F.A. (1983) *Journal of Dairy Research*, **50**, 341.

Gutfield, M. and Rosenfeld, P.P. (1975) *Dairy Industries International*, **40**(2), 52.

Hansson, E. (1979) *Lebensmittel Industrie*, **26**(7), 306.

Humbert, G., Driou, A., Guerin, J. and Alais, c. (1980) *Lait*, **60**, 599.

Hyslop, D.B., Swanson, A.M. and Lund, D.P. (1979) *Journal of Dairy Science* **62**(8), 1227.

IDF (1987) *Determination of Chymosin and Bovine Pepsin Contents Standard 110A: 1987.* International Dairy Federation, Brussels.

Kosikowski, F.V. (1977) *Cheese and Fermented Milk Foods*, F.V. Kosikowski and Associates, Box 139, Brooktondale, New York.

Kosikowski, F.V. (1982) *Journal of Dairy Science*, **65**, Suppl. (1), 73.

Lynggaard-Jensen, B. and Petersen, N. (1982) *Proceedings XXI International Dairy Congress*, Moscow, Vol.1, Book l, p. 429.

Mabbit, L.A., Chapman, H.R. and Berridge, N.J. (1955) *Journal of Dairy Research*, **22**, 365.

Marshall, R.J. (1982) *Journal of Dairy Research*, **49**, 329.

Mehta, R., Sharma, H.S., Marshall, J.T. and Bassette, R. (1980) *Journal of Food Protection*, **34**(5), 390–391.

Metz, F.L. (1980) US Patent 4 199 609.
Munya, J.K. and Larsson-Raznikiewicz, M. (1980) *Milchwissenschaft*, **35**(12), 748.
Olsen, N.F. and Kowalchyk, A.W. (1978) *Dairy and Ice Cream Field*, **161**(8), 68F.
Quist, K.B. (1979) *Milchwissenschaft*, **34**(10), 600.
Ramet, J.P and Weber. F. (1980) *Lait*, **60**, 1
Reuter, H., Hisserich, D. and Prokopek, D. (1981) *Milchwissenschaft*, **36**, 3.
Scott, R. (1979) In *Topics in Enzyme Fermentation Biotechnology* (ed. A. Wiseman), Ellis Horwood, Chichester 109–169.
Shahata, A.E., Iyer, M., Olson, N.F. and Richardson, T. (1967) *Journal of Dairy Science*, **50**(6), 824–827.
Stanley, D.W., Emmons, D.B., Modler, H.W. and Irvine, D.M. (1980) *Canadian Institute of Food Science and Technology Journal* **13**(2), 97.
Storry, J.E. and Ford, J.D. (1982) *Journal of Dairy Research*, **49**(2), 343.
Storry, J.E., Grandison, A.S. Millard, D., *et al.* (1983) *Journal of Dairy Research*, **50**, 215.
Teuber, M. (1991) *IDF Bulletin* No. 251, International Dairy Federation, Brussels.
Thomsen, D. and Birkkjaer, H.E. (1980) *Beretning Statens Forsoegsmejeri*, No. 24, p.18.
Tofte-Jespersen, N.J.T. and Dinesen, V. (1979) *Journal of the Society of Dairy Technology*, **32**(4), 194.
Vanderheiden, G. (1976) *CSIRO Food Research Quarterly*, **36**(2), 45.
Van Leeuwen, H.J. (1982) *XXI International Dairy Congress, Moscow*, Vol. 1, Book 1, p. 488.
Walstra, P. and Jenness, R. (1984) *Dairy Chemistry and Physics*, John Wiley, New York.
Zannoni, M. and Annibaldi, S. (1981) *Scienza Technica Lattiero Casearia*, **32**(2), 79.

12 Cheesemaking operations

12.1 Basic cheesemaking operations

All food manufacturing processes should have a standard written procedure, programme or method of operation. In cheesemaking, the term recipe has been used to cover aspects of the manufacture of a specific variety of cheese, particularly when operating on a small scale.

Within the range of recipes available for varieties of cheese, there are certain operations which occur repeatedly, i.e. coagulation of the milk, cutting the coagulum, scalding the curd, manipulating the curd to develop texture, salting and pressing the curd. It is convenient to discuss these operations broadly before discussing recipes for individual cheeses.

12.2 Standardization of cheese milks

Adjustment of the composition of cheese milks has been practised for many years by the simple process of adding skim milk or by skimming some cream off the milk. There are several reasons for standardization of cheese milks:

(a) To compensate for seasonal variation in the compositional quality of raw milk to produce a consistent raw material (see Chapter 6).

(b) For economic reasons cheeses of different fat standards have been required. Thus, 'full fat', 'half fat', and 'quarter fat' standards have been normal for some semi-hard cheeses (e.g. Edam) and soft cheeses (e.g. Camembert).

(c) To satisfy a growing market demand for reduced and low fat cheese, cheesemakers have introduced standardization and adapted cheesemaking techniques to produce such cheese.

(d) Preconcentration by ultrafiltration (UF) is normally carried out on skim milk and this UF concentrate is then blended with homogenized cream to the desired protein/fat ratio, e.g. for St. Paulin cheese (Ramet *et al.*, 1982). Low fat semi-hard cheese made from UF concentrated skim milk (35.9% total solids (TS)) and 80% fat cream to give 41.3% TS is reported by de Boer and Mooy (1980).

(e) Where the cheese milk is prepared by recombination, either through choice or lack of a local milk supply. Sometimes anhydrous milk fat is used instead of cream, for instance in making Herve cheese (Mottar and Delbeke, 1978).

A number of studies of standardization of Cheddar have been made (Banks *et al.*, 1984a,b; Banks and Muir, 1985). Ideally, the standardization of milk should be in terms of the casein/fat ratio, typically 0.70:1 for Cheddar cheese. However, since crude protein is more easily estimated (formerly by dye binding, now more often by infra-red), a crude protein/fat ratio of 0.9:1 can be used since this approximates to the desired casein/fat ratio. Developments in FTIR measurement have enabled casein to be estimated directly.

It is essential that final heat treatment should follow standardization of milk for cheese even if by skim milk or cream addition.

12.3 Homogenization of cheese milks

Cheese milk does not normally require homogenization since homogenized cheese milk produces a finer network of protein which does not fuse easily in a curd structure and such curds tend to retain excess moisture (Green *et al.*, 1983). However, lightly homogenized concentrated (UF) cheese milks give curds which tend to retain fat and moisture more easily and so produce higher cheese yields (Green *et al.*, 1983). Cheddar cheese from milk in which only the cream is homogenized give softer, smoother and more elastic curds (Emmons *et al.*, 1980).

Mahaut and Maubois (1978) reported that 45–60 day blue cheese made by the MMV (Maubois, Mocquot, Vassal) method from pasteurized, homogenized, concentrated milk were good organoleptically. (See section 10.3 with regard to the homogenization of milk for blue-veined cheese.)

Abd-El-Salam and El-Shibiny (1982) reported on homogenization pressures for recombined milk for cheese, finding that the yield of cheese was greater when pressures of 20 MPa rather than 10 MPa were used. (See homogenization, section 10.3 and also Chapter 14.)

Pasteurization must follow homogenization of cheese milks unless a special aseptic machine is used, to minimize post-pasteurization contamination.

12.4 Heat treatment of cheese milks

The heat treatment of milk for cheesemaking is an attempt to standardize the biological quality of the milk by destroying unwanted or harmful bacteria and some enzymes. In the process of heat treatment most of the bacteria useful in the cheese process are destroyed (e.g. lactic acid bacteria) along with some of the milk enzymes such as lipases.

The temperatures and holding times used for heat treatment of the milk depend on the desire of the cheesemaker to achieve certain objectives. Thus lower temperatures, e.g. 65°C, may inactivate some coliforms while keeping the lipase enzymes intact, but cannot be relied upon to destroy vegetative pathogens. The aim should be to avoid the presence of pathogenic bacteria in the cheese milk. The minimum thermal treatment to achieve this objective is a treatment of 71.7°C for 15 s, or equivalent.

Many cheesemaking procedures are very sensitive to the heat treatment of the milk. Denaturation of β-lactoglobulin increases with heat treatment and denatured β-lactoglobulin will complex with κ-casein, inhibiting coagulation and syneresis from the curd. Adverse effects on the cheese quality can occur, particularly with cohesive cheeses such as Cheddar. In practice, the pasteurization conditions should be kept below 75°C for 15 s, preferably at 72–73°C for 15 s or equivalent.

It is unfortunate that some groups of bacteria normally found in milk are destroyed by heat, because they provide a system of enzymes active in the production of flavour and aroma in cheese which is not provided by the normal starter bacteria. Some starter organisms, e.g. *Lactococcus lactis* sub-sp. *lactis* biovar. *diacetylactis* and *Leuconostoc mesenteroides* sub-sp. *cremoris*, do provide flavour compounds, including diacetyl. Commercial cultures are being developed to provide additional flavour components.

12.5 Starter addition

The heat-treated milk is normally cooled to a temperature of ≈ 30°C, the exact temperature depending upon the cheese recipe. This temperature is necessary for both the growth of the inoculated starter bacteria and the subsequent coagulation process if rennet is used.

The bulk starter culture is added according to the recipe in quantities from 0.05 to 4% or even 5%. The quantity of starter used is often controlled by the activity of the starter and by the preference of the cheesemaker to allow the bacteria to grow in the cheese milk slowly in order to produce lactic acid required for the next stage. This 'ripening' of the milk by starter may be quite prolonged, up to 2 h, whereas inoculation of larger quantities of starter (2–4%) shortens the ripening time to 5–20 min.

During this ripening period, however, micro-organisms other than the starter bacteria may grow. Thus, there is the risk that coliform organisms may proliferate in the less acid conditions, though later they are held in check by more rapid lactic acid production. This is an example of where the recipe must be interpreted in the light of experience.

The types of bacteria in the starter are often dictated by the recipe (i.e. rapid or slow acid producers). The activity of the starter bacteria, individually and in mass, controls the rate of production of lactic acid in the milk and, later, in the curd. Larger rates of starter inoculation, while achieving the acid necessary for rennet action, may, in the later stages, produce excessive amounts of acid. Some starter cultures are slow to produce acid in the early stages, but acid production speeds up later.

Direct vat (DVS) cultures are now available in frozen or freeze-dried forms. Addition rates to the cheese milk are much lower, typically 0.01–0.02%. These starters normally exhibit a longer lag phase than well-made bulk starter and this needs to be taken into account, often by increasing the ripening and coagulation temperature slightly, e.g. from 29 to 31°C, and by a small increase in the ripening time. Acid production tends to be much slower initially but is at least as fast as for bulk starter in the latter stages of the cheese make. Previous experience is a guide for the cheesemaker in the interpretation of the recipes when using unfamiliar cultures.

Some cheesemakers add starter into the vat whilst filling proceeds, and this normally saves time. However, much time can be lost if the incoming milk is too cold for starter growth, since this produces a check in the growth of bulk starters.

Bulk starter which is granular does not release its bacteria into the milk. Such granules are often hardened by the scalding process and, therefore, persist in the cheese as white specks. The action of pumping bulk cultures into the vat or into the milk filling pipelines is normally sufficient to break up granular starter.

While most modern larger installations inoculate the starter into the vat filling lines or into a pre-ripening tank, there are some small dairies and artisanal producers using open rectangular vats which inoculate the vat milk with starter from cans just before the rennet is added. This practice may produce several qualities of coagulum in the length of the vat. Some parts of the coagulum at cutting are granular and tough, while some parts are softer. Although curds produced are mixed up during stirring and scalding, some patches of the curds produce acid too rapidly, without ‘mellowing’. The uneven curds produced as a result have portions of crumbly curds, acid in flavour, discoloured and tainted. If the cheese has been coloured, the colour may be bleached in patches and uneven. Wet patches of leaking whey may also form in the ripening cheese.

12.6 Colour and additives

Colour and any chemicals (sodium nitrate or calcium chloride) are added to the milk before it is ripened or while the temperature is raised to rennetting temperature, e.g. 29–30°C according to the recipe. Colour need not be diluted before adding to the vat, but must be evenly distributed and stirred into the milk. Chemicals must be added in solution and quantities are usually specified by recipe.

12.7 Rennet addition

Testing of the milk for suitability to receive rennet differs widely from cheesemaker to cheesemaker. Mostly, the acidity or pH at which rennet is added is given in the recipe, since this governs the type of coagulum and the speed at which the coagulum forms. With less variable bulk milk supplies and DVS cultures then a fixed time may be allowed for ripening.

The test cup (Marshall) is used by many cheesemakers. The cup is a vessel holding 500 ml with an accurately sized hole in the base. The sides of the cup are straight, white and marked into five divisions. In use, the cup is filled with vat milk (at vat temperature) and 1 ml of rennet is stirred into the milk. The milk runs from the base until it sets. The level of coagulum on the side divisions indicates the 'coagulability' of the milk rather than the acidity. The cheesemaker uses a particular division (e.g. No. 3) as indicating the suitability of the cheese milk for rennetting. This device should only be used by experienced cheesemakers.

The second method uses a vessel holding 112 ml of vat milk into which is stirred 3.5 ml of rennet. A straw or piece of charcoal floating on the milk is made to rotate by the stirring action. The time from adding the rennet until the rotation ceases is measured in seconds. Thus, 22 s indicates normal coagulation and equates with an acidity of 0.21%. Milk taking more than 22 s is judged not to be ready for rennet. This method also should only be used by the experienced cheesemaker.

The choice of type of rennet to be used is a decision which the cheesemaker has to make. For some of the factors involved in this choice see Chapter 11. The quantity of rennet and the rennetting temperature are fixed by reference to the specific cheese recipe or through experience with the type of rennet and local milk supplies in the manufacture of the cheese.

12.8 Coagulation

The coagulation of milk occurs in two phases. The first phase, which is almost independent of temperature, is the enzyme splitting of the κ-casein amino acid chain at the bond between the 105th amino acid, phenylalanine, and the 106th, methionine. This splitting produces para-κ-casein and a macro-peptide which diffuses away from the micelle and will subsequently be lost in the whey. The splitting can take place at cold temperatures, $< 10°C$, although in traditional cheesemaking they are usually higher. At cold temperatures this first phase is used to advantage in some curd making methods: for example, the Hutin-Stenne method, where concentrated milk is rennetted while cold, and then warm dilution water is added to effect coagulation (see Chapter 15).

The second, or clotting, phase is more temperature dependent and does not take place until heat is available and even then only in the presence of available calcium ions.

As already stated, the κ-casein exerts a stabilizing influence on the casein micelle. Once approximately 90% of the κ-casein has been converted into para-κ-casein this influence is lost and, in the presence of sufficient calcium ions and sufficient temperature (> 20°C), the casein micelles combine to form a coagulum enclosing the remainder of the milk components. If the coagulation is only partial, the curds formed will not reach the degree of firmness required by the recipe for the cheese. Some of the other components of milk (i.e. fat, non-casein proteins, degradation products) will be lost in the whey.

Ten factors influence the firmness or complete coagulation (i.e. curd tension) as follows:

1. The curd tension rises when the quantity of rennet increases from 0.006% to 0.03%, but does not increase with more rennet.
2. The curd tension increases with temperature up to 40°C and then declines (see Figure 11.2; see also the temperatures quoted for various cheese recipes). Thus, curd formed at 40°C is rubbery and will not cut cleanly if left too long.
3. Low curd tension and longer clotting times occur if the milk has been cold stored before rennetting. Curd tension increases with added calcium chloride up to a maximum (0.07% calcium chloride) (see Figure 11.5).
4. The ratio of fat to protein can influence the curd firmness. High fat milk gives softer curds.
5. Decreased pH (i.e. more acid) increases curd tension up to pH 5.8 when the tension begins to decrease. In traditional cheesemaking, most milks are rennetted between pH 6.5 and 6.35, but higher values will pertain with DVS cultures.
6. Coating of the κ-casein by denatured protein or free fatty acids released by proteolysis and lipolysis affects the completeness of 'clotting'.
7. Proteolysis or interaction with the amino acids on either side of the bond Phe_{105}–Met_{106} in the κ-casein may result in such damage that the cleavage is incomplete and a soft curd results.
8. The amount of whey proteins in the curd affects the firmness of the curds.
9. Dilution with water affects the curd tension.
10. Some coagulants produce softer curds initially.

These factors are sometimes included in the recipe of particular types of cheese, but they are very often part of that experience without which interpretation of a recipe cannot be made.

Rennet needs to be diluted up to 10 times with clean water before it is distributed over the vat milk, the milk having been well stirred just prior to the addition. It is usually safe to stir the milk for up to 5 min after addition of rennet to ensure good mixing. Inadequate stirring, apart from incomplete dispersion of the rennet, and hence patchy coagulation, allows fat to rise to the surface of the milk. This increases the loss of fat at cutting, which gives floating fat on the whey surface. Too vigorous, particularly too prolonged stirring of the milk (i.e. over-stirring) causes the newly forming coagulum to break up; these curds 'whey off' quickly and do not reunite but lose fat in the whey. Vibration from heavy machinery in motion can damage the coagulum and cause broken curds.

When the milk is about to set, the point of setting can be detected by dropping one or two drops of the milk gently into clear water. Curd which is 'set' causes strings and granules of curd to settle gently in the water.

It is essential during the secondary clotting phase that the milk remains stationary. The casein micelles coalesce or bond together during this stage and if disturbed do not join up again later.

The firmness of the curd within the period specified by the recipe fixes the type of curd needed for that cheese. During this period of coagulation lactic acid is being produced, thus causing the curd tension to increase as well as the enzymatic breakdown of some components to give that 'mellowing' of curd so often described by cheesemakers.

Approximately 6% of the chymosin originally added to the milk remains in the cheese curd where it is still active. The secondary phase of rennet activity occurs in the curd, with less specific proteolysis of the α and β caseins contributing to the maturation of the cheese.

12.9 Cutting the coagulum

The coagulum is ready to cut after a period of from 25 min to 2 h, as defined by the recipe. However, some cheesemakers calculate the time for cutting by multiplying the time taken to set by three. For example, if the time taken from adding rennet until the curd first sets is 12 min then the time for cutting is 36 min from the time rennet was added.

The point at which the coagulum is cut is a vexed question amongst cheesemakers, who do not have common agreement (Figure 11.4) and hence definition of the desired characteristics.

Many attempts to measure this point have been made. One apparatus measures the pressure or tension needed to cut the coagulum or to turn a spade electrode in the coagulum. Another apparatus measures the transmission of vibrations through the coagulum. Since these instruments are measuring different characteristics then correlation of data is difficult. One of the difficulties is that some of the equipment needs to be operated on milk outside the cheese vat. The point at which the cheese coagulum needs to be cut is, therefore, extrapolated from a graph or chart not necessarily drawn from data relating to the cheese milk in the vat (Figure 11.4).

Few mechanical devices measure more than resistances to cutting; they do not measure the 'quality' of the curd. Non-destructive tests such as by visco-elastic rheometry are not suitable for a dairy environment. Methods of measurement of coagulum formation have been reviewed by Green and Grandison (1993).

Cheesemakers' attempts to judge the point of cutting are fraught with difficulties. The surface layer of a coagulum, particularly in open vats, is often some degrees colder than the coagulum underneath and is, therefore, softer. To judge firmness of curd on the surface, therefore, has little meaning.

The main method employed by the traditional cheesemakers is to plunge the hand, rod or thermometer stem below the surface layer and to lift the coagulum causing it to break in a cleavage line. A clear cleavage with green whey at the base of the cleft indicates that the curd is ready to cut. A soft irregular cleavage with white whey indicates that the curd is too soft. The sides of the cleft show the quality of the curd. Thus, granular curds indicate that the curd is too firm. A rule used by some cheesemakers is that the curd should be cut early rather than later, and once cut the curd should be left to complete its forming process in the warmer whey which rises over it. If the coagulum becomes too firm, knives or curd breakers crush the curd rather than cutting it cleanly. Curds which have been cut cleanly will 'heal' or join up the cut fibrils on the new curd surfaces and thus prevent loss of fat and other milk components.

The size of the curds after cutting is defined in the recipe. Thus, curds which need to be scalded to higher temperatures are cut into smaller pieces to aid both heat transfer and dehydration, while those curds which are scalded to lower temperatures can be left in larger pieces unless the curds are very acid.

During the cutting or breaking of the coagulum and in the subsequent scalding (cooking) process, the cut surface surrounding each curd particle assumes some importance in the cheese curd maturation, during which it can retain some membrane characteristics.

Fat globules are held in the matrix of the casein network, partly by physical enclosure and partly by a loose bonding of the globule membrane and protein. Fat globules near the cut surfaces leak away. Such fat, although only 0.2–0.3% in the whey, is more than 5% of the original fat in the milk and constitutes a loss of cheese yield. The finer the cut then the greater the surface area and the greater the potential for fat loss, so curd is seldom cut into smaller than 6-mm cubes. The whey from the cut curds carries water-soluble components including lactose, whey proteins, salts, peptides and other non-protein nitrogenous substances.

Smaller open vat systems may use multi-bladed, hand-held steel knives. The blades vary from 6 to 18 mm apart and the assembly can be 150–200 mm wide and 700–800 mm long. The blades may be mounted to give either a vertical or horizontal cut. Some cutters are composed of wires strung on steel frames. Stainless steel has virtually replaced tinned steel.

Mechanically operated curd knives are larger than the manual knives and use either blades or, occasionally, wires. It is very important that the edges of blades are kept sharp enough to cut cleanly. Heavy gauge wires tend to tear the curd more than steel knives, and some cheesemakers cut slightly earlier with wires than with knives. Wires may be used for weak curds such as with Cottage cheese coagula. For small-scale cheesemakers, cutting the coagulum lengthways once manually in the long rectangular vats normally cut by mechanically operated knives may prevent crushing of the soft curds during the first mechanical cutting.

With modern enclosed vats the curd must be cut mechanically. The knives cannot be removed from the vats so these are designed to have a dual function: the angle of the blade presented to the curd is such that if the rotation of the knife is reversed it stirs rather than cuts the curd. The rotating knives in round or oval vats do not crush the curd against the vat sides. Even so, the speed of rotation in some equipment can be controlled to minimize damage.

In the production of soft cheeses, like Brie, Coulommiers or Camembert, the cutting or breaking or curds (sometimes ladling of curds) is followed by hooping or moulding the curds for drainage of whey without stirring, as in semi-hard or hard cheese. The curds retain the shape of the mould as the whey drains away and the curd shrinks.

In some cheeses, such as Lactic, where the curd is made into a paste, the coagulum is ladled or poured into cloths suspended on hooks or poles so that the whey drains from the curds. The 'Berge' process, for instance, uses vibration of the cloths to assist whey drainage. Depending on the consistency of curd required, it may be pressed to force out the free whey remaining, to leave a firm paste to be moulded to shape for packaging. Alternatively, curds like the Quark may be separated from the whey by a special type of rotary separator in which the size of nozzle and throughput of the machine determines the consistency of the curds. Since the fermented product is separated when warm, a cooler is also required (see Chapter 13, and recipes in Chapter 19). Ultrafiltration provides an alternative concentration process for such products, especially Fromage Frais.

12.10 Stirring and scalding

The curd, when first cut, is soft and the coat surrounding the particles is open. Stirring the curd gently until the first flush of whey has left the curd particles is necessary to prevent undue crushing and loss of fat and curd dust. Once the curd coat becomes more membrane-like the rate of stirring can be increased.

Scalding or cooking the curd promotes contraction of thc protcin matrix, causing the curd to shrink and expel more whey. The increase in temperature also speeds up the metabolism of bacteria enclosed within the curd. Lactic acid production increases, the pH declines, and this acidity assists in shrinking the particles to expel more whey.

Since the whey has, in solution, lactose and salts, the amount of these substances retained in the cheese is proportional to the amount of moisture in the curd. The calcium phosphate associated with the casein and in a colloidal state will gradually become solubilized as the pH falls. Thus, high acid curds (e.g. blue-veined cheese curds) lose more calcium (92%) than low acid curds (e.g. Edam), which lose only 35%.

Lactose is the main substrate for the lactic bacteria in the curd and is essential for the production of lactic acid. The lactose concentration in the curd is a controlling factor in the metabolism of the bacteria. Thus, once the lactose concentration has decreased to a critical point, then much smaller decreases have greater effect on the growth of bacteria and on the production of lactic acid. The cheesemaker has control over lactose in the curd, and therefore the amount of lactic acid formed, through the size of the curd particles, scald temperature and rate of rise of temperature of the curds.

The curd particle can be regarded as an immobilized bacterial fermenter bound by a permeable membrane. Within the particle the starter bacteria are converting lactose to lactic acid. This lactic acid then diffuses out of the particle into the bulk of the whey down a concentration gradient, making the whey steadily more acid. This process depletes the lactose in the curd particle so lactose diffuses in along a similar concentration gradient to provide additional substrate to support the microbial metabolism. Diffusion of lactose into the curd will be opposed by the expulsion of whey from the curd particle as the casein matrix shrinks.

There are two methods of reducing the amount of lactose in the cheese curds.

1. Shrinkage of the curds brought about by heat and lowering of the pH (development of lactic acid in the curd).
2. The addition of water to the whey, usually following removal of part of the whey. This addition decreases the level of lactose in the now diluted whey and hence reduces the lactose level in the curd particle. The addition of hot water to the whey/curd mixture may also be used as a method of scalding (heating) the curd in the washed curd cheese processes.

The aim of scalding the curds is to shrink the curd to expel moisture and so firm up the curd to a state ready for texture formation, pressing or salting.

This state provides the dividing point between four main groups of cheese (excluding soft cheeses, some of which may be scalded).

1. The textured cheeses like Cheddar, or Cheshire.
2. The Pasta filata types or kneaded cheese.
3. The cheeses untextured in the vat stage, like Edam and Gouda cheese, and also those which acquire texture later, like Tilsiter, Emmental, etc.
4. Blue-veined cheese.

The variation of scalding rates is carried out according to the acidity produced and is under the control of the cheesemaker. This is a further point in the recipe where previous experience aids in interpretation.

Figure 7.2 shows the difference between the effects of scalding rates on acid production in three separate cheeses.

A high rate of scald will shrink the coat of the curd particles so much that the membrane is so firm that moisture is locked in the curd, giving a high moisture cheese that will lose whey as an exudate on maturation. The resultant cheese is often of acid, hard, harsh texture, crumbly and eventually dry, with off flavours associated with the retained whey.

Low rates of scald may be necessary for curds in which the bacteria are slow to produce lactic acid. Alternatively, curd shrinkage may be by acid alone without the use of scald. The maximum scald temperature is determined by the recipe but it is important to note that the normal lactic starter bacteria will be inhibited, if not destroyed, if the scald temperature is too high (i.e. temperatures above 40°C). Scald temperatures higher than normal need the inclusion of high-temperature-tolerant, i.e. thermophilic, starter bacteria, e.g. *Streptococcus thermophilus*, *Lactobacillus delbrueckii* sub-sp. *bulgaricus*).

The cheesemaker has a decision to make in respect of when to cease stirring the whey–curd mixture; this is not often included in the recipe. The cessation of stirring is called the 'pitching' point when the curd sinks down to the bottom of the vat.

Normally, fast acidifying curds are stirred until the whey is removed. With very slowly developing curds, the stirring ceases altogether and the curd is 'pitched'; sometimes, to prevent excessive 'matting' of the curd into lumps, the curds are stirred at intervals. The effect of the pitching will be to reduce the loss of lactic acid from the curd into the bulk of the whey so that the pH of the curd particle drops more rapidly, bringing the curd to the point where it can be taken on to the next stage in the process.

The acidity testing at these stages needs interpretation, as already noted, since the acidities of the curd and the whey are different. Normally, the curd acidity is greater, due to the larger numbers of bacteria enclosed in it. Occasionally, the reverse happens and a slow, inhibited curd is less acid than the whey.

Since the whey provides a protective influence on the curd, temperature-wise and for provision of metabolites, the point at which the whey is removed from the curd is a major guide stage in the recipe.

Thc curds of cheeses not textured in the vat, e.g. the Edam, Gouda or Emmental curds, are taken directly from the whey into the press moulds at this stage. The textured cheese curds, like Cheddar or Cheshire curds, at this stage are left in the vat, or removed for further manipulation on a draining tray, and the whey is removed from the curd. In both types of cheesemaking this constitutes the 'whey off' stage in the recipe.

It is the rate of acid development per unit of time which largely determines the future development of acid in the curds. It is usual to find that the lactose content of the curds is exhausted within a short period after the curds have been pressed and, therefore, this is the period when the acidity in the curd is highest (i.e. the pH is the lowest).

A pH of 5.0 is regarded as too acid for many hard-pressed cheeses. For Cheddar cheese the optimum is nearer to pH 5.2–5.25. Hard-pressed cheeses with pH values of 4.9–5.0 are usually acid, of short texture, crumbly and may have harsh curds. However, the pH is low (pH 4.2–4.8) at this stage in certain groups of cheese like Feta cheese and the white brined cheese of the Middle East.

12.11 Washed curd cheese

The addition of water to a curd–whey mixture causes removal of lactose and other solubles from the curd as well as the water being re-absorbed into the curd. When the water is hot it will also scald the curd and so expel moisture. The effect of this activity is to hold the pH of the curd at a higher level than would otherwise be the case, since the activity of the lactic bacteria is reduced by depletion of the reserves of lactose. A few examples of the use of water in this way are given below.

1. With Edam cheese one-third of the whey is removed and then scalding is achieved by adding hot water at 50–60°C. This raises the temperature to 36–37°C and the effect is to keep the pressed curd at pH 5.2–5 4, otherwise it might reach pH 4.6–4.8.
2. Similarly, Gouda cheese curds are scalded with hot water at 60°C to raise the temperature from 30 to 40°C. The cheese is held to pH 5.2–5 4.
3. In the processing of St Paulin cheese one-half of the whey is removed and is replaced with weak brine. The brine assists in firming up the curd to ensure adequate drainage of the whey.
4. In the making of Samsoe, Maribo, Havarti and similar cheese one-third of the whey is removed and replaced with hot water in order to scald the curd to 37–38°C and to dilute the whey.

Table 12.1 Changes in the acidity and pH of whey or curd during the manufacture of washed curd cheese

Stage	Elapsed time (h:min)	Temperature (°C)	Acidity (%)	pH
Milk		4	0.16[1]	6.60[1]
Starter	0:30	31	0.17[1]	6.50[1]
Rennet	1:30	31	0.18[1]	6.40[1]
Curd cut	2:10	31	0.11[2]	6.35[3]
Wash water added	2:30			
Pitching	3:00	39	0.09[2]	6.30[3]
Whey off	4:00	38	0.11[2]	6.10[3]
Curd moulded	5:10	37	0.25[2]	5.80[3]
Cheese after brine	72:00	7	–	5.15[3]

[1] Milk.
[2] Whey.
[3] Curd.

The hot water used to form the curds into a plastic fibrous structure in the Pasta filata group of cheeses, Mozzarella, Provolone, etc., cannot be regarded as wash water, but since it is at a higher temperature, in effect it 'pasteurizes' the curd. Therefore, pasteurization of the milk for these cheeses is not so necessary.

Table 12.1 shows the changes in acidity and pH during the manufacture of a washed curd cheese.

12.12 Salting of cheese

Untextured cheeses are salted mainly by immersion in brine (18–27% sodium chloride) at temperatures dependent on the type of cheese, varying from 8 to 16°C. The immersion time varies from 15 min up to 5 days depending on size of cheese and type of curd.

The aim is to produce a salt content of 1.5–2.5%, but in some types this may be different (Emmental may have 0.6%, and Pecorino from 5% up to 7%). The presence of significant quantities of salt will inhibit the metabolism of the starter bacteria – in the case of Cheddar, a 1.95% salt addition will give 5% salt in moisture (S/M). Thus, the salting stage halts the acid production, and so the pH of the curd does not decrease further after salting.

Dry salting is also used, whereby salt is rubbed on to the outside of the cheese after pressing. The salt is slowly absorbed into the curd over a period of some days.

Varietal differences in the percentage of salt in cheese are quite wide as shown in Table 12.2.

The rate and pH at which salt is added to the curd are controlling factors in the further ripening of the cheese. In some varieties, e.g. Emmental, the intensity of the scald combined with the washing and earlier production of

Table 12.2 Levels of salt and salt in moisture for various cheeses

	% Salt	Salt/moisture
Cottage cheese	0.25–1.0	0.3–1.25
Emmental cheese	0.5–1.2	1.5–3.5
Gouda cheese	1.5–2.2	3.5–>5
Cheddar cheese	1.75–1.95	4.5–>5
Limburger cheese	2.5–3.5	4.0–>6
Feta cheese	3.5–5.0	6.0–>10
Gorgonzola	3.5–5.5	6.0–>10
Other blue cheeses	4.0–7.0	

acid denudes the curd of lactose so that less salt is required to hold the pH steady in the earlier stages of maturation. With Swiss style cheeses, the upper level of salt is limited by the sensitivity of the secondary ripening starter, *Propionibacterium freundruechii* sub-sp. *shermanii*, to S/M levels above 6.7%.

In some textured cheeses and in semi-soft cheeses using lower scald temperatures, the lactose persists in the curds for 15–20 days after pressing, so that more salt is required to limit the acid production by lactic acid bacteria. This prevents a lowering of the pH of the curd nearer to the isoelectric point of casein (pH 4.6) which would result in an undesirable, brittle, short textured cheese.

The level of S/M has a significant effect on sensory attributes. While chymosin-mediated hydrolysis of α_{s1}-casein increases with rising S/M levels, that of β-casein is inhibited above 5, thus reducing the incidence of bitter off-flavours. S/M levels much above 5 can, however, lead to the cheese being downgraded on other grounds.

Some cheesemakers have used sea salt but there may be legal restrictions of 2% maximum by weight of calcium and magnesium carbonate for such salts. The salt may be of standard food grade crystalline form rather than the finely ground salt used for buttermaking.

12.12.1 Methods of salting

The dry salting methods may be divided into three main groups:

1. Textured cheeses are salted with dry salt after milling. It is usual to allow the salt to become dissolved in the curd (mellowing) before the cheese is pressed. Dark lines are produced, especially in mature Cheddar cheese curds, if the salt has not fully dissolved in the curd before pressing. Undissolved salt is harsh and gritty in the curd. Salt, by dissolving the casein surface, helps to produce early matting of the cheddared curd into a close texture in the press. However, the salt helps to release more free whey in the press, so that a very wet curd will lose a quantity of salt in the whey released during pressing. For this reason it is usual to allow extra salt to be applied to wet cheese. For very dry curds slightly less salt is

required. Most recipes do not allow for these variations and the cheese-maker must decide from experience how much salt to use.

2. Blue-veined cheeses are usually salted on the curd before moulding (e.g. Stilton), but in some cheeses the salt is applied either on the curd while in its mould or after the cheese has been removed from the mould. Different varieties of blue-veined cheese have different salt contents, varying from 2% up to 5%. Moist cheese curds will absorb more salt by the dry salting method than the drier curds.
3. Soft cheeses are dry salted over the whole surface by rubbing the cheese with salt. Sometimes a further salting is carried out after the first salt has dissolved. A soft cheese of 150 g weight requires 6–9 g salt. Some soft cheeses are brine salted for 20-80 min at 18-20°C in a 22% brine (20° Baumé).

The amount of salt used depends on the type of cheese or dryness of the curd and is usually specified in the recipe. Salt inhibits some bacteria but is a stimulant to others. For this reason salt induces the growth of a different flora in the salted curd. A strength of 2% salt in a cheese of 40% moisture content means that the cheese moisture is a 5% salt solution. While this will inhibit most starter bacteria, coliforms are not inhibited until the salt solution is 12%; indeed, they are stimulated by a 3–4% salt solution. Until the salt is completely distributed in the cheese curd the pH of the curd may be unstable. The even distribution of salt in a large dry-salted cheese may take 3 days, but in the smaller soft cheese, particularly where the curd has been finely milled, the salt is well distributed in 16 h.

Brine salting has been used extensively with washed curd cheeses, since the lactose depletion from the washing limits the drop in cheese pH. While the rate of uptake of salt from the brine increases with temperature, this is limited by release of fat from the curd and by growth of undesirable organisms, so temperatures of $\leqslant$15°C are used in practice, giving a maximum solubility of $\approx$26% salt. The higher the salt level then the more rapid the salt uptake and the greater the dehydration of the cheese due to the difference in osmotic pressure between the cheese and the brine. The brine will thus become more dilute and accumulate whey components from the cheese. Periodic addition of salt should be made to the brine, and the brine either pasteurized and filtered or regenerated by ultrafiltration or similar membrane filtration processes. The cheeses should be held in racks during brining to ensure brine contact with the maximum surface area of the cheese, with more even uptake promoted by brine circulation.

When brine salted, hard-pressed cheeses may hold 16–18% of salt in the rind for a time. The rind becomes horny and hard and the cheese can be handled safely. Dry salting for a short time gives up to 20% salt in the rind, and some cheese, which depend on *B. linens* for ripening, have 18–20% of salt in the rind. This salt does not inhibit growth of *B. linens* but does prevent proteolytic organisms growing and 'pitting' the surface of the cheese.

12.13 Cheese pressing

The pressing of cheese has often been the least understood of all the cheesemaking operations. The main aim of the cheese pressing is to form the loose curd particles into a shape which is compact enough to be handled, and to expel any free whey. A brief look at the structure and components of raw curd highlights some of the problems of this apparently simple operation.

The curd is composed of a matrix of protein enclosing fat globules, moisture, lactose, salt, non-protein nitrogenous substances as well as peptides, etc. The curd also contains some air and some gas (CO_2) so that while it is still warm it is springy, elastic and soft. The fat is also mainly in the liquid state. Salt (NaCl) may or may not have been applied and salt will dissolve some of the casein surfaces, but it also retains water. The surface layer of casein may be rendered hard and horny if the salt is not allowed to dissolve freely into the warm curd.

Pressing the curd should, therefore, be gradual at first, because sudden high pressure compresses the surface layer of the cheese creating an impermeable layer which will lead to moisture being retained within pockets in the body of the cheese. The temperature of the curd before pressing should be below the liquid fat temperature, i.e. $\leqslant 24°C$ in summertime, 26°C in wintertime. Otherwise, fat will leak from the curd and be lost in the whey, or will fill spaces in between the curds and give a greasy cheese.

The pressure applied to the cheese should be expressed as force per unit area of the cheese and not per cheese, which may vary in size. Table 12.3 shows the pressures traditionally applied and a comparison with pressures applied to an 18 kg block of cheese.

Since the cheese curd holds a volume of air before pressing, those cheeses requiring very close curds (e.g. Cheddar) have been pressed under a vacuum of minus 85–95 kPa (5–28 inch Hg). (Note: the unit of pressure is the pascal (Pa), equal to one newton per square metre (N/m^2)). The vacuum is applied for a short time (2–3 h) to deaerate the curd but also assists in cooling the curd by evaporation as well as improving the heat transfer coefficient. Pressures have been traditionally applied for 2–3 days to Cheddar cheese, but the more recent 'block cheese' pressing has been limited to 24–36 h, and with vacuum pressing, 10–15 h. This has enabled the cheese moulds to be washed and reused the following day.

Cheese presses have been either spring, dead weight, pneumatically or hydraulically operated and many cheesemakers have had their own preferred type. Spring and dead-weight presses are usually operated vertically

Table 12.3 Pressures applied to cheese of various sizes

Cheese	Size	Area	Traditional application per cheese		Equivalent applied to 18 kg block
	(m)	(m^2)	(Tons)	(kPa)	(kPa)
Cheddar	0.41	0.13	0.5	38	65
			1	76	130
			2	153	260
			3	230	391
			5	384	652
Cheshire	0.30	0.07	0.5	68	36
			1	137	73
			2	274	146
			3	411	292
Leicester	0.51	0.20	0.25	12	50
			0.5	24	101
			1	49	203
			2	98	407
Truckle mould	0.10	0.03	0.25	77	16
			0.5	154	32
			1	308	65
			2	617	130
Block mould	0.36 × 0.28	0.1	0.5	50	50
			1	100	100
			2	200	200
			3	300	300

Note: 1 lbf/inch^2 = 6.89 kPa; 1 inch = 0.0254 m.

for a single stack of cheeses with the applied force expressed in terms of weight (Table 12.3). With the pneumatically and hydraulically operated presses there is potential for errors in estimating pressures if standard gauges calibrated for the line pressures are used, as these will need correction for the cross-sectional area of the piston. This is illustrated in Figure 12.1 where, ignoring frictional losses, the product of the cross-sectional area of the piston (A_p) and the pressure of the service (air or hydraulic fluid) (P_p) will equal the product of the cross-sectional area of the cheese mould (A_c) and the pressure exerted on the cheese curd (P_c), i.e.

$$A_p.P_p = A_c.P_c$$

One of the requirements of a pressed cheese is that the surface or rind is close, smooth and has no crevices to aid mould penetration. The traditional methods used coarse hessian cloths in the initial stages of pressing to assist in closing up the holes in the curd. Sometimes the cheese was immersed in hot water at ≈50°C to plasticize the coat and the cheese was then re-pressed in a stiff calico cloth to obtain a smooth finish. These methods were very labour intensive, and textured synthetic films have replaced the cloths previously

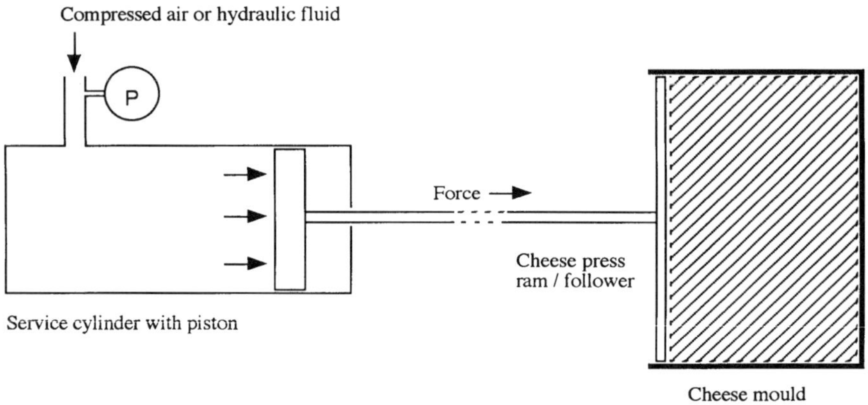

Figure 12.1 Illustration of the principle of a pneumatic or hydraulic cheese press.

used. Traditional cheesemakers and certain cheese buyers still prefer the older methods of cheese preparation, especially for the textured cheese varieties, e.g. Cheddar or Cheshire.

Block moulds replaced most of the traditional shapes and sizes of moulds for most large-scale production, giving standard cheeses approximately 18 kg in weight and $0.36 \times 0.28 \times 0.18$ m in size These moulds have been produced in tinned steel, aluminium and in stainless steel and are used in conjunction with single use perforated plastic films.

Larger 'Ton' or 'box' presses have been also been used. The presses have vacuum cylinders so that the curd can be pressed under vacuum. The press holding $\approx$ 900 kg of curd is vertical and allows a single large block to be produced. Large blocks made by this technique did not mature evenly, so the freshly pressed blocks are cut into 18-kg blocks for wrapping and maturation.

Development of cheese moulds made up of finely perforated stainless steel (e.g. Perfora™ moulds) has enabled cheesemakers to dispense with cheese cloths. At the same time the advent of perforated plastic materials for moulds (e.g. Laude™ moulds) has enabled automatic filling of curds into clothless, plastic moulds to be used on automatic cheese lines. The elimination of manual labour in the handling of moulds and cheese on such lines, with the use of vacuum presses, has enabled the pressing time to be reduced to approximately 2 h (Hansen, 1984).

More recently, the block formers have been developed to replace the presses completely (Chapter 14).

Whatever the system adopted, the freshly pressed curd, or 'green' cheese, is still relatively warm and plastic. Such cheeses must be supported or they will flow and distort, giving shapes that are unappealing in traditional cheeses and result in excessive cutting losses with block cheese.

12.14 Cheese rind coating, bandaging and wrapping

The softer types of cheese acquire a rind during ripening, often as a result of the growth of moulds and bacteria. Later, the evaporation of moisture hardens the rind so that it is more rigid to handle. In some cheeses, the rind is dried off and coated, by the use of ashes (e.g. Cendre D'Aisy), or by the use of grape must or marc (e.g. Tomme au Marc), or by wrapping in leaves (e.g. Aromes de Lyon).

In many instances, the rind is kept clean by repeated washing with a salt impregnated cloth (e.g. Emmental), or by repeated brushing to remove mould growth (e.g. Cantal). When these cheeses are ripe and ready for sale, the rind is simply coated with vegetable (olive) oil, which may be coloured brown or black (e.g. Parmesan, Pecorino Romano).

Smoking of cheese also gives the coat a fatty layer, can cause some surface dehydration and has a further preservative effect due to phenolic compounds from the smoke. Spices are also used on the coats of some cheese to impart a flavour to the curd, but mainly the spices are included in the curd. Resinous, softwood bands used for packing Vacherin Mont d'Or cheese impart a spicy flavour to the cheese. Feta and similar white cheese are packed in casks or drums filled with brine or salted whey.

For many years, cheeses have been pressed into ceramic or pottery jars in order to preserve them for longer periods. Blue cheeses are preserved in this way. Slightly damaged cheese stripped of any defect can be satisfactorily potted for sale during festival seasons (e.g. Christmas or Easter). The pots or jars are sealed with wax or plaster of Paris. Gorgonzola has also been coated with plaster of Paris as a protective coat inside a woven basket. The plaster is not completely air tight and allows the cheese to 'breathe' and the mould to remain blue. Packing in sealed glass, ceramic or pottery jars, which are air tight, does not allow the mould to remain blue; but when such cheeses are exposed to the atmosphere for a short time the blue colour returns to the whitish brown bleached moulds. The flavour does not seem to be affected.

The larger hard-pressed cheeses like Cheddar or Emmental have in recent years been produced in block shapes for two main reasons:

1. The packaging of cheese in retail markets for consumer sale in small portions has accelerated the use of block cheese shapes. These cheeses, 4.5–9 or 18 kg in size and rectangular in shape, can be cut mechanically without waste into consumer portions.
2. The use of impermeable films with blocks avoids mould and cheese mite damage which has caused serious loss in traditional round-shaped cheeses.

Attempts to overcome these defects at first employed, and still use, chemical treatments, i.e. sorbic acid and its salts, or pimaricin to stop mould growth, and/or waxing or resinous coating of the cheese rind to prevent both mould growth and mite infestation. Dipping the cheese in hot wax effectively pasteurizes the surface of the cheese as well as severely reducing the availability of oxygen. There is a tendency for the sorbates used as anti-mould treatments to be absorbed into the curd. Mould growth may be more prolific after such treatments when the sorbate has been absorbed than if no treatment had been given. Therefore, retreatment needs to be applied for long-term storage.

The waxing or resinous coat was applied to many cheeses, even including ripe blue-veined cheeses (Stiltons). Coloured wax (red or yellow) is now reserved for the sweeter types of cheese (Edam, Danbo, Elbo) mainly for the export trade.

The waxing of cheeses like Cheddar or Cheshire over the bandage requires that the bandage is completely dry (2–3 days drying). If the bandage is not dry, the wax coat peels away and is not effective as a mould preventative. The waxes used are available with different melting points, from 49°C to 82°C, for either temperate or tropical usage. The application of the wax is usually by dipping the cheese in a bath of melted wax for up to 30 s and then allowing it to cool quickly. It may be necessary to dip twice if the cheese is not totally covered on the first occasion. Paraffin wax for a first coat has been used with a second coat of a petroleum micro-crystalline wax. This double coating prevents cracking of the first wax coat during storage. An example of a wax coating machine is shown in Figure 12.2.

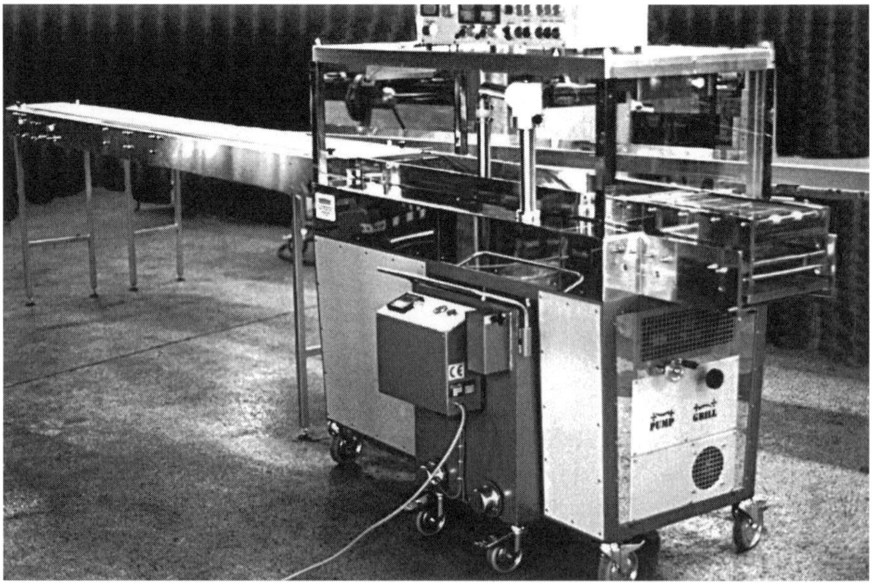

Figure 12.2 Wax coating machine. (Courtesy of Vantage House)

The treatment of cheeses at 104°C and over, even for short periods, is likely to cause steam pockets to develop under the bandage or skin. Later, moulds may grow in these pockets where the wax has not adhered to the cheese coat.

Earlier cheesemakers used whey butter and lard to coat the bandages and rinds of cheese. These greases, however, tended to give rancid odours to the cheese.

Resins developed for application to the cheese were applied in thin coats, and during long-term ripening the cheese had to be recoated. Solvents for the resins have caused off-flavours in the rind and care in choosing the resin is necessary. Anti-mould additives are used in the resin mixture or sprayed onto the rind before coating.

Losses of cheese from mite attack (*Tyroglypus siro* and other species) became serious in long-hold cheese stores for cylindrical cheeses, and often caused losses of up to 15%. The use of film wrapping materials, not easy to use with the round cheese, became common when the block shape cheese came into use and this prevented mite attack on the cheese.

Since cheese respires for 8–15 days, this may create problems. Furthermore, the cheese wrapped in film does not lose water by evaporation and the cheese has to be made with less water than traditional cheese so that loss of water is unnecessary. This, of course, leads to a slightly different form of ripening in that the evolution of gas, carbon dioxide, propionic acid, etc. must be suppressed by the use of storage temperatures not more than 8°C. Temperatures of 10°C allow gas to form and lift the wrapping of the rind of the cheese and this is discussed further in Chapter 17. The block cheeses have been termed 'rindless' because, since there is no evaporation, a rind does not form and thus there is an economic saving of cheese. The first of the wrapping materials were cellulose acetate and its derivatives, but as experience was gained, other materials came into use, and Table 12.4 shows some of the range of materials used. Wrapping materials for cheese are required to have a number of properties as follows:

1. Low permeability to oxygen.
2. Low permeability to carbon dioxide.
3. Low permeability to water vapour.

Table 12.4 Typical permeabilities for some films

	Water vapour[1]	Oxygen[2]	Carbon dioxide
Polythene	130	10.6	35
Nylon 6	7000	0.38	16
Cellulose acetate	75 000	7.8	68
Polyvinylidene chloride copolymer	14	0.053	0.29

[1] 10^{10} cc/cm^2/mm/s/cm Hg/25°C/90% RH.
[2] 10^{10} cc/cm^2/mm/s/cm Hg/30°C.

4. Strength and thickness of film.
5. Stability under cold or warm conditions.
6. Stability to fats and lactic acid.
7. Resistance to light, especially ultraviolet.
8. Ease of application, stiffness, elasticity.
9. Ability to seal and accept adhesives.
10. Laminated films to remain laminated.
11. Low shrinkage or ageing unless shrinkage is a requisite.
12. Ability to take printed matter.
13. Should not impart odours to the cheese.
14. Suitability for mechanization of packing.
15. Hygienic considerations in storage and use.
16. Cost effectiveness as a protective wrapping.

There are new laminations being developed from time to time, and new synthetics for new sources of raw materials, but the range of packaging materials includes:

Paper, waxed, laminated or parchment
Polythene, low or high density
Cellulose acetate
Nylon (polyamide) in various forms
Polyester
Polystyrene
Polypropylene
Polyvinylidene chloride copolymer (PVDC)

Table 12.5 gives some properties of typical wrapping and rigid films. The British Standards Institution has provided the following standard: BS 1133 Section 22 (1991) Wrapping materials, foils, films and laminates.

The vapour permeability of films is sometimes quoted as the grams of water passing through a square metre of film 0.0254 mm (0.001 inch) in thickness in 24 h at 25°C with a relative humidity (RH) of 75% (i.e. g/m^2/0.0254 mm/24 h/25°C/75% RH). There are many other expressions (e.g. per square yard or per 1000 inch^2 and temperatures from 20 to 100°C. Relative humidities from 50 to 100% RH affect the results when using films like cellulose. The oxygen permeability is often quoted as 10^{10} cc/cm^2/mm/s/cm/atmosphere at temperatures varying from 20 to 30°C as illustrated in Table 12.4; care is necessary in the reading of the different expressions for permeabilities, since only if the expression is the same can the figures for each material be compared.

BS 1133 (1989) gave methods for measuring the permeability of films to gases but there are many methods described in the literature. The difficulty in comparing the permeabilities of films is that they vary from sample to

Table 12.5 Properties of materials used for cheese packaging

Material	Permeability		Normal usage
	Water vapour	Oxygen	
Single films			
Cellulose PT	Poor	Poor if wet, good if dry	Outer wrappings
Cellulose MXXT	Very good	Very good	Wrappings (Saran coated)
Nylon 6	Fairly good	Good	Wrappings
Polyester	Fairly good	Good	Wrappings
Polyethylene	Very good	Very poor	Laminates
Polypropylene	Very good	Fair	Rigid, containers and wrappings
Polystyrene	Fair	Very poor	Rigid containers
Polyvinylidene chloride copolymer	Very good	Very good	Wrappings
Laminates			
Polyethylene-cellulose	Fair	Fair to good	Wrappings
Polyethylene-cellulose acetate	Fair	Moderately good	Wrappings
Polyethylene-paper	Fair	Fair to poor	Wrappings
Waxed cellulose	Fair	Fair	Wrappings

sample and on ageing in storage. Therefore, only an indication of the permeabilities is given in Table 12.5.

The laminated films are used where one of the laminates provides the strength and only some of the other qualities (e.g. cost effectiveness). The other laminating materials are mainly concerned with permeability to water, oxygen and carbon dioxide as well as saleability or acceptance of adhesives.

Waxes coated on films give better qualities to the film (polyisobutylene 30–50%, paraffin wax 30–50%, and polyethylene 15% is typical of a mixture used). The wax side of the film is used next to the cheese rind and may be heat sealed to the cheese in a hot press. Indeed, if the cheese fat is brought to the surface in contact with the film this alone often conveys moisture-proof qualities to the film. Lacquered films are often used in place of laminates to confer certain properties to the film (e.g. printability). Cellulose films in the moist conditions next to the cheese allow water vapour to pass through easily, whereas in dry conditions on the outside or central laminate the water vapour permeability is low. Polyethylene is not always resistant to solvents and oils unless coated on the side affected. Cellulose films have retained their popularity in packaging systems for soft cheeses (Sweetman, 1991).

Some Films may be stretched during manufacture to orientate the molecules in one direction; this gives added resistance and lower permeability to the film. Films like the polyvinylidene chloride copolymers are shrunk onto the cheese coat after packaging. This seals the film to the cheese rind and so gives extra resistance to permeability (e.g. the use of PVDC shrinkable barrier bags for large-scale packaging). The following materials are laminated in either of two, three or even four thicknesses (some of these laminates are lacquers or very thin coatings): paper, cellulose acetate, polyethylene, polyester, polyvinylidene chloride copolymer and aluminium foil.

Aluminium foil is used for laminates, the foil being coated with an adhesive film (polyethylene) to prevent excessive bending of the foil and pinholing in the bends and crevices. The intact foil is impermeable to water, gases, solvents and light, but if pinholes occur the foil is permeable. Tin foil has been used for processed cheese but is not used for general natural cheese packaging. Information on the types and usage of film available in any country or area can be obtained from local suppliers of cheese packaging materials.

Basic types of film are as follows:

1. Cellulose film is a carbohydrate material from wood, treated to give certain properties, used as outside packaging and for strength. It is more useful when laminated with polyethylene (inside laminate) or wax coated (requires heat sealing to the cheese).
2. Polystyrene and polypropylene are often used for rigid containers for soft cheese or curds, rather than flexible packaging. Both are resistant to chemicals.
3. PVDC (Saran™) is a copolymer film, sometimes used for coating cheaper materials. Alone it is not easy to seal due to its limited melting temperature range, and metal 'clip on' seals are used. The film is used for vacuum packaging cheese. Cryovac™ is an alternative PVDC film made by W. G. Grace Ltd, London.
4. Polyethylene film is usually laminated and is used as the inside film, next to the cheese, since it is impermeable to water vapour. Its almost transparency to oxygen (in the low density form) makes it unsuitable for single film packing of cheese. Excessive mould growth occurs on polyethylene packed cheese and so an oxygen-tight membrane is used as a laminate on the outside. Waxed cellulose or aluminium foil has been used as the outer lamination.
5. Polyamide (nylon 6 or 11) has been used for both ripening films and consumer packs, although in some areas it may be expensive.

Where cheeses are mass produced and are subject to long-term storage, PVDC films (e.g. Cryovac™), made up into bags, have been successfully used. The cheese is filled into the bag which is then subjected to high vacuum. The bag, under vacuum, -91 to $-98\,kN/m^2$ (27–29 inch Hg), is sealed with a metal clip. The sealed bag is then shrunk by heat onto the cheese. Both block and round-shaped cheese can be packed by this method. Replacement of the film by a heat-sealable PVDC laminate improves productivity and avoids clips as a source of metallic foreign bodies at the subsequent cutting operations.

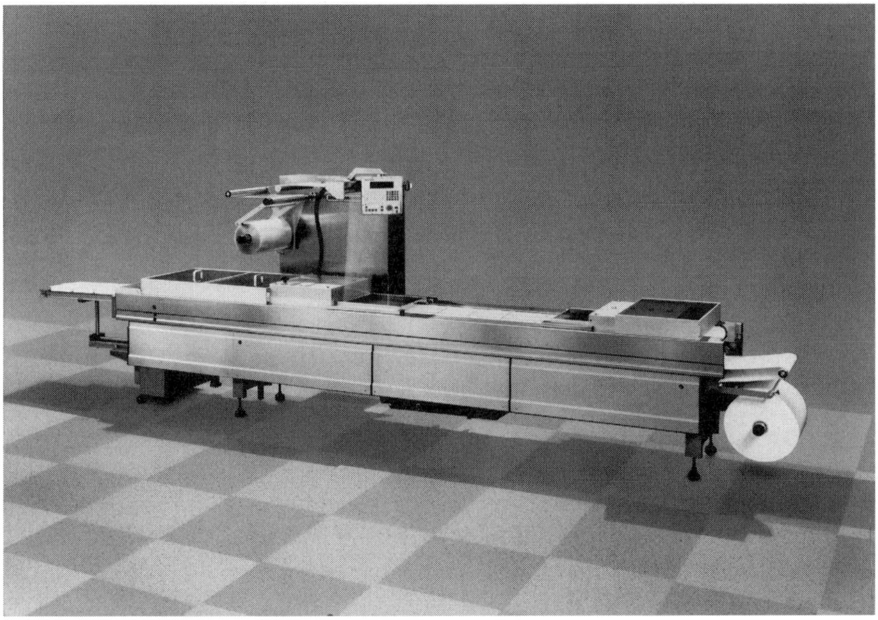

Figure 12.3 Thermoforming and vacuum packing machine for retail cheese. (Courtesy of Multivac UK Ltd)

Since the cheese taken from moulds or from block formers is still fairly warm, and therefore, likely to deform on storage, film wrapping is not sufficient in itself and a rigid container is required for the early storage periods. Refrigerated storage for initial cooling of the cheese is needed, not just to help firm up the cheese but to inhibit growth of heterofermentative secondary microflora. A number of methods have been used to support the cheese.

Traditional cheeses have been supported by bandaging, the bandages being reinforced by the flour paste or latterly carboxy methyl cellulose based adhesives which may contain sorbate to inhibit mould growth. The bandaging or coating of traditional cheese must withstand rougher usage during turning of the cheese, which is essential on a daily, then alternate day, basis in the early stages of maturation to minimize distortion and give the desired cylindrical/slightly barrelled shape.

Vacuum packed block cheeses (typically $36 \times 28 \times 18$ cm) are normally packed into various types of boxes. Cardboard boxes have been used for some textured cheeses, particularly those maturing quickly, e.g. Cheshire, but give little mechanical support to the warm cheese without some additional support such as a palletized crate. The restricted ventilation with this approach slows the cooling of the cheese.

Rigid plastic or wooden crates have been widely used, giving reasonable support with easy handling and ventilation (plastic crates being better in the

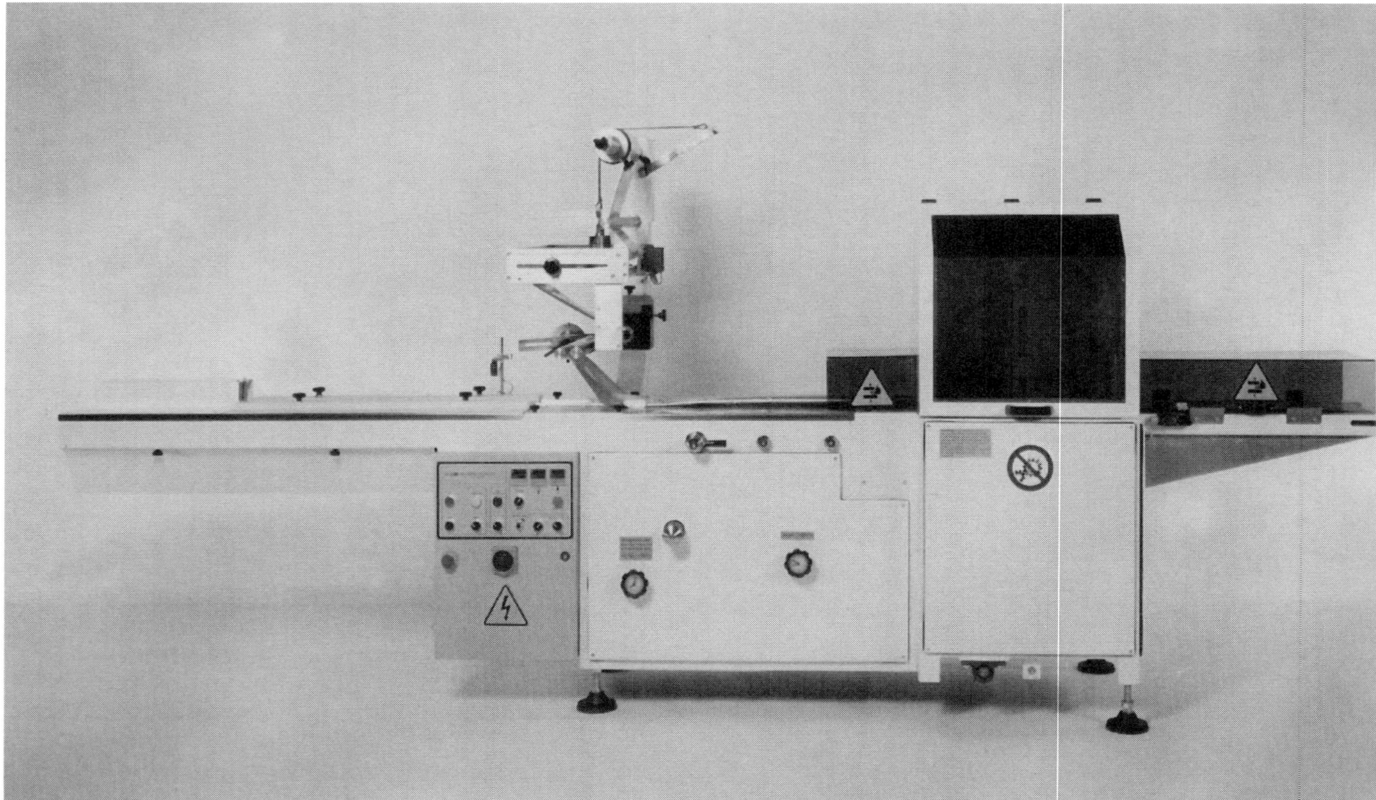

Figure 12.4 Flow-wrap modified atmosphere packing machine for retail cheese. (Courtesy of Wrapid Packaging Systems Ltd)

latter respect). The limited height of the block and the minimal clearances between the cheese and the crate do not require any turning of the cheese. The Unibox system used six plywood boards, tightly bound to the cheese surfaces by metal bands, giving a strong stackable structure and sharp edges on the cheese upon opening. This system however is relatively labour intensive and has been largely replaced by crates.

There are two types of film or wrapping materials needed for cheese packing: (1) those wrappings which are to be used for ripening cheese in store and (2) those films used for retail consumer packs, when heavier quality, printable materials may be required. However, both packs need to be kept in temperatures of less than 10°C in order to inhibit gas production in the cheese curd. Cheese stored at 15°C has been known to develop 100 kPa (an atmosphere) pressure in the gas-tight packaging materials.

The lack of rind adherence to the packaging materials, or the use of wrong materials, assists in the growth of moulds on the surface of cheese. Moulds produce *inter alia* proteolytic enzymes which penetrate and digest the curds and this may cause heavy wastage.

Cheeses which have been ripened at low temperature in gas-tight packings often exhibit aromas and flavours not characteristic of the cheese. However, on exposure to the atmosphere for some time (30–60 min), those locked-in volatile substances are lost to the air and the cheese assumes its normal flavour and aroma. The flavour and aroma of hermetically sealed cheese cannot be expected to be the same as traditional cheese. The volatiles which would normally leave traditional unwrapped cheese undergo secondary reactions and so affect the final flavour and aroma of some of the hermetically sealed cheese, especially if ripened at higher than normal temperatures.

A wide range of packaging systems have been developed for retail packs, the most popular being vacuum packaging and modified atmosphere packaging, examples of which are shown in Figures 12.3 and 12.4. Cheese packing has become a highly specialized sector of the cheese industry, and is frequently carried out independently of the manufacturing.

Warm storage in consumer retail stores has given rise to problems in respect of flavour, aroma and whey leakage in some of the hermetically sealed cheeses. The increase in temperature alters and speeds the process of ripening, and the more rapid change in the state of bound moisture results in free moisture (whey) which is released.

References

Abd-El-Salam, M.H. and El-Shibiny, S. (1982) XXI *International Dairy Congress, Moscow*, Vol. 1, Book 1, p. 388.

Banks, J.M. and Muir, D.D. (1985) *Journal of the Society of Dairy Technology*, **38**(1), 27.

Banks J.M., Muir, D.D. and Tamine, A.Y. (1984a) *Journal of the Society of Dairy Technology*, **37**(3), 83.

Banks, J.M., Muir, D.D. and Tamine, A.Y. (1984b) *Journal of the Society of Dairy Technology*, **37**(3), 88.
British Standards Institution (1991) BS 1133 Section 22: 1989. *Packaging in Plastic Containers*. British Standards Institution, London.
De Boer, R. and Mooy, P.C.F. (1980) *North European Dairy Journal*, **46**(3), 52.
Emmons, D.B., Kalab, M., Larmond E. and Lowrie, R.J. (1980) *Journal of Texture Studies*, **11**, 15.
Green, M.L. and Grandison, A.S. (1993) In *Cheese: Chemistry, Physics and Microbiology*, Vol. 1, 2nd edn (ed. P.F. Fox), Chapman & Hall, London, pp. 101–140.
Green, M.L., Marshall, R.J. and Glover, F.A. (1983) *Journal of Dairy Research*, **50**, 341.
Hansen, R. (1984) *North European Dairy Journal*, **8**, 3.
Mahaut, F. and Maubois, J.L. (1978) *XX International Dairy Congress*, Vol. E, p. 793.
Mottar, J. and Delbeke, R. (1978) *XX International Dairy Congress*, Vol. E, p. 797.
Ramet, J.P., El-Mayda, E. and Weber. F. (1982) *XXI International Dairy Congress, Moscow*, Vol. 1, Book 2, p. 546.
Sweetman, A. (1991) *Dairy Industries International*, **56**(11), 28–29.

13 Cheese manufacture

13.1 Varietal differences

Although the list of cheese presented by Birkhalter (1981) can be divided into 29 groups, in this section it is convenient to consider cheese in the groups shown in Table 3.2.

Chapter 19 presents a number of recipes for making various types of cheese, but a recipe can only be regarded as a skeleton for practical application. Indeed, due to differences in milk composition, starter activity and the local cheesemaker's aim to produce cheese to satisfy customer demand, there is obviously a need for variation within a skeleton recipe (see recipes for Cheshire cheese). It is, therefore, prudent for a cheesemaker to gain practical experience in so far as deviations from recipes are necessary.

The techniques of cheesemaking change from time to time, especially according to the degree of mechanization, but even with mechanization the general principles laid down for any specific variety still apply even though some stages are enclosed and combined within the mechanics of the process.

Cheesemaking is a fermentation process within which the following play a role.

(1) Time.
(2) Temperature.
(3) pH (often measured as acidity).
(4) Concentration of reactants (or metabolites in relation to bacteria).
(5) Water activity (A_w) in the curds.
(6) Concentration of inhibitors (e.g. salt, nitrate).

Apart from the growth of micro-organisms, the cheesemaker has physical control of some of these parameters, e.g. ingredients, times and temperatures. It is the succession of activities of different micro-organisms and enzyme systems which contribute to differences in flavour, aroma and sometimes texture between varieties of cheese.

The water activity (A_w) of a curd affects the growth of micro-organisms (Stadhouders and Langeveld, 1966; Stadhouders, 1975). The A_w of cheese, while dependent on the water present, is also dependent on salt and other components (Hardy, 1979; Marcos *et al.*, 1981). The water holding capacity of casein increases with A_w over the range A_w 0.7–1.0 (Hardy and Steinberg, 1983). In this respect the use of wash water for scalding curds (as in Samsoe,

Danbo, Gouda and Edam) results in softer, moister curds, but it also results in loss of lactose, lactic acid, calcium salts and other solubles from the curds.

Scald temperatures above rennetting temperature shrink the protein networks and thus expel whey from the curd. Acidity, developed by lactic bacteria from lactose, lowers the pH of the curd, which also causes curd shrinkage. The released whey contains lactose and other solubles. Indeed, in some cheese like Lancashire and Feta, acid alone and not temperature is responsible for curd shrinkage.

Salt also assists in the removal of whey from the curd, but has another role in the curd: limiting or even inhibiting the activity of micro-organisms (see Salting of cheese, section 12.12). The final pH of the curd before salting is an important stage in cheesemaking. The nearer the pH value of the curd to the isoelectric point of casein (pH 4.6), the more brittle and short textured the cheese curd becomes. While there is lactose available the lactic bacteria can produce lactic acid, unless they are inhibited by salt concentration.

Cheddar curds, because of the scald temperature and the rate of acid production throughout the process, do not lose all lactose at the time of salting. Instead, it persists for 15 to 20 days during early ripening. In unpressed Cheddar curds with a 40% moisture content, the application of 2% w/w of salt to the curds represents a 5% concentration of salt in the curd moisture, and this is sufficient to inhibit or slow down the production of acid, the pH of the curd falling only slightly during the first few weeks of ripening (e.g. Cheddar cheese pH values falling from 5.25 down to 5.10).

In some cheeses, like Emmental and Gruyère during manufacture, high scald temperatures, together with acidity, denude the curds of lactose by the time the curd is pressed. Brine salting is satisfactory for these cheeses and the pH remains steady (pH 5.2–5 4). Final pH values outside this range tend towards the formation of 'blind' cheese. Cheese showing the characteristic 'eyeholes' (e.g. Emmental) are not textured during manufacture but are subject to a second ripening at a higher temperature (22–25°C). It is during this time that gas generated by *Propionibacterium* forms holes in the curds.

Edam, Gouda and other cheeses developing smaller eyeholes are not subjected to a secondary ripening stage, but are stored at temperatures from 15 down to 2°C according to the speed of ripening required.

Textured cheeses receive manipulation of the curds before salting to cause the texture formation. In Cheddar cheese pressure, temperature, and pH cause the particles of curd to spread out to five to six times their normal size and to mat together into a firm mass. This produces the chicken breast or cheddared structure; in Cheshire cheese, another textured variety, the object of manipulation is to keep the particles of curd separate to give an open-structured curd.

Blue-veined cheeses need an open-textured curd to aid ingress of oxygen from the air and for the hyphae of the mould to penetrate through the curds.

Pasta filata cheeses have their curds immersed in hot water or hot whey causing them to plasticize and become pliable and elastic, after which the curds are moulded into shape and cooled before being brine salted.

Different quantities of starter and ripening times and different amounts of rennet and rennetting temperatures, all affect the firmness of curds when cut. These features have an effect on the varietal differences between cheeses, but the different species of micro-organisms present in the curds are also responsible for some of the varietal differences such as flavour, aroma and texture.

The rate of acid production within the curd, together with the temperature employed in scalding it, are both important in controlling the quality of the curd to go forward to ripening. In the final stage, salt also plays a role by slowing down the rate at which acid is produced. The amount of salt needed depends on the moisture content of the curd, since a wet curd loses more salt in the whey, which is expelled. The cheesemaker, therefore, needs to apply more salt to wet curds and less to dry curds.

Even though most stages in cheesemaking are longer than 1 h, the hourly rate of acid increase is an indication of the activity of micro-organisms in the curd. Table 13.1 shows some of the hourly rates of acidity increase in some textured cheese. In textured cheeses, the acidity increase between 'whey off' and moulding the curd should have caused expulsion of the whey, leaving very little lactose in the curd for further acid increase. If the curd is left covered with whey, which contains lactose, for too long then an excessively acid cheese results which will become dry and hard during ripening. This problem is shown in Table 13.2 which indicates the acidities in two experimental Cheddar cheeses. At 5 months the normal cheese (A) was of good flavour, texture and body, whereas cheese B was very dry, hard and cracked with an acid and unpleasant flavour. The whey from cheese B became very acid and its pH was lower than that of the curd.

Comments on the various cheese groupings are given below and recipes of representative types of cheeses are given in Chapter 19.

Table 13.1 Some typical hourly rates of acidity increase during manufacture of textured cheese

Cheese	Rate per hour (as % lactic acid)		Curd cut to cube size	Scald temperature (°C)
	Curd cutting to whey off	Whey off to moulding		
Cheddar	0.044	0.235	Rice	40
Cheshire	0.046	0.255	Pea	33.5
Derby	0.038	0.176	Beans	35
Leicester	0.035	0.170	Rice	36
Double Gloucester	0.048	0.250	Pea	38
Caerphilly	0.003	0.082	Pea	34.5
Lancashire	0.006	0.096 (0.032[1])	Pea	–

[1] After 24 h.

Table 13.2 Acidity development in two experimental Cheddar cheeses

Stage	Whey		Curd	
	acidity (%)	pH	pH	moisture (%)
Normal cheese (A)				
After cutting	0.12	6.49	6.56	76
After scald	0.16	6.32	6.35	66
Whey off	0.21	6.00	6.15	61
At milling	0.69	5.22	5.34	41
At 1 month	–	–	5.15	36.8
Experimental cheese (B)[1]				
After cutting	0.12	6.50	6.58	77
After scald	0.16	6.24	6.38	67
Whey off	0.35	5.40	5.75	49.5
At milling	0.67	4.95	5.45	39
At 1 month	–	–	4.92	33.5

[1] Curd pitched in whey for 2 h longer than for cheese A.

13.2 Grating cheese

Parmesan cheese is made in many countries and is often called Grana cheese; it originates from Italy. The cheese is a grating cheese, very hard in body and has a shelf life of 3–5 years. A typical analysis is:

fat-in-dry-matter	35–38%
dry matter	70–74%
water in fat-free substance	40%.

The apparent low fat percentage is because the cheese is often made from reduced fat milk (2.8–3.2%). Leaking cheese often arises from too high a fat content in the milk. The curd is hard and dry, the flavour pleasant, agreeable, spicy but not piquant or with a strong 'tang'. The 28–34% moisture in young cheese declines in the older cheese to 18–27%, and the salt content rises to 3.2–3.4%.

Since the cheese is scalded to high temperatures, 54–58°C, it has been thought unnecessary to heat treat the milk. Parmigiano-Reggiano and Grana Padano are the two better known types in Italy. The rind is often coated with oil and coloured brown, umber or sienna, the cheese being marked on the surface with its identity. Bottazzi (1964) studied the production of Grana cheese, and Besana (1962) wrote about the manufacture of Grana Lodigiano cheese. Delforno and Losi (1980) discussed Sbrinz cheese making. Messner (1968) wrote on Parmesan cheese and Dieci *et al.* (1970) considered the effect of whey starters on Parmesan cheese ripening. More recently, these cheeses have been included in a review by Battistotti and Corradini (1993).

13.3 Textured cheese

This group of cheeses includes many of the English varieties of cheese, i.e. Cheddar, Cheshire, Derbyshire, Leicestershire, Double Gloucester and Lancashire. Even the blue-veined cheeses are textured, i.e. Wensleydale and Stilton. These textured cheeses differ in manufacture from other groups of cheese in that after the whey has been drained away the curds are manipulated to give a texture before being moulded and pressed.

13.3.1 Cheddar cheese

Cheddar cheese is made in many countries but due to local influences and market preferences a number of slightly different Cheddars have evolved, e.g. traditional English, New Zealand, American and Canadian Cheddars. All these cheeses have in common the **Cheddaring** of curd during manufacture, which is a distinguishing feature of this cheese. There are also differences in manufacturing procedure, especially where mechanization has been introduced.

The aim of the cheesemaker is to produce a buttery but firm body with close texture and a clean nutty-flavoured cheese with no eyeholes and a keeping quality of up to 1 year (Bryan, 1957). The following description of the manufacture of Cheddar cheese is given in some detail since many of the features involved in the process can be applied to other cheese varieties. This description is based on traditional Cheddar; where there are possible differences in technique, then references are given in parentheses to previous sections or chapters dealing with these differences. The traditional method for the manufacture of Cheddar cheese was a mainly manually operated process. Small- and medium-scale production methods still require a large, skilled manual input, which will be evident from Figures 13.1 to 13.3.

For many years, it was considered that cheese made from raw clean milk, free from disease and other contaminating micro-organisms, acquired an entirely distinct and characteristic aroma, flavour and texture from cheese made from heat-treated milk. Today the collection of milk from many sources and the greater awareness of the risks of food poisoning place an emphasis on heat treatment. Table 13.3 gives some details of the acidities, temperatures and operational times for Cheddar cheese manufacture.

Unfortunately cheese milk stored between farm and cheese vat sometimes presents difficulties (Ali *et al.*, 1980) (section 8.1). Addition of calcium chloride up to 0.2% may be required to overcome difficulties brought about by cold storage of the milk, although heat treatment at 60–70°C for 15 s avoids the effect of low temperature storage (Johnson *et al.*, 1981).

The raw milk should be of good chemical quality, clean and fresh, with an acidity of approximately 0.16% before heat treatment. The most suitable heat treatment was considered to be 68–70°C for 15 s in order to preserve

Figure 13.1 Open cheddaring table. (Courtesy of APV Nordic Cheese AS)

Figure 13.2 Curd mill for textured dry salted cheese. (Courtesy of APV Nordic Cheese AS)

lipases in the milk (section 10.6), but treatment at 72°C is now regarded as safer. After heat treatment the vat milk is left at starter temperatures of not less than 21°C. At this stage a virile starter in the 'soft set' state (i.e. not over acid) is added to the milk at the rate of 1% of the milk volume. Since starter activity varies in different milk supplies or media, previous cheesemaking experience is a guide to any possible variations in quantities used. The ripening time for the milk may be from 45 min to 1 h, at which stage the acidity should have increased by 0.02%. In view of the possibility of bacteriophage attack (section 9.6) the starter addition may be increased to 2– 5% and the ripening time reduced to 5–10 min.

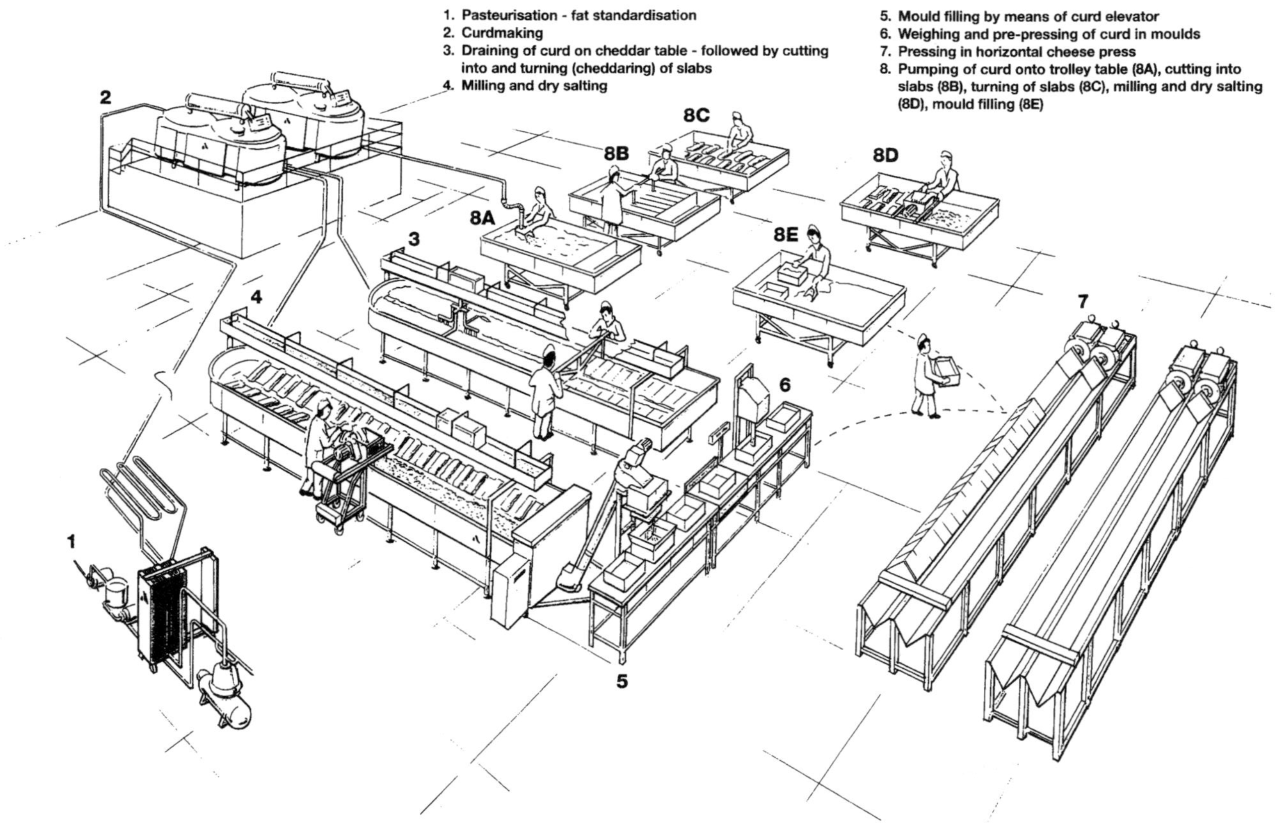

Figure 13.3 Process line for small- and medium-scale Cheddar production. (Courtesy of APV Nordic Cheese AS)

Table 13.3 Summary of acidities, temperatures and operational times of traditional Cheddar cheese, as taken from cheese logs

Process stage	Acidity at the end of stage % or (pH)	Temperatures at the end of stage (°C)	Stage length (min)	Elapsed time (h:min)
1. Raw milk	0.16	< 8	–	–
2. Heat treatment	0.16	72	0	0
3. Starter	0.18	30	20	0:20
4. Rennet (Coagulant)	0.20	30	45	1:05
5. Cut	0.115	30	15	1:20
6. Stirring and scalding	0.135	38	40	2:00
7. Pitching	0.155	38	55	2:55
8. Whey off and draining	0.25	35	25	3:20
9. Cheddaring	0.62	32	65	4:25
10. Milling	0.65	30	15	4:40
11. Salting	–	29	20	5:00
12. Hooping	–	28	10	5:10
13. Pre-press	0.75 (whey)	26	40	5:50
14. Dressing	–	–	10	6:0
15. Pressing	1.15 (whey) (5.2, curd)	24	13–18 h	
16. Storage and grading		–	7–15	(21+ days)

If coloured cheese is required then annatto may be added to the milk at the rate of 0.015–0.03%. The colour should be well stirred into the ripening milk while at the same time the temperature is raised to 29–29.5°C. To ensure a curd firm enough to cut in a reasonable time (e.g. 45 min), a rennetting temperature of up to 31°C may be required with some milks, or if DVS cultures are being used. Milk supplies from some localities may require additional calcium chloride to achieve a satisfactory, firm curd. Rennet at the rate of 0.022–0.025% of milk, diluted with cold water six to ten times, is added into the ripened milk. When using bulk starter, the acidity should be 0.19–0.22% before rennetting, but some cheesemakers prefer to use the rennet test for assessing the readiness of the milk for rennetting. To avoid an 'over stir', i.e. broken coagulum (section 12.5), the rennet should not be stirred into the milk for more than 5 min.

The coagulum is cut, when firm enough, in about 45 min (section 12.9). Multi-bladed steel knives or nylon-strung frames are used to cut the curd to the size of small peas. The curd–whey mixture is stirred slowly at first to allow the cut surfaces of the curd to heal. Then the temperature of the mixture is raised at the rate of 0.5°C per 5 min, rising to 1°C per 5 min. After 45 min the temperature should be 39°C. Stirring of the mixture should continue for a further 45 min before the curd is 'pitched' (i.e. allowed to sink in the whey). The acidity at this stage should be 0.15–0.175%. When the acidity of the whey reaches 0.20–0.25%, i.e. about 2.25 h–2.75 h after rennetting, the whey is removed from the vat and the curds collected along the sides of the vat to leave a drainage channel down the centre. The next stage is one unique to Cheddar cheese. The curds, whilst still draining, are cut into large blocks which are turned over and after 10–15 min piled one on another. This process is repeated every 15 min. The piling of the curd squashes the individual curd particles as well as releasing more whey. It is this repeated piling of the blocks which produces the 'chicken breast' structure of the curd (i.e. cheddared curd) which is typical of this cheese. When the acidity of whey draining reaches 0.68–0.85% and cheddaring is complete the curd is milled into walnut-size pieces by a tearing or cutting chip mill. After milling the curd, salt is sprinkled over it at the rate of 2% w/w and the salted curds are stirred whilst allowing them to 'mellow', during which the salt largely dissolves in the curd and the temperature drops further to 25–27°C.

The well-mixed curds are moulded and then pressed for 2 h at 19 kPa (250 kg per cheese), after which the cheeses are removed and 'dressed' (i.e. the rough edges are removed). The cheeses are returned to press for 24–30 h at higher pressure. The advent of mechanization has altered the traditional methods of cheesemaking by removing the laborious tasks of handling individual cheeses.

Various forms of curd-making vats are available (Chapter 14) but the coagulation is still a batch process with the curd from a succession of vats going forward to a continuous system of draining, texturing, milling, salting and moulding machines. Since such devices are mechanically connected, the curd entering the system must be of consistent quality, both chemically and bacteriologically. The introduction of large silo tanks for raw milks and the use of milk-ripening tanks with automatic starter injection have assisted in providing a reasonable level of quality of milk, including the rate of acid development for further treatment in the curd-making vats.

Various forms of mechanisms have been devised for dealing with the curd from the vats (Chapter 14) and there are a number of curd-moulding and pressing arrangements available. Vacuum pressing and the block formers using vacuum have shortened the pressing time from the traditional 2–3 days to only 1–6 h in the vacuum press. While the smaller cheeses have time to cool and 'set', the larger cheeses need support until they are cool enough to 'set'. 'Hubert' Cheesemaking Systems have devised a press using Laude plastic moulds for block Cheddar cheese in which the cheeses are vacuum pressed for only 40 min, after which the cheeses are transferred to another open press for a further 30 min. The cheeses need careful handling since they are not fully set until they have been allowed to cool.

Some of the problems that can occur in making Cheddar cheese are illustrated in Tables 13.4 and 13.5. Although milk composition does cause problems, the rates of acid development are the main cause of many problems. Table 13.4 contains analyses of a selection of Cheddar cheese made in September/October, while Table 13.5 includes those made in February.

Table 13.4 Analyses of cheese logs showing effect of cheese composition and rate of acid development on the quality of the final cheese as judged by grading

Composition and rate of acid development, etc.	Cheese make no.				
	1	2	3	4	5
Fat (%)	33.0	33.0	34.0	33.8	34.0
SNF (%)	33.4	28.9	29.1	28.6	28.6
Moisture (%)	33.6	38.1	36.9	37.6	37.4
FDM (%)	49.6	53.3	53.8	54.1	53.4
WFFS (%)	50.1	56.9	55.8	56.7	56.6
pH of cheese at Grading	5.3	5.35	5.0	5.3	5.45
Acidity at salting (%)	0.65	0.68	0.66	0.66	0.64
S/M (%)	5.0	3.5	4.5	4.1	4.5
Hourly rate of acidity development					
Starter to rennet	0.03	0.06	0.00	0.015	0.045
Cut to pitch	0.06	0.02	0.04	0.01	0.035
Pitch to cheddaring	0.27	0.27	0.17	0.18	0.19
Cheddaring to salting	0.32	0.27	0.36	0.31	0.06
Time starter to salting (h:min)	4:20	4:30	5:20	5:40	5:30
Grading points awarded					
Flavour and aroma	41	41	41	41	41
Texture and body	33	32	33	34	35
Total points	89	88	89	90	91
Grade awarded	Selected	Selected	Selected	Selected	Selected

Comments: 1, fairly firm body, mealy texture; 2, soft weak body, high moisture cheese; 3, fairly firm body, open texture; 4, good body, fair texture; 5, fair body, slightly rubbery close texture.
Notes: FDM, fat in dry matter; S/M, salt in cheese moisture; SNF, solids not fat; WFFS, water in fat-free solids (also known as MNFS, moisture in non-fat solids).

Table 13.5 Analyses of cheese logs showing effect of cheese composition and rate of acid development on the quality of the final cheese as judged by grading

Composition and rate of acid development, etc.	Cheese make no.				
	6	7	8	9	10
Fat (%)	33.9	34.0	32.7	33.8	33.3
SNF (%)	27.9	27.7	27.5	27.7	28.2
Moisture (%)	38.2	38.3	39.8	38.5	38.5
FDM (%)	47.0	55.0	54.3	54.9	54.1
WFFS (%)	57.6	58.0	59.1	58.1	57.7
pH of cheese at Grading	5.2	5.2	5.15	5.35	5.35
Acidity at salting (%)	0.79	0.72	0.76	0.75	0.73
S/M (%)	5.3	5.3	3.6	4.3	4.4
Hourly rate of acidity development					
Starter to rennet	0.005	0.012	0.018	0.00	0.00
Pitch to cheddaring (end)	0.07	0.07	0.06	0.006	0.04
Cheddaring to salting	0.48	0.24	0.26	0.06	0.40
Time starter to salting (h:min)	4:50	4:30	4:30	5:30	5:40

Table 13.5 (*Contd.*)

Composition and rate of acid development, etc.	Cheese make no.				
	6	7	8	9	10
Grading points awarded					
Flavour and aroma	37	38	37	39	40
Texture and body	26	30	31	28	31
Total points	75	83	84	82	86
Grade awarded	Graded	Selected	Graded	Graded	Selected

Comments: 1, off-flavour, crumbly harsh body; 2, off-flavour, crumbly open texture; 3, off-flavour, soft sticky body; 4, off-flavour, soft sticky body, open texture; 5, lacks flavour, open mealy crumbly texture.

Comments on the individual cheese analyses given in Tables 13.4 and 13.5 are as follows:

Cheese 1 The FDM at 49.6% is lower than normal, indicating a firmer cheese. The WFFS at 50.1% also confirms this assessment. The high cheese pH at 5.3 shows that the curd was not acid but that the acid development in the last stage and the high salt content were responsible for the mealy texture of the cheese.

Cheese 2 The fat and solids contents of the cheese were fairly normal but the moisture content was higher than is consistent with good quality (WFFS @ 56.9%). The acid development was slow in the early stages and the cheese was not acid (pH 5.35). In view of the low salt content (3.5% S/M) and high moisture, off-flavours may develop later.

Cheese 3 The moisture content of the cheese was higher than might have been expected in an acid cheese (pH 5.0). The rate of acid development in the last stage could have resulted in the open texture in the final cheese.

Cheese 4 The composition of the cheese was normal, but the WFFS @ 56.7% showed that the cheese was not as firm as might be expected. The longer manufacturing time of 5 h 40 min assisted the final texture formation.

Cheese 5 The rate of acid development slowed down after cheddaring and the cheese was fairly sweet with pH 5.45. This contributed to the rubbery curd. Although graded higher than Cheese 4, this cheese would not keep its quality on further ripening but would deteriorate.

In Table 13.5, Cheeses 6–9 all had high moisture and off-flavours.

Cheese 6 The high WFFS value, 57.6%, shows the high moisture in relation to the solid content and hence the off-flavours of the final cheese. The cheese acidity, pH 5.2, was normal but the high salt in moisture content could have contributed to the harsh crumbly body and texture.

Cheese 7 The high WFFS value @ 58% shows the high moisture in relation to the solids content. The rate of acid development was slow in the early stages and contributed to the open, crumbly texture of the cheese. Although on grade points it was of 'Selected' grade the 38 points for flavour and aroma did not reflect the off-flavour comment.

Cheese 8 The WFFS value, 59.1%, showed the high moisture content and hence the off-flavour development later. The lower S/M @ 3.6% also contributed to the off-flavour development. The points for flavour and aroma (37) were sufficient to keep the cheese in the 'Graded' group.

Cheese 9 The WFFS value, 58.1%, showed the high moisture content in relation to the solids in the cheese. The low rate of acid development is reflected in the high pH value (5.35) of the curd. Thus the cheese had a soft sticky body and open texture and was in the 'Graded' group on account of these defects.

Cheese 10 This cheese developed acid slowly at first and then more quickly after cheddaring. This led eventually to a mealy open texture although the cheese at pH 5.35 was not acid. In spite of the comment that the cheese 'lacks flavour' it received 40 points for flavour and aroma and this brought it into the 'Selected' grade group.

Thus, Cheese No. 1 (Table 13.4) was short of fat as against the other cheeses, while all of the Cheese Nos. 6–10 (Table 13.5) had more moisture per unit of solids than usual. In such instances the cheesemaker could have cut the curds smaller than the usual 13-mm cubes to reduce the moisture content more quickly.

The manufacture of Cheddar cheese by traditional and by the earlier mechanized methods has been described by Maddever (1955), Bryan (1957), Ministry of Agriculture, Fisheries and Food (MAFF; 1959), Scott (1967a,b,c), Davis (1976), the Australian Society of Dairy Technology (1970), Kosikowski (1982) and Irvine (1978). Cheddar cheesemaking by the Cheddarmaster equipment is described by West (1980).

13.3.2 Cheshire cheese

Cheshire cheese is an open-textured cheese with a firm body and a mildly lactic flavour. The colour varies from white to a light straw colour to a light red. It originated around the banks of the River Dee and, because of the fluctuating quality of the milk produced from the water meadows near to the river, three recipes were needed, for:

Figure 13.4 2000 kg Wincanton/Stoelting end unloading finishing vat for Cheshire and other traditional UK cheeses. (Courtesy of Wincanton Engineering Ltd)

(a) a quick ripening cheese in spring,
(b) a medium ripening cheese in summer,
(c) a longer ripening cheese in late summer and autumn.

The cheese takes only 3.25–3.5 h to make (compared to 5.5 h for Cheddar). The cheesemakers' aim is to obtain an enhanced rate of starter activity so that acid helps to dry out the curds rather than a scald temperature of only 31–34°C. The drained curd is kept free, usually by stirring as illustrated in Figure 13.4, till at 0.65–0.75% acid the curds are milled through a fine peg mill to produce a free-flowing curd. Salt is sprinkled on the curd which may sometimes be re-milled. This produces the open-textured fairly acid curd typical of Cheshire cheese, but continued acid production in the curd before pressing is achieved by keeping the moulded curd warm at 21°C overnight (i.e. in a 'cheese oven').

Cheshire cheese is often coloured and is marketed in 2 weeks to 10 months. Blue-veined Cheshire cheeses are becoming more widely produced from moist open curds inoculated with *Penicillium roqueforti* spores. The 'blueing' is often very irregular (MAFF, 1959). Cheshire can be made on systems similar to those for cheddar if the texturing stage is shortened, but there is a tendency for the cheese to be rather closer textured than the traditional cheese. However, the retail trade requires a closer textured cheese to cut into consumer portions and the 'industrial' cheese is more suitable for this.

13.3.3 Dunlop cheese

Dunlop cheese is a close-textured waxy cheese with a pleasant, mild, creamy flavour and is quicker maturing than Cheddar, although older cheeses develop a sharp flavour (Smillie, 1955). This cheese is often used for toasting. It is said to have originated in Ayrshire, based on an Irish method (Smith, 1995) but has been modified *inter alia* by nineteenth century Somerset Cheddar cheesemaking practices.

The cheesemakers' aim is to use less starter than normal (0.5–1%) and to cut the curd in 45 min to maize-size cubes. Scald temperatures of not more than 36.5°C are used to produce a moist curd of 0.22% acid after whey draining. The curd is milled through a peg type mill at 0.65% acid to give a friable, free curd before salting (1.85%). This process produces a moist curd which matures quickly in 6–8 weeks. Dunlop cheese must have less than 44% moisture.

13.3.4 Derby cheese

Derby cheese is softer and less acid than Cheddar and is therefore a quick ripening cheese (Scott, 1954a; MAFF, 1959). The flavour is bland and the texture fairly close with few mechanical holes, but the body must not be pasty or chalky, i.e. lacking acidity or excessively acid. Derby cheese is sometimes produced with added sage leaves for extra flavour. Since it is a low acid cheese it requires clean milk, which is usually pasteurized. Moderate amounts of starter are used (0.75–1.0%) to produce a mildly acid curd. The coagulum is cut to the size of beans and is very fragile. The scald, to not more than 36°C, is achieved slowly. The acidity at salting the curd is from 0.45 to 0.50%. The cheese is marketed from 1 to 8 months old.

13.3.5 Leicester cheese

Leicester cheese is one of the most highly coloured of cheeses. It should have a clean buttery cheese flavour and firm mellow body with a short texture but be free of gas holes (Scott, 1954b; MAFF, 1959). The cheese is made from low acid curds. Only 0.5–1.5% of starter is used and the milk is rennetted at low acidity (0.14–0.15%). The coagulum is cut finely (size of rice) and the scald to 35–37°C is achieved slowly to produce a firm curd. After the whey has drained away, the curds are cut into blocks and further drainage is obtained by re-cutting the blocks smaller until the acidity reaches 0.45–0.55%. Finally, the curd is milled with a fine-toothed peg mill and then after salting and moulding the cheese is pressed.

13.3.6 Gloucester cheese

Gloucester cheese was originally made in the Vales near the River Severn (UK) as Single and Double Gloucester cheeses differing in shape. Now the Double Gloucester is almost a hybrid cheese with a character between Cheddar and Cheshire cheese (Colnett, 1955; MAFF, 1959). Double Gloucester cheese is of medium red colour and is made from curds which develop acidity slowly. The coagulum is cut into wheat-sized cubes which are scalded to 37.5°C to produce a firm rubbery curd, only slightly Cheddared when ready for milling at acidity 0.60–0.65%. A peg mill is used to produce a fine friable curd which is salted at the rate of 1.75–1.9% before moulding and pressing. Unlike traditional Cheshire curds, the Double Gloucester curd is not ovened overnight. The cheese is ready for sale in 6–8 weeks but will be mature at 4–6 months with a stronger sharp flavour.

13.3.7 Caerphilly cheese

Caerphilly cheese is a quick ripening cheese often sold at 10–14 days old. The cheese has a firm springy body with a flaky but short texture (Monie, 1955; MAFF, 1959). The flavour is mildly acid but when old the flavour becomes sharp and strong. Normally the acid production is slow at first but faster later. Since the scald temperature is low with little more than a degree Celsius rise in temperature, the curd is dried out by the acid produced. The curd is milled through a coarse mill at an acidity of 0.28–0.40% (according to season) and one-half the salt is mixed with the curd. The remaining salt is rubbed onto the surface of the pressed cheese and brining may be used.

13.3.8 Lancashire cheese

Lancashire cheese is sometimes called the 'Leigh Toaster' because of its toasting character. The cheese falls between the hard-pressed and the semi-hard group, with a maximum permitted moisture of 48% (UK Regulations, 1996). It is of a soft crumbly texture with a clean, mellow, mildly acid flavour becoming strong if over 6 months old (Haley, 1954; MAFF, 1959). Sometimes sage leaves are mixed in with the curds. Lancashire cheese should be made by mixing curds of 2 or 3 days' manufacture to give a final acidity of 0.99% (Haley, 1954). The mixed curds are milled and then salted at the rate of 2%. The very soft curd needs 2 days pressing before being bandaged in order to keep its shape. The curds are often made daily from raw milk ripened with a small amount of starter (0.15%) to give a low rate of acid production. When the curds, without scalding, reach an acidity of 0.20% they are divided into two or three parts, one part being used for that day's mixed curds and the other part or parts being kept warm overnight. These latter curds produce a high acidity (1.2–1.6%) ready to be mixed with new low acid (0.2%) curd. The cheeses are marketed at 3–12 months old. In order to hasten the process and produce a firmer but non-typical cheese the curds have been scalded.

13.4 Untextured cheese

In untextured cheeses, the curds are taken from the whey directly into the cheese moulds and then into the presses. Pressing expels the free moisture from the curd and also removes the lactose and other substances in solution with it. Since the acidity is lower than for textured cheese the curd particles mat together into a solid shape during pressing. However, further acidity still develops in the curd during this period and until the curd is salted. During the pressing the curd cools and when removed from its moulds it retains its shape.

After pressing, the cheese curd is allowed to develop to its optimum pH (e.g. from pH 5.15 to 5.4 according to type of cheese), which is usually a stage in the recipe. At this stage the cheese is brine salted and a little further lowering of the pH value occurs (see Textured cheese, section 13.3).

The Romano cheeses are typical of this group and are made in many areas. When made from sheep's milk, in Italy, the prefix Pecorino is added (i.e. Pecorino Romano) (Carta, 1963). The Romanian Tismana and Bulgarian Pekorino cheeses are also made from sheep's milk; cow's, sheep's and goat's milk alone or mixed are used in other countries. The sheep's milk, according to breed, has 6.5–8.5% fat and so the cheese is a high fat variety. Typically, it is also a high salt variety with 4–7% salt. The aroma and flavour are characteristically piquant and sharp, with the 'bite' or tang associated with sheep's milk cheese. The cheeses are coated with vegetable oil, or with tallow mixed with yellow clay, which becomes dark with age; the cheese is normally ripe in 8–10 months. The main faults arise from blown curds and rough rinds. Mould growth (red) on the rind may produce bitter flavours, although retained pockets of whey in the curd also give rise to off-flavours.

In 1967, Balducci wrote about the Sordi system for the manufacture of Pecorino Romano cheese (Balducci, 1967). Ledda and Arrizzo (1969) also described Pecorino Romano cheese.

The Pressato, another Italian cheese, is allied to the Asiago and Fontina cheeses. It is a soft sweet cheese (pH 5.2–5.3) which develops a piquant flavour with a strong but not unpleasant aftertaste. Salt at 2–3% is kneaded into the soft curds before moulding, to enhance the keeping quality. Normal cheeses ripen in 6-8 months but acid cheeses are hard and unsuitable for long ripening. Delforno (1970a) described the manufacture of Pressato cheese, and also (Delforno, 1967) the manufacture of Taleggio cheese. Demeter (1966) described Fontina cheese and Delforno (1970b) described Asiago cheese. Kammerlehner (1967) described equipment and methods for Romadur and Munster cheese. Olson (1981) wrote on the trends in Italian cheesemaking and technology; a more recent review has been made by Battistotti and Corradini (1993).

13.5 Cheese with eyeholes

There are two groups of cheese which develop eyeholes in the ripe curd.

The first group (dating from AD 1104) is headed by the older cheese Gruyère and includes Emmental, Greyerzerkäse, Gruyère de Comte, Gruyère de Beaufort and the smaller Danish Samsoe. Even so, there are distinct divisions in the group. Emmental is the larger cheese, approximately 153 cm in diameter with large and shiny eyeholes. The curd is ivory white rather than yellow, the texture is close and the body is firm and yet elastic, so that no slits occur at the eyeholes. Swiss cheeses were traditionally made in open copper vats. Copper-lined stainless vats are still available, as illustrated in Figure 13.5. Larger sized cheeses may require special handling

Figure 13.5 Copper coated Alpvat for Swiss-style cheesemaking. (Courtesy of APV Nordic Cheese AS)

Figure 13.6 Brining and handling of large Swiss-style cheese. (Courtesy of APV Nordic Cheese AS)

(Figure 13.6). Other members of the Gruyère group though employing similar manufacturing methods develop different characteristics. The curds may be more golden yellow; firm and fatty rather than moist. The eyeholes are small (1–3 mm) and the cheese develops acid more quickly than Emmental and the flavour is stronger, aromatic and with a distinct savour of ammonia. Smear coats of *Brevibacterium linens* are often allowed to develop on these smaller cheeses. Gruyère is ripened at 17–18°C for 2–3 months for the secondary ripening (eyehole formation) by *Propionibacterium freundreichii* sub-sp. *shermanii,* as against Emmental, which is ripened at 20–24°C for 6–8 weeks. Thus, the influence of the rate of acid development and different ripening temperatures assist in producing quite different cheeses while employing similar manufacturing procedures. Sbrinz cheese is very similar to Gruyère, but does not have the secondary ripening to develop the eyehole texture.

The second group of cheese with eyeholes is rather more diverse than the first group. The eyeholes are much smaller, and in some instances less numerous. The cheeses in this group are more likely to have a smear coat (*B. linens* growth). Perhaps one of the better known in this group is Tilsiter cheese. The cheese quality is according to the fat content of the cheese, which means that the milk used has to be standardized as in

Table 13.6 Standardization values for milk to be made into different cheese qualities

Cheese quality	FDM (%)	Fat in cheese milk (%)
Half fat	20	0.95–1.0
Three-quarters fat	30	1.65–1.75
Normal	40	2.5–2.55
Full fat	50	3.6–3.65
High fat	60	5.0–5.1

Table 13.6. These figures may be compared to the figures for soft cheese (Camembert) given earlier.

Peter and Zollikofer (1959) produced a book on Emmental, Greyerzer and Sbrinz cheese. Kursteiner and Staub (1950) wrote on Emmental cheese. Steffen *et al.* (1993) have reviewed Swiss-type varieties. Schällibaum (1965) wrote on Tilsit(er) cheese.

13.6 Semi-hard cheese with few or no eyeholes (including washed curd cheese)

The semi-hard group of cheeses are produced in many countries. Although the curds initially are very similar, the differences between named varieties of cheeses are caused by differences in milk, starter and the temperature regimes used. Saint Paulin (Port du Salut) and the Trappist cheeses are fairly typical of at least 26 other varieties in this group.

Saint Paulin, first registered at the Stresa Convention in 1957, was further defined by the French Ministry of Agriculture in 1960. The cheese is made from cows' milk which is normally pasteurized. Raw milk cheeses have a stronger flavour. Faults in the cheeses are often due to cracked or split rinds into which moulds penetrate. When the air in the store is too dry, or the cheese has been too long in the salt bath, the rind is hard and cracks easily.

Just as the monks at the Notre Dame Monastery at Port du Salut developed Saint Paulin cheese, so the monks at the Maria Stern Monastery at Banja Luka in Bosnia developed the Trappist cheese in 1885. The Trappist cheese curds are fairly similar to Saint Nectaire (France), Saint Paulin (France), Carrick (Scotland) and Esrom (Denmark). The cheeses differ in curd consistency and in the type of eyeholes in the curds. Some cheeses, e.g. Trappistenkäse (Germany), have a smear coat which dries up when ripe and this produces a curd with stronger flavour. The Hungarian Trappista and the Bosnian Trappist cheese may be made from goat's or sheep's milk and this gives them a stronger and more aromatic flavour.

The differences between the various Trappist cheeses arise mainly from the type of milk and the temperatures employed, although the smear coating of some cheeses gives a stronger flavour. Dilution of the whey with hot water, however, produces a less acid type of cheese with a fairly bland, mild and pleasant flavour. Slit-type eyeholes may form in some varieties rather than the few round pea-sized holes usually associated with this group of curds.

The cheeses are of the washed or diluted whey type which dilutes the whey lactose, limiting the fermentation and hence restricting the acid development. The pH of the whey before salting varies from 6.0 to 6.1. However, the curd before salting may have a pH of 5.5–5.7 and after salting the pH is 5.2–5.3. The goblet-shaped traditional wooden moulds have almost all been replaced by plastic moulds, and the pressures, normally 80 kPa, reduced to 30–50 kPa.

The washed curd cheese group typified by Edam and Gouda cheese have been made and copied in many countries. The fat in dry matter (FDM) varies from 20 to 60% according to the class of cheese. The factory cheeses are made from standardized milk as illustrated in Table 13.7 and Figure 13.7. In some trades, 20 and 50% FDM contents are required. While the Edam and the Gouda cheeses are typical of this group, different countries have adapted the recipe to suit local conditions. Blarney cheese produced in Ireland belongs to this group. The use of sodium nitrate to control coliform organisms in the sweeter curds of these cheeses is a normal but fading practice. In some instances, 'blown' curds result from excessive citrate utilization by some starters and, therefore, starter selection is essential. Traditionally the Edam is coated with red wax for export, while the Gouda has a yellow or a sienna coloured coating. The wax is almost essential to prevent moisture loss in the moist types of cheese, although plastic resin coatings as well as synthetic film coatings are also used. The moisture contents of this group of cheeses determine the rate of ripening – typically from 3 to 8 weeks.

The Swedish cheeses Kadett, Skandia and the smaller Halsingost and Slatte have a smear coating and belong to the Steppenoste group of cheeses, which also includes Esrom (Denmark). This group also have in common hot water scalding and the later development of pea-sized eyeholes during a period of warmer storage at 15–18°C as against the usual 8–12°C.

When the curd is washed (e.g. Reading Yellow cheese, Butterkase, Edam, Gouda cheese) to remove some of the lactose and to scald the curd, the resulting whey from the curd is reduced in acidity even though the curd maintains its pH value. Typical values for Gouda cheese are as shown in Table 13.8. The acidity of the curd influences the moisture expulsion from

Table 13.7 Fat in milk used for Edam cheese of various qualities

Cheese FDM (%)	Fat in cheese milk (%)
30	1.5–1.6
40	2.5–2.6
45	2.9–3.1

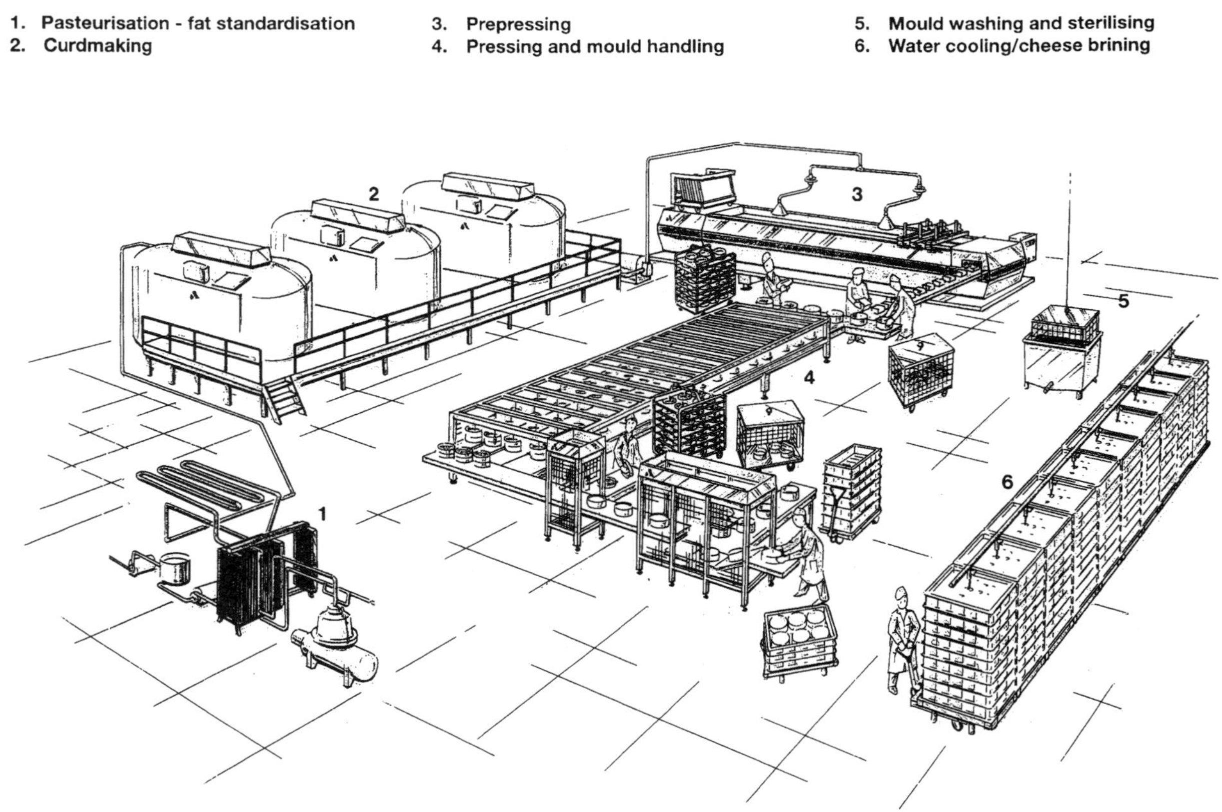

Figure 13.7 Process for medium-scale production of Gouda, Tilsit, Swiss, etc. cheeses. (Courtesy of APV Nordic Cheese AS)

Table 13.8 Typical pH and acidity values for Gouda cheese

	pH	Acidity (%)
Whey, after cutting	–	0.13
Whey, after scalding and water added	–	0.09
Curd, after scald	6.45	–
Curd, after salting	5.20	–

the curd either alone or together with an increase in temperature (i.e. scalding the curd), as in most hard cheese types. The Netherlands' influence is shown in the Queso do Bola cheese made in Mexico which has similar curds to Edam.

A number of fresh white cheeses are made in most Latin American countries. Queso Blanco is a popular variety which must be eaten fresh. A method of making Queso Blanco cheese which was creamed and brined but rennetted rather than coagulated by acid has been described by Siapantas and Kosikowski (1973).

The Tilsiter curds are ivory white becoming more yellow in ripe cheese, although some cheeses may be coloured. The body is close and elastic rather than short or crumbly. The texture is close with slits, oval or round eyeholes. The flavour is slightly 'dry', piquant and mildly acid but not sour. While the modern tendency is for milder flavours the use of added flavours is quite common. For example, cumin has been used for flavouring the German Erntekäse, a skim milk harvest cheese, while caraway seeds are used for other varieties. Stangenkäse (Austria) and Brick cheese (USA) are similar types of curd although softer in body. On the other hand, Steppenkäse (Germany) has more regular eyeholes and is a firmer bodied cheese. It is more piquant but milder in flavour than the Tilsiter when ripe. This group of non-textured cheeses is widely represented in Bulgaria, Austria, Hungary, Poland and countries of the former Soviet Union.

The type of milk and slight divergences in manufacturing methods contribute to the differences between named varieties. Most of the groups are dependent on starter activity and on the temperatures during manufacture for the early character of the curds but the subsequent ripening temperatures are of vital importance to the quality of the final product. Tiersma and Edam (1968) described the differences in techniques in the manufacture of some cheese varieties. Palosy (1965) described Trappist cheese. Delbeke and Naudts (1970) described the manufacture of Herve cheese from raw milk. Mechanization of Gouda cheesemaking was described in 1970 (Anonymous, 1970) and more recently Dutch-type cheeses have been reviewed by Walstra *et al.* (1993). Lampenauer (1952) produced a booklet on the making of semi-soft and soft cheese. Torres and Chandan (1981) wrote on Latin American white cheese.

13.7 Blue-veined cheese

Blue-veined cheeses, e.g. Stilton, are characterized by the growth of a *Penicillium roqueforti* in the air spaces between incompletely fused curd particles within the cheese. Mould spores may be added to the milk or to the curd during draining rather than relying on environmental contamination.

The aim of curd production for blue-veined cheese is to produce an acid type of curd with little or no scalding (rarely more than a 2°C rise). The manipulation of the curd after breaking or cutting the coagulum is simply to allow slow drainage of the whey after the curd has become firm enough and acid enough in the vat. The draining curds are moved (stirred) sufficiently to prevent pools of free whey forming. When the curd is acid and dry enough it is put into moulds to give it shape. The curds are not generally pressed (Wensleydale is lightly pressed) in order to allow spaces to form between the curd particles in which the blue mould eventually grows. The curd in its mould is turned repeatedly to assist drainage. Salt is spread on the surface of the partially dried out curd also to assist drainage of whey. When the cheese is firm and well drained, it is removed to the ripening store.

Most blue-veined cheeses belong to the semi-hard group and are attractive to consumers because of the extra piquancy of flavour brought about by the breakdown of the milk components due to enzymes from the mould growth inside the curds. The flavour, especially when goat's or sheep's milk is used in part, arises from the degradation of fatty acids to ketones as a byproduct of the mould metabolism. The methods of manufacture of blue-veined cheese vary from variety to variety but all aim to produce a curd which has an open texture, which is white to creamy in colour (due to curd structure, bleaching or complementary colour additives), and in which the protein breakdown provides a medium in which moulds will grow. The quicker ripening blue-veined cheeses often have extra salt. When the protein breakdown reaches a certain stage the cheese is pierced by needles to admit more oxygen and provide outlets for carbon dioxide. This piercing is an essential feature of the ripening process of the blue-veined varieties. The colour of the mould growth varies from green through blue–green to a blue colour.

Formerly, most blue-veined cheese was manufactured from a mixture of curds made on consecutive days but the modern systems use one day's curd only. Individual cheesemakers use their own slightly variable processing procedures. The types of cheese curd produced are influenced by the rate of starter growth, salt content, the strain of moulds used and type of milk, whether sheep's, goat's or cow's. To aid the earlier production of more piquant flavours, modern processing resorts to homogenization of the fats so that the lipolytic enzymes can act more easily.

France, with its premier blue-veined cheese, Roquefort, made from sheep's milk, has approximately 30 other blue-veined cheese varieties. Many other countries also produce blue-veined cheese, some examples of which are listed below:

Austria	Grunschimmelkäse
Bulgaria	Magura
Finland	Aura or Adelost
Germany	Edelpilzkäse
Hungary	Marvany Sajt or Merinofort
Poland	Rokpol
Romania	Bucegi
Sweden	Adelost
Bosnia–Hercegovina	Rokfor

The Danish Danablu (Roquefort type) and Mycella (Gorgonzola type) have been developed for the home and export trade. Other countries have also developed their own blue cheese.

The UK has its own blue-veined cheeses, Stilton, Wensleydale and Blue Vinny. The 'Blue Cheshire' (sometimes also coloured with annatto), Shropshire Blue and Caledonian blue (Scotland) are fairly distinctive types of blue cheese of more recent origin. Most UK blue-veined cheeses are drier, less salty and longer maturing than many of the other varieties. However, the high cost of labour and high interest charges are forcing the cheesemakers to produce quicker blueing and maturing types of these cheeses, which for this reason do not have quite the same flavours as the traditionally produced, longer ripening cheese.

The Italian Gorgonzola cheese has been copied (e.g. Danish Mycella), and as with other blue cheese, it also has a white counterpart variety 'Pannerone'. The rind of the Gorgonzola is reddish with numbers of stab holes where the cheese has been pricked. The curds are white to creamy in colour, with a stronger blue veining than some other blue cheeses. The flavour is spicy but pleasant and piquant. In earlier days the cheese was made by the two-curd system but today most milk is pasteurized and only one day's curd is used.

Faults in blue-veined cheeses are common to many varieties. Curds which are too close-textured do not blue well or evenly. Large numbers of coliforms will cause gas holes in the curd. Antibiotics in the milk prevent the growth of lactic bacteria and soft, close-textured curds result. The excessive growth of lactic bacteria cause high-acid curds to develop, causing brittle cracked rinds and loss of flavour in the curds. Historically, cheesemakers have not objected to a small number of gas holes since the coliform organisms seem to assist in the preparation of the curd as a medium for the growth of moulds. However, excessive numbers of coliforms tend to produce off-flavours in the cheese as well as discoloured curds, and should they survive the acid conditions generated in the curd mat subsequently pose a health threat.

The culturing of *Penicillium roqueforti* was reported by Morris (1970). Demeter (1963) described the salting, ripening and pricking of Roquefort cheese. Shehata (1967) described the making of blue cheese by direct acidification. Stilton cheese making has been described by Scott (1954b). Bleu d'Auvergne manufacture was described in 1980 (Anonymous, 1980). Delforno (1970c) wrote on the technology of a single curd system for Gorgonzola cheese.

13.8 Pasta Filata cheese (kneaded curds)

The Pasta Filata cheese types have been developed separately in different countries. The cheeses are first produced as normal enzyme coagulated curds and are then kept warm in hot whey at 63–65°C and are consolidated into a solid mass. Acidity is allowed to develop in the curd mass (from pH 5.2 down to pH 4.9). The acidified mass is now plasticized in whey or hot water at 70–80°C and from this mass two curd types are formed. Figure 13.8 shows some of these varieties and the manipulation required to create them. Although the curds can be manipulated by hand, the kneading process is now carried out by machine (Figures 13.9 and 13.10).

This group of cheeses falls into two sub-groups. The kneaded cheese (e.g. Kashkaval cheese), and the plasticized and formed cheese (e.g. Provolone, Caciocavallo and Mozzarella). The Kashkaval or Kaschkawal cheeses are highly scalded and kneaded cheeses made widely in Central and Eastern Europe under a variety of names – Kaskavalsajt (Hungary), Gilad (Israel), Cascaval dalia and Cascaval dobrogea (Romania), Kashaval (Bulgaria), and Kachavalj (Bosnia–Hercegovina). According to the language used, some 14 variations in the spelling of the names are to be found. The cheeses have a high salt content (2.5–3.5%). The flavour is piquant, salty and when made from sheep's milk, sharp and aromatic. The chief faults are typical of many cheese varieties, i.e. lack of acidity, off-flavours, and rancid flavours due to butyric acid bacteria.

The kneading or 'cheddaring' formerly performed by hand is mainly done by a type of mixing machine with a worm extruder (Figure 13.9). Since the cheese is kneaded hot at about 72°C, it has to be cooled to its de-aired shape in a mould or hoop before final transfer to the ripening store.

The Provolone cheese belongs to the plastic 'spun' curd cheese made in a number of different sizes and shapes, e.g. bottle shaped, cylindrical, square or oblong. Each shape has its own name, e.g. Salami or Giganta, the long cylindrical cheese. The flavour is typically sweet, but when fully ripe, sharp and piquant. This is especially noticed when the cheese is made by using kid's rennet in paste form or when lipolytic enzymes are

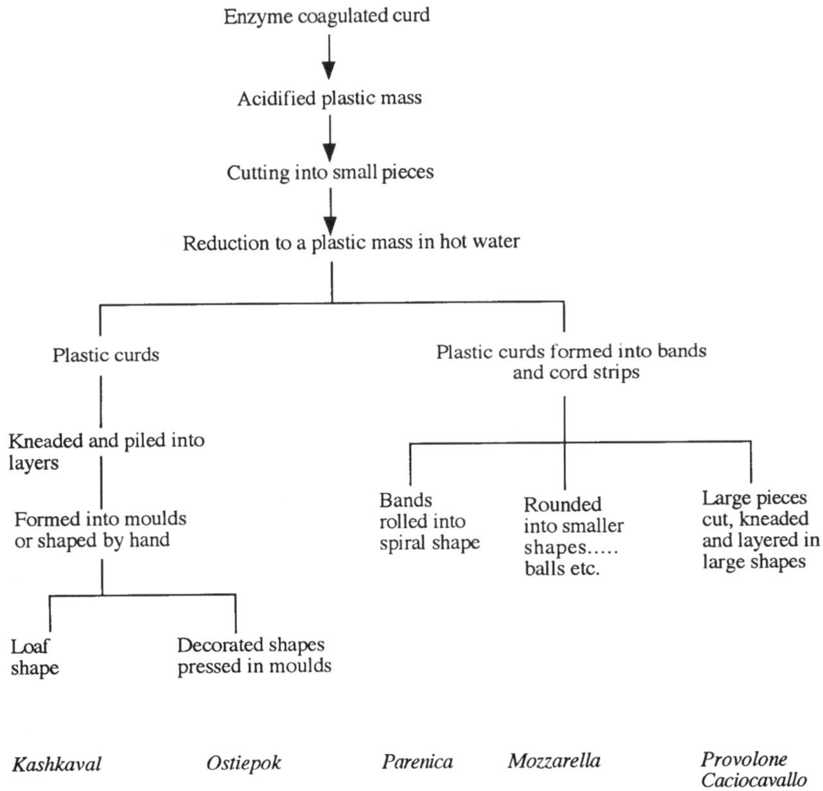

Figure 13.8 Schematic diagram of the manufacture of varieties of the Pasta Filata group of cheeses.

added. The dry matter content is from 58 to 65% and the fat in dry matter content is 45%. The manufacture of Mozzarella cheese has been described by Kosikowski (1982) and Friis (1981) and that of Provolone cheese by Emaldi (1968).

The object in the manufacture is to produce a curd of pH 5.1 when ready for ripening. High temperature starters are used, typically *Lactobacillus delbrueckii* sub-sp. *bulgaricus* and *Streptococcus thermophilus*; other organisms reported include *Micrococcus* and *Enterococcus* spp.

The faults found in the cheese include blown curds, due to the use of 'silage' milk (i.e. *Clostridium butricum* contamination), bitter flavours due to propionic bacteria, discoloured curds, cracked rinds and large holes in the curds. Certain trade outlets favour smoked Provolone cheese.

Caciocavallo cheese is Italian in origin but is made where Italian communities exist in other countries, e.g. USA, Argentina, Canada and Australia.

Figure 13.9 Cooker/stretcher for Pasta Filata cheese plus moulding machine. (Courtesy of APV Nordic Cheese AS)

The cheese curds are similar to the Provolone curds but smaller and the cheeses are hung up in pairs. There are no eyeholes and the flavour is piquant when sheep's milk has been used. It is not usually smoked, but can be kept for 12 months as a cheese for grating.

Mozzarella cheese belongs also to the Pasta Filata group and is made in many countries from cow's milk, buffalo milk and even milk powder. The cheese exists in a large number of forms and sizes from 50 g to 500 g weight. When made in Italy the cheese from buffalo milk must have 45% fat in dry matter but the cow's milk cheese need only have 44%. Typical analyses are as shown in Table 13.9. The acidity must be high (0.7% as lactic acid, pH 5.3) before being 'pulled' to give a fine texture at 85°C. Benzoyl peroxide has been used to bleach the colour of the curds, but the treatment leads to loss of vitamin A.

Pizza cheese is a low moisture cheese made in a similar manner to Mozzarella. Both Mozzarella and Pizza cheese need protective packaging, i.e. Saran or multicoat or multilayer films, and have to be stored at low temperature, 4°C, until used. A large number of Italian cheeses are reviewed by Lembo and Spedicato (1992).

The Italian influence in the Southern American is shown by the Mexican Asadero, a plastic cheese like the Provolone. In Honduras, the Queso Emredo is a pulled and twisted curd cheese. Queso de Pera, a cheese made on farms in Colombia, is a Provolone-type curd cheese.

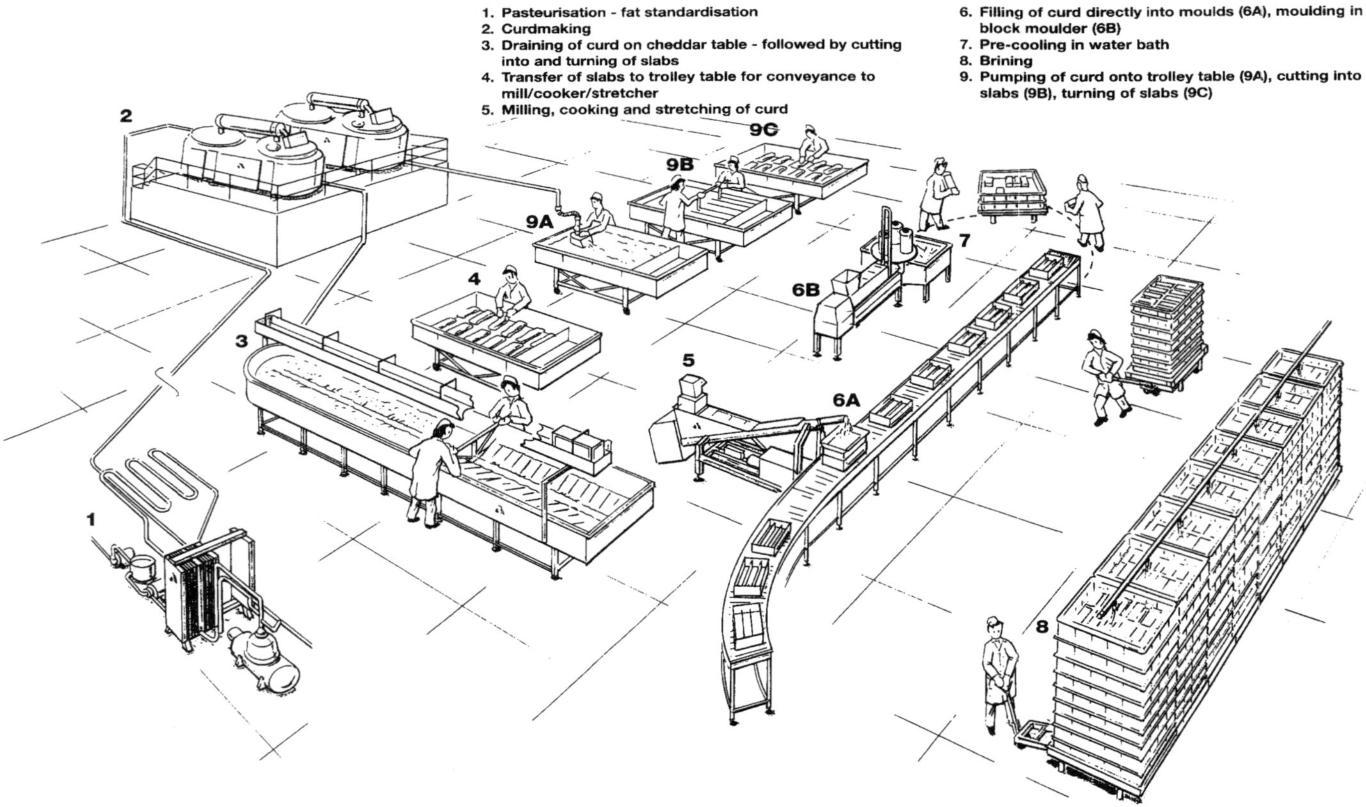

Figure 13.10 Process line for small- and medium-scale production of Pasta Filata cheese. (Courtesy of APV Nordic Cheese AS)

Table 13.9 Some figures from analysis of Italian cheese from different milks

	Buffalo milk cheese	Cow's milk cheese
Fat content (%)	25–30	15–18
WFFS (%)	56–58	68–70
Water content (%)	40–45	52–66

13.9 Soft cheese

There are many different curd types amongst the group of cheeses often classed as 'soft' cheese. The curds vary from the very soft Petit Suisse and Gervais, which border on the lactic cheese, through the firmer Brie, Coulommiers, Carreé de l'Est and Camembert cheese, and on to the firmer (but still semi-soft) cheese like Herve, Limburger and Romadur. The differences amongst this group of cheeses are due to the method of ripening as well as to the moisture content.

Camembert cheese is most popular amongst the surface mould ripened cheeses, both in France and other countries. The traditional home of the Camembert cheese is in the Orne Département in Normandy in France. The cheese was made popular by Mme Marie Harel in 1791, brought under the control of a Syndicate of Manufacturers in 1909 and granted its AOC in 1983.

Up to 1909, the Camembert cheese had a blueish mould growth due to mixtures of *Penicillium roqueforti* (formerly *P. glaucum*) and *P. camemberti* (formerly *P. albus* or *P. candidum*). Isolation and use of pure cultures of *P. camemberti* eradicated the blue colour. Pure cultures of the moulds are available commercially. Saint Marcellin is another soft cheese with surface mould growth, made from goat's milk or a mixture of goat's and cow's milk.

Brie de Coulommiers is reputed to be the oldest of a group of cheeses from the Département of Isle de France, east of Paris. This group includes Brie de Meaux, Brie de Montreau and Brie de Valois cheeses which are surface ripened in the first instance by *P. camemberti*, which produces a white mat on the surface of the cheese. Later the mat becomes dry and other moulds or bacterial smear cultures may grow. The variants of Brie cheeses are mainly due to size (varying from 12.5–15 cm for Coulommiers to 36–37 cm for Brie de Meaux) and texture but the Cendre de Brie is coated with ashes which impart a dark grey colour to the surface.

Livarot, another soft cheese, has a washed rind coated with annatto and is kept free of surface mould growth. The flavour is strong when ripe (after 3 months) due to growth of *Brevibacterium linens* and there is a distinctive binding of the cheese with sedge leaves (*Typha latifolia*) to give it a characteristic appearance. Pont l'Evêque is also a soft cheese, which, after being drained on straw mats is matured in cellars and acquires mould growths of

Geotrichum candidum on the surface. When the growths dry up, a red smear of *B. linens* develops. The surface growths contribute to the development of flavour in the mature cheese.

The still firmer curds of the cheeses like the Herve, Limburger, Romadur and Munster are drained curd cheese, unscalded and unpressed. The salting regimes vary a little from variety to variety to produce salt contents of from 2.1% to 3.2%. The cheeses are made in several fat qualities, from 20% to 60% fat in dry matter of the cheese. The group all have smear coats due to the growth of *Brevibacterium linens* which contributes to the mature cheese flavour.

The Butterkäse (Germany) and the Bel Paese (Italy) cheeses are often grouped along with the semi-hard types, but are more properly grouped with the semi-soft. The Butterkäse has an elastic soft body but is not runny or pasty. Eyeholes may rarely be present. The taste and aroma are mild and faintly acid. Different fat qualities are made, for instance 45% and 50% FDM.

Mocquot (1955) reviewed the manufacture of soft cheese. Anonymous (1952) dealt with the manufacture of Brie cheese. Alba (1954) published a book on *The Camembert of Normandy*. Nikolaev (1980) and Lampenauer (1952) wrote on soft and semi-hard cheese. Eck (1987) discusses some aspects of the manufacture of French cheeses, and Masui and Yamada (1996) have written an illustrated guide to French cheese types.

The Italico cheese originally set at high temperatures 43–46°C and gave a coagulum in 8–10 min. The cheese was drained in hot moulds and developed an elastic soft silky texture and body. It is now marketed under the trade name Bel-Paese.

Many of the soft cheeses are small in size, and the problems of handling large numbers of small cheeses on a daily basis have meant that mechanization of the soft cheese industry has received earlier attention than that for larger and harder varieties of cheese.

13.10 White brined cheese

The Middle Eastern countries, including Egypt, enjoy a hot climate which is not so suitable for the storage of hard cheeses. For many years, the answer to the problem of storage of cheese has been to store the cheese in strong brine in airtight packages, e.g. tinned steel or plastic boxes or drums, which have replaced the older animal skin bags.

Feta cheese is made in Bosnia–Hercegovina, Greece, Israel and nearby countries, and is a rindless, highly salted, drained-curd cheese of low pH (4.8–4.2), which is stored in a 16% brine (15° Bé) at temperatures between 2 and 8°C according to the cheese usage and trade. The ripening time may extend from 15 days to 2 months or longer. The cheese is normally made from sheep's milk but goat's and cow's milk may be used alone or in a mixture.

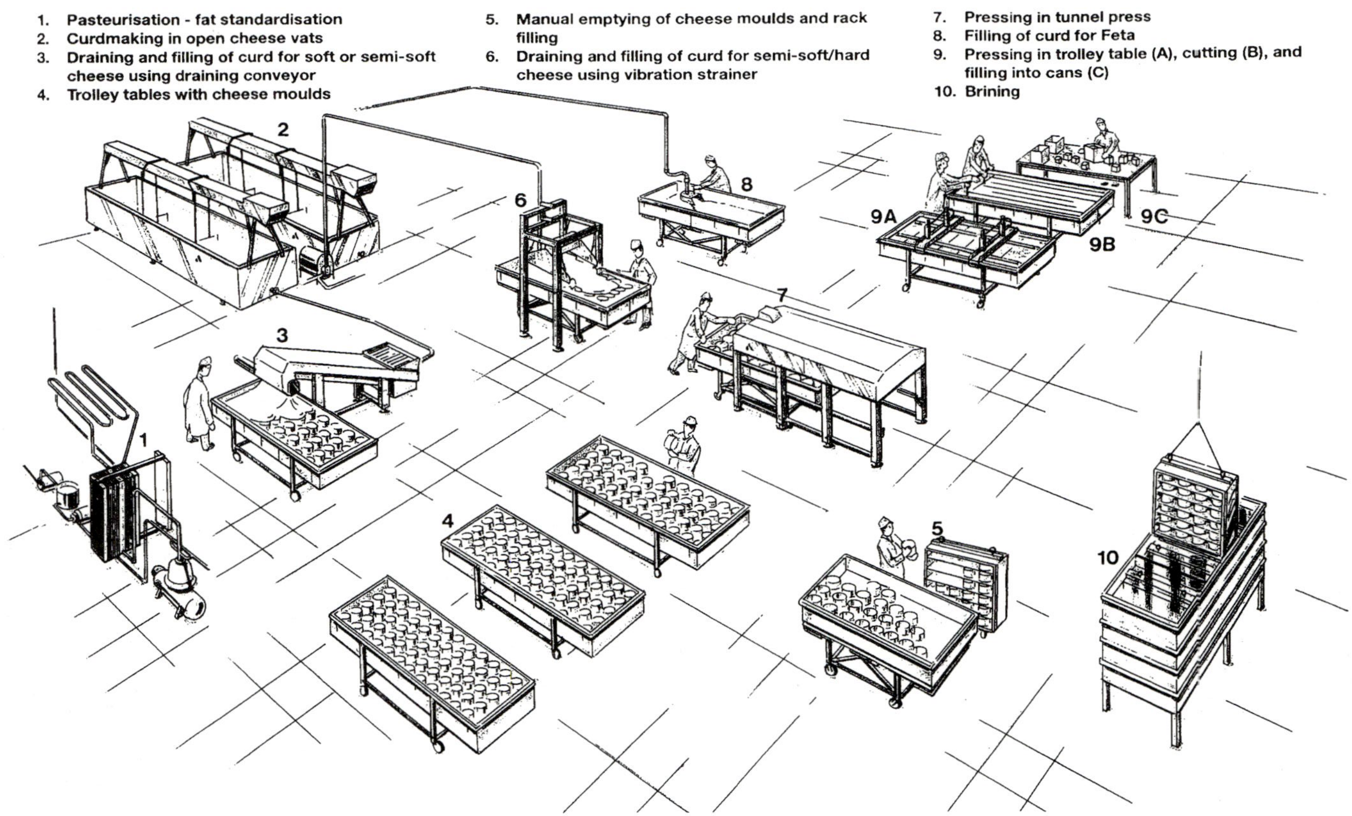

Figure 13.11 Illustration of small-scale manufacture of Feta and similar specialist cheese. (Courtesy of APV Nordic Cheese AS)

Occasionally, a decolorant of the chlorophyll type is used to keep the curds white. A process line for small-scale manufacture of Feta and similar cheeses is illustrated in Figure 13.11.

Halloumi is a similar cheese made in Cyprus. Domiati or Beda are Egyptian white cheese made in square, oblong or cylindrical shapes which may be preserved in brine for 3–6 months at 7°C, especially when the ambient temperature is 30°C or more.

El-Sokkary *et al.* (1957) described the making of Domiati cheese. Delforno and Tsarouchas (1968) discussed the production of Feta cheese in Greece, and Beinoglou (1969) wrote on Feta cheese from ewe's milk. El-Safty and El-Shibiny (1980) described a continuous process for Domiati cheese. Robinson and Tamime (1991) have produced a more recent review of brined cheeses.

13.11 Lactic, Cottage and similar cheese

A brief survey of the literature shows that there are at least 30 different types of curd based on skim milk, whole milk and powdered milks which are variously classed as lactic curds, Cottage cheese, Quark, or cheese bases for incorporation of added non-cheese flavours.

Lactic cheese must conform to the appropriate analytical legal limits for moisture and fat. Thus, there are skim-milk cheeses and cheeses with varying fat contents. The market for the cheese determines the texture, which varies from an homogenized, smooth curd to one that is rougher and/or more granular. The keeping quality of the cheese may be enhanced by the inclusion of potassium sorbate at up to 0.1%, to give a shelf-life of 3–4 weeks when stored at 4–5°C. Alternative methods giving a less acid cheese use rennet as coagulant (3 ml per 100 litres of milk). The resultant curd needs to be pressed in woven nylon or similar bags to reduce the moisture content to the required level.

The equipment originally required for the lactic type of curd was simple. Milk would be soured then strained through a cloth. On a larger scale the coagulation/fermentation vats may be of the rectangular square-ended type or the vertical enclosed type. Normally temperature is controlled through the jacket of the vat. Lactic curds are usually drained through linen, nylon or polyester cloth covered frames, before they are salted and packed. Sometimes the curds are lightly pressed to aid drainage. Berge (1956) patented a system consisting of a number of flexible troughs lined with whey-permeable material, e.g. nylon cloth. These troughs were linked at each end to cams which caused the troughs to vibrate slowly. This enhanced the drainage of whey by disturbing the curds. Pressure could be applied for final drainage (Vassal *et al.*, 1962). Gurjanov and Bogdanova (1965) described a horizontal semi-circular vat which was used for curd making. A second similarly shaped vat could be lowered onto the curd to improve final drainage of the curd.

Lactic cheeses can be packed by extruder-type packing machines, some of which impart an extra smoothness to the curd.

A typical analysis of a full fat lactic cheese would be:

Fat (%)	29
Moisture (%)	55
Fat in dry matter (%)	67
Salt (%)	1
pH	4.48

Quark is a fermented milk curd produced primarily in Europe (cf. the Hungarian Taho) under many names – Koarg, Kwark, Qwark, Quarg, Twarog, Sauermilchquark and Speisequark. The division between lactic cheese, Cottage cheese, Quark and the very soft cheese like Gervais, Neufchâtel or Petit Suisse is sometimes quite narrow, depending on the manufacturer of the product. The chief faults in these products arise from excess moisture, off-flavours due to contamination, high proteolytic activity in the curd, and uneven quality of production from batch to batch.

Industrial preparation of Quark starts with separation of skim milk, usually by a self-sludging centrifugal separator. The skim is pasteurized at 72°C for 15 s, then cooled to 25–30°C (a higher temperature for a faster fermentation) and a mesophilic starter added at about 1%. Rennet is usually added but at less than 0.001%. The fermentation is carried out in insulated, sometimes jacketed, vertical tanks fitted with two-speed mixers. When the pH reaches 4.6 the acid coagulum is broken up by stirring at the higher speed, the mixer then being run at the slower speed to maintain the casein suspension. This suspension is then pumped via a duplex filter to the nozzle bowl separator where the casein is concentrated and discharged through the nozzles as a viscous white paste which may then be pumped away for cooling and blending with other ingredients. The type of plant used is illustrated in Figure 13.12.

The mechanical separation technique still results in loss of whey proteins. This economic problem was addressed by invention of a 'Thermo process' whereby the skim milk was subjected to a severe heat treatment of 90–95°C for 2–3 min to denature whey proteins. Fermentation is similar to the standard process except that gas-producing cultures should not be used (cf. Cottage cheese production). The slurry of fermented curd is then subjected to a further heat treatment at up to 60°C for 1–2 min to promote further complexing of whey proteins and casein, great care being taken in the design of the plate heat exchanger to minimize temperature differentials (less than 3°C between the hot water and the fermented milk) before cooling to 40°C and feeding to the separator. Protein yields may be increased by up to 10% by this method.

Mechanical separation of Quark depends on the curd being denser than the whey. This factor and the tendency of fat to be lost from acid curds limits

Figure 13.12 Equipment for the production of Quark: two nozzle bowl centrifuges with the necessary tubular coolers, tanks and pumps. (Courtesy of Westfalia Separator Ltd)

the fat content of the milk and hence of the Quark. Milk fat may then be added back via a high fat cream or by blending with a cream cheese, which may be produced by traditional processes or via a cream cheese separator which operates on a similar principle to that of a cream separator but with modifications to cope with the viscous cheese. The fat in dry matter of the different Quark-type curds varies from virtually zero to 60% in different countries.

The development of membrane filtration has introduced the possibility of simplifying production of lactic curds and further increasing protein recovery. However, it is still necessary to culture the milk before concentration so that the calcium level in the curd may be kept low and bitter flavours avoided. Ultrafiltration systems also have difficulty in handling the viscous pastes produced so that while ultrafiltration has proved popular for production for the lower solids Fromage Frais type products, the centrifugation route remains the most popular for production of those Quark products where the solids would normally be in excess of 20%.

Cottage cheese, while similar to lactic cheese in fermentation, is expected to have a particulate curd structure. Pasteurized skim milk is fermented with non-gas producing mesophilic cultures, typically in open rectangular vats as illustrated in Figure 13.13. Low levels of rennet may be added ($\leqslant$0.0002%) and also calcium chloride at 0.01–0.02% to help form a less fragile curd and improve drainage. Fermentation continues till coagulation

Figure 13.13 Rectangular 9000 litre Wincanton/Stoelting Cottage Cheese vats. (Courtesy of Wincanton Engineering Ltd)

at pH 4.5–4.7, at which point the weak coagulum is cut. The knives may consist of stainless wires on a frame. Unlike most soft cheese, the curds and whey are stirred gently and scalded, the temperature being raised slowly to about 49°C over a period of about 30 min then held for a further 30–60 min to shrink and firm up the curd particles as well as to kill off the starter and most contaminant micro-organisms. The whey is then drained off and the curd washed at least three times with water, the last wash being with chilled water to complete the cooling of the curd. This water must be of good bacteriological quality, free of psychrotrophic contaminants, in order to ensure a satisfactory shelf life.

Ernstrom (1963) evolved a process (patented) for making Cottage cheese by direct acidification of the cold milk to pH 4.7. The milk, warmed in a quiescent stage, was coagulated and the curds were treated as normal Cottage cheese by hot water scalding and cold water washing. The Creamery Package Manufacturing Co., USA, evolved a tubular system in which milk could be heated to 38°C after acidification according to the Ernstrom patent.

The curd particles at this stage of the process are very bland, being essentially pieces of moist insoluble casein. The size of the curd particles is determined primarily by the market requirements. The flavour and mouth-feel are modified by addition of creams and dressings, e.g. fruits, vegetables or savouries. These ingredients may be added at the rate of about 15% of the total. The final pH is from 4.5 to 4.7 and so cream dressing is normally acidified with starter before adding to the Cottage cheese curds. The cream fat and dosage levels are often adjusted to give about 4% milk fat in the final product, but the increased popularity of Cottage cheese as a slimmer's cheese has led to the development of lower fat cream dressings for the curd.

The historical development of Cottage cheese may be followed via review articles by Emmons (1963), Hinds (1972), Bassette (1980), Brown (1982) and Shaw (1993).

13.12 Cheese made in mountainous areas or by nomadic tribesmen

There are at least 40 varieties of cheese made by the nomadic tribes, from Mongolia through Tibet to Afghanistan, countries of the former Soviet Union, Iran, Arabia and North Africa (FAO, 1992).

Milks from the cow, sheep, goat, yak, mare and camel are used to produce either fresh curds or dry, hard cubes or balls of curd. These hard balls need to be soaked in water before use. The methods in use include:

(a) Naturally soured milk drained in sacks or bags and pressed into shape, often by hand.
(b) Inoculation of milk by cultures, i.e. yoghurt, koumiss, etc. and drainage of the coagulated mass.
(c) Use of rennets from young, slaughtered animals to coagulate milk which is then drained and shaped into cheese curds.
(d) Buttermilk soured and drained as for milk in (a).
(e) Use of whey as a souring agent for milk. The coagulum is drained as in (a).
(f) Cheese made by warming colostrum.
(g) Mixtures of whey and milk, buttermilk and milk, and buttermilk and whey are made into curds by acid development, evaporation, etc.

This economic usage of all the milk constituents by the nomadic people provides cheese either for immediate food (as a supplement to meat) or for storage to provide food in wintertime.

The base for making many of these cheeses is provided from yoghurt, koumiss or similar fermented milks, or by curdling milk with strips of lamb's stomach (rennet). In both types the curds so produced are drained and salted; sometimes they are sun dried for long-term storage. A typical analysis of a dried cheese (Table 13.10) illustrates the potential nutritive value of the cheese.

Krut or Qurut is an Afghan cheese made from buttermilk. The cheese, rolled into balls and then dried, is called 'kurd', and is made in many areas

Table 13.10 Typical values for undried and dried curd cheese

	Undried	Dried
Dry matter (%)	19	93
Fat (%)	5	27
Protein (%)	5.4	36
Lactose (%)	6	23

inhabited by nomadic herdsmen. Drained sour milk Quark is made into Angharaghul in Mongolia.

The Bedouin have made Gibna cheese from goat's or sheep's milk curdled with either lamb rennet or the juice of the fig tree, but a firmer cheese is made by immersing a strip of lamb's stomach (i.e. rennet) in milk. The milk is heated until curds form. These curds are collected and pressed between wooden boards; afterwards they are salted. In Iraq, a similar curd is produced and the curds are pressed in bags. The dry curds are cut into strips, salted and then stored in sheepskin bags to produce Meira cheese. Panir Kusei cheese is made in Iran by farmers. The curds, flocculated on the top of heated milk, are pressed into earthenware pots. The pots are sealed with paper and earth.

Pots of nomads' cheese, Shur Panir, are made in the nomadic areas from Saudi Arabia through Iran and Afghanistan to Tibet. The milk is curdled with strips cut from the stomach of a newly slaughtered lamb. The curds, flocculated on top of the whey, are pressed in cotton sacks. When dry the curds are sliced with a knife and salted down for future use. The Tibetans make a drained curd cheese called Chura, which is mixed with tea and flour and made up into rolls.

Dixmerias *et al.* (1979) described the goat's milk cheeses of the mountainous areas of Savoie in France, and included Grataron, Cherotin and Persille type cheese. Mocquot and Bejambes (1960) reviewed the technology of ewe's and goat's milk products. Delforno (1981) described various Alpine cheeses.

References

Alba, F. (1954) *The Camembert of Normandy*, Jean le Moulec, Caen, France.
Ali, A.E., Andrews, A.T. and Cheeseman, G.C. (1980) *Journal of Dairy Research*, **47**(3), 371.
Anonymous (1952) *Zuievel en Handel*, **58**, 655.
Anonymous (1970) *Svenska Mejeritidningen*, **62**, 20, 365.
Anonymous (1980) *Technique Laitiere*, No. 947, 53, 55.
Australian Society of Dairy Technology (1970) *Pocket Book of Cheddar Cheese Manufacture*, Highett, Australia.
Balducci, A. (1967) *Scienza Technica Lattiero Casearia*, **18**(2), 117.
Bassette, R. (1980) *Dairy Field*, **163**(8) 4, 6.

Battistotti, B. and Corradini, C. (1993) Italian Cheese. In *Cheese: Chemistry, Physics and Microbiology*, Vol. 2, 2nd edn (ed. P.F. Fox), Chapman & Hall, London, pp. 221–244.
Beinoglou, B.K. (1969) *Delt. Agrot. Trapezés*, **168**, 37.
Berge, J.M.A. Ch. (1956) French Patent No. 1 111 562.
Besana, C. (1962) *La Rivista del Latte*, **18**, 2.
Birkhalter, G. (1981) *Catalogue of Cheeses*. IDF Bulletin Doc. 141, Brussels.
Bottazzi, V. (1964) *Il Latte*, **38**, 97.
Brown, G.D.L. (1982) *Dairy Industries International*, **17**(7), 3.
Bryan, R. (1957) *Quality of Scottish Cheese*, Report to The Company of Scottish Cheesemakers Ltd, 127 Candleriggs, Glasgow.
Carta, M.G. (1963) *La Rivista del Latte*, **19**(4), 15.
Colnett, A. (1955) *Dairy Engineering*, **72**(11), 339.
Davis, J.G. (1976) *Cheese*, Vol. III, Manufacturing methods, Churchill Livingstone, London.
Delbeke, R. and Naudts, M. (1970) *Revue de l'Agriculture (Brussels)*, **23**, 333.
Delforno, G. (1967) *L'Industria del Latte*, **3**, 227.
Delforno, G. (1970a) *Mondo Latte*, **24**, 35.
Delforno, G. (1970b) *Mondo Latte*, **24**, 439.
Delforno, G. (1970c) *Mondo Latte*, **24**, 799.
Delforno, G. (1981) *Mondo Latte*, **35**(2), 93.
Delforno, G. and Losi, A. (1980) *Mondo Latte*, **34**(8), 516.
Delforno, G. and Tsarouchas, F. (1968) *Industria Latte*, **4**, 189.
Demeter, K.J. (1963) *Deutsche Molkereizeitung*, **84**(49), 1763.
Demeter, K.J. (1966) *Deutsche Molkereizeitung*, **87**, 1124.
Dieci, E., Resmini, P. and Saraccii. S. (1970) *Scienza Technica Lattiero Casearia*, **21**(3), 145.
Dixmerias, J., Magnan, M., Touchebuef, J.C. and Valleix, Y. (1979) *Chevre*, **110**, 9.
Eck, A. (1987) *Cheesemaking – Science and Technology*, Levoisier Publishing Inc., New York.
El-Safty, M.S. and El-Shibiny, S. (1980) *Egyptian Journal of Dairy Science*, **8**(1), 41.
El-Sokkary, A.M., El-Sadek, G.M. and Hamed, M.G. (1957) *Annals of Agricultural Science Cairo*, **2**, 123.
Emaldi, G.C. (1968) *La Rivista del Latte*, **24**(6), 1.
Emmons, D.B. (1963) *Dairy Science Abstracts*, Review Article 110, Parts 1 and 2, **25**(4), 129 and **25**(5), 174.
Ernstrom, C.A. (1963) US Patent No. 3 098 776.
FAO (1992) *Fermented Milks and Cheeses*. FAO, Rome.
Friis, Torben, L. (1981) *Nordeuropaeisk Mejeritidsskrift*, **7**, 3.
Gurjanov, A. and Bogdanova, E. (1965) *Molochnaya Promyshlennost*, **26**(10), 14.
Haley, M.M. (1954) *Dairyman*, **71**(9), 266, 276.
Hardy, J. (1979) *Revue Latiere Francaise*, **337**, 19.
Hardy, J. and Steinberg, M.P. (1983) *3rd International Congress on Engineering and Food*, Dublin, Doc. 97.
Hinds, H. (1972) *Cultured Products Journal*, **7**, 21.
Irvine, D.M. (1978) *Cheddar Cheese*, Bulletin of the Ministry of Agriculture and Food, Ontario, Canada.
Johnson, D.E., Murphy, R.J., Gilmour, A. and Macelhiny, R.S. (1981) *Milchwissenschaft*, **36**, 449.
Kammerlehner, J. (1967) *Deutsche Molkereizeitung*, **88**(12), 429.
Kosikowski, F.V. (1982) *Cheese and Fermented Milk Foods*, 2nd edn, 3rd printing, F.V. Kosikowski and Associates, Box 139, Brooktondale, New York.
Kursteiner, J. and Staub, W. (1950) *Merkblätter für den Emmentalerkäser*, Thun, Switzerland.
Lampenauer, H. (1952) *Die Herstellung von Weichkasen*, Volkswirtschaftlicher Verlag GmbH, Kempten, Allgau.
Ledda, A. and Arrizzo, S. (1969) *Scienza Technica Lattiero Casearia*, **20**(3), 143.
Lembo. P. and Spedicato, E. (eds.) (1992) *l Prodotti caseari del Mezzogiorno coordinamento scientifico*, Consiglio nazionale delle ricerche, Rome.
Maddever, K.D. (1955) *Dairy Engineering*, **72**(2), 41.
Marcos, A., Alcala, M., Leon, F., Ferandez-Salquero, J. and Esteban, M.A. (1981) *Journal of Dairy Science*, **64**(4), 622.
Masui, K. and Yamada, T. (1996) *French Cheeses*, Dorling Kindersley, London.

Messner, E. (1968) *Ostmaerkische Milchwirtschaftliche*, **23**, 381.
Ministry of Agriculture, Fisheries and Food (1959) *Cheesemaking*, 4th edn, Bull. 43, HMSO, London.
Mocquot, G. (1955) *Journal of the Society of Dairy Technology*, **8**, 17.
Mocquot, G. and Bejambes, M. (1960) *Dairy Science Abstracts*, Review Article 84, **22**, 1.
Monie, E.M. (1955) *Dairy Engineering*, **72**(5), 149.
Morris, T.A. (1970) *XVIII International Dairy Congress*, Vol. IE, 382.
Nikolaev, A.M. (1980) *Technology of Soft Cheese*, Pishchevaya Promyshlennost, Moscow, 212 pp.
Olson, N.F. (1981) *Journal of Dairy Science*, **64**(6), 1063.
Palosy, L. (1965) *Industria Alimentara (Bucharest)*, **16**, 310.
Peter, A. and Zollikofer, E. (1959) *Lehrbuch der Emmentalarkäserei*, K.J. Wyss Erben AG, Bern, Switzerland.
Robinson, R.K. and Tamime, A.Y. (1991) *Feta and Related Cheeses*, Woodhead Publishers Ltd., Cambridge, UK.
Schällibaum, U. (1965) *Schweiz Milchzeitung*, **91**, 691, 695, 701.
Scott, R. (1954a) *Dairyman*, **71**(8), 229.
Scott, R. (1954b) *Dairyman*, **71**(9), 261.
Scott, R. (1967a) *Process Biochemistry*, **2**(2), 5.
Scott, R. (1967b) *Process Biochemistry*, **2**(3), 23
Scott, R. (1967c) *Process Biochemistry*, **2**(5), 49.
Shaw, M.B. (1993) In *Modern Dairy Technology*, Vol. 2 (ed. R.K. Robinson), Chapman & Hall, London, pp. 221–280.
Shehata, A.E. (1967) *Dissertation Abstracts Section B*, **28**(2), 401.
Siapantas, L.G. and Kosikowski, F.V. (1973) *Journal of Dairy Science*, **56**(5), 631.
Smillie, M.D. (1955) *Dairy Engineering*, **72**(3), 77.
Smith, J. (1995) *Cheesemaking in Scotland, a History*. Scottish Dairy Association, Clydebank.
Stadhouders, J. (1975) *Netherlands Milk and Dairy Journal*, **29**(2/3), 104, 126.
Stadhouders, J. and Langeveld, L.P.M. (1966) *XVII Internationa; Dairy Congress*, Vol. D, 577.
Steffen, C., Eberhard, P., Bossett, J.O. and Rüegg, M. (1993) Swiss-Type Varieties. In *Cheese: Chemistry, Physics and Microbiology*, Vol. 2, 2nd edn (ed. P.F. Fox), Chapman & Hall, London, pp. 83–110.
Tiersma, P. and Edam, Y. (1968) *Veeteelt-en Zuivelberichten*, **11**(11), 497.
Torres, N. and Chandan, R.C. (1981) *Journal of Dairy Science*, **64**(3), 552.
UK Regulations (1996) *The Food Labelling Regulations*, 1996 No. 1499. HMSO, London.
Vassal, L., Ducret, P. and Mocquot, G. (1962) *L'industrie Laitiere*, **185**, 39.
Walstra, P., Noomen, A. and Guerts, T.J. (1993) Dutch-Type Varieties. In *Cheese: Chemistry, Physics and Microbiology*, Vol. 2, 2nd edn (ed. P.F. Fox), Chapman & Hall, London, pp. 39–82.
West, A. (1980) *Dairy Industries International*, **45**(9), 21.

14 Mechanization of cheesemaking

Attempts at the mechanization of cheesemaking operations began in earnest in the 19th Century. The first of the irksome tasks performed by the cheesemaker was the stirring of curds in the vat, while the turning of cheese on the shelves in the store was a task assigned to farm labourers. Small circular vats were convenient for installation of a revolving paddle for stirring the curd, and the Edam and Emmental cheesemakers were amongst the first to benefit.

Turning of cheese in the store, especially the larger Cheddar and Cheshire types, was aided by joining two shelves together at the ends and, by making them rotate 180°, many cheeses could be turned together. In the UK the 'revolving shelves' ('Turning Dales') existed until the 1960s when the block-type cheeses were introduced, which did not require turning over as did the taller cylindrical cheeses.

In both farmhouse and factory practice it was the heat treatment of the cheese milk which received first attention, particularly after the introduction of pure culture starters in the early 1900s.

The Danish Kettle rotary heater was amongst the earlier heat treatment devices used for pasteurizing milk. In 1923 the introduction of the 'plate heat exchanger' by Dr Seligman of the Aluminium Plant and Vessel Company Ltd, UK (subsequently APV) provided the means for large-scale, automatic, temperature-controlled, heat treatment equipment for milk.

The collection and storage of cheese milks has reached a high degree of efficiency in many dairying areas, but the change from can collection to tanker collection from refrigerated farm tanks has added a new problem in respect of the growth of psychrotrophic organisms in the cold milks. Thermization before cold storage of the milk is not widespread. The hygiene of some milk supplies is often not good and sieves or filter cloths do not improve the bacteriological quality of the milk.

Centrifugal clarifiers, a development from the cream separators, while removing suspended matter from the milk, did not remove the bacteria. Indeed, the clarifier by breaking up clumps of bacteria seemed to make matters worse and contaminating organisms proliferated with more vigour. Subsequently, the bactofuge was developed to reduce the numbers of both vegetative bacteria and spores. The difficulty that the bactofuge also slightly reduced the protein content of the cheese milk was overcome by sterilizing the bactofugate and returning it to the cheese milk. Microfiltration of skimmed milk may be used in a similar manner (Chapter 6).

14.1 Electrodialysis and ultrafiltration equipment

Electrodialysis is covered in connection with whey (Chapter 18) and is used for the removal of salts from liquids.

Ultrafiltration and reverse osmosis are coming to be regarded as similar processes which are operated under different conditions (Chapter 18). Ultrafiltration is used for concentrating cheese milks by removal of water and lactose from the protein solution (retentate), while reverse osmosis is useful for removing the water from whey, thus concentrating the lactose and whey proteins in solution. Indeed, by using ultrafiltration for concentrating the proteins (including whey proteins) in cheese milks, followed by reverse osmosis of the permeate (whey), a lactose concentrate almost free of protein is achieved. The concentrate still contains salts which can be removed by electrodialysis.

Both ultrafiltration and reverse osmosis equipment use membranes for separation of the effluent (permeate) from the retentate, and the pressures used and porosity of the membranes define the process.

There are many plants used for separation of solubles from liquids, such as salt from sea water and solvents from paints. The ultrafiltration equipment used commercially for the treatment of dairy produce varies widely in design. Both Abcor and Patterson Candy International (PCI) have developed membranes of the tubular type. These are arranged in modules, the numbers of which can be increased for higher volumetric throughput.

The Danish Sugar Corporation (DDS) have used flat sheets of membrane tightly sandwiched between conductor plates. These modules are arranged in vertical tiers. Dorr Oliver and Rhone-Poulenc use horizontal frames and flat membranes in the equipment, while Alfa-Laval use modules containing bundles of microbore tubes as the membrane separation material.

From time to time new equipment made by other firms appears. However, it is the membrane material which is of the greatest importance in any equipment for ultrafiltration or reverse osmosis. The earlier membranes, based on cellulose acetates, have served a useful purpose, but operating characteristics (e.g. pressure, pH, temperature, etc.) have been improved upon by the use of newer synthetic membrane materials.

The pumps used for pumping liquids through the membranes are often those already used for other purposes. High-pressure piston and multistage centrifugal pumps have been brought into use, especially for reverse osmosis equipment.

Design of equipment, as well as membrane compositions in the future, depends largely on research into use of the equipment for milk concentration. The research has already led to the production of commercially viable equipment for soft cheese. Extension of the investigations will expand the types of cheese which can be made from milk concentrated by this method, and overcome some of the defects which arise in such cheese. The Tetra

Laval 'Camatic' equipment has been developed to take advantage of the new technology in soft cheese production.

A complete plant consists of the normal equipment for milk treatment, which is followed by the ultrafiltration of milk to concentrate the liquid to a predetermined level (i.e. 1:4.7 or 1:5.1), the concentration being determined by refractometer. The concentrated milk (i.e. 'pre-cheese') is ripened and stored cold until it can be used. The pre-cheese is then heated to rennetting temperature, 30°C, and rennet is injected. The rennetted pre-cheese is filled into stainless steel moulds arranged in rows on a plate on a conveyor system. The cheese curd sets while the multi-mould trays are stacked to allow coagulation. When the curd is firm, the stainless steel moulds are inverted and emptied into plastic moulds on trays by ejecting the curd by air from the inverted mould. Thereafter, the cheeses are treated in the traditional way to brining, inoculation of mould and ripening.

When introducing mechanization and when only one section of a process is changed at one time, e.g. from open vats to enclosed vats, the change is fairly readily absorbed into the routine. Even so, it has often been necessary to have a 'settling in' period to get the staff acquainted with ancillary operations such as the cleaning-in-place (CIP) routines. Where the whole processing system is changed there is a greater problem in training staff to operate the new techniques. More difficult is the change in mental attitudes towards the new techniques.

There is a reduction in the numbers of staff required to operate the mechanized systems as against those required for traditional processes; but there is a need for more highly trained and, therefore, more highly paid personnel to staff the new techniques. This change in emphasis towards training of personnel is not always appreciated by non-technical managements.

Reception of milk, heat treatment and starter preparation have already been discussed (Chapters 9 and 10). The importance of consistent, active starter free of contaminants cannot be overemphasized. All starter rooms must be constructed to and run at the highest level of hygiene with tanks and ancillary equipment designed to avoid contamination (Figure 14.1). The use of additives, according to the recipe in use, has been discussed in Chapter 8.

It is important to know something of chemistry, microbiology and physics when discussing mechanization. The cheese industry has suffered greatly in the past simply because the engineer, while proficient in engineering skills, has not had the background knowledge in chemistry and biology to appreciate the requirements of fermentation processes or the practical difficulties experienced by the cheesemaker. Many new plants have been installed which have had to be modified, often at great cost, because of lack of appreciation of the process requirements or the practical difficulties. Closer co-operation between engineers and practical cheesemakers has eliminated some of these problems. On the other hand, some cheesemakers brought up on

Figure 14.1 Starter preparation room. (Courtesy of APV Nordic Dairy AS)

long-standing traditional ideas have objected to change even when the same end result is achieved, for instance, in the UF concentration of milk for soft cheese production.

Today, commercial units for washing drying, waxing and wrapping cheese have become available. In large stores, particularly for traditionally ripened Edam, Gouda and the larger Emmental cheese, units are available for conveying the cheese into and out of store and for turning the cheese while in store.

Because the different varieties have evolved with different methods of cheese curd manipulation, the mechanization of cheesemaking has been developed along different lines. Figure 14.2 shows schemes for making cheese of the different groups; the process involves almost 20 stages, each of which may need different types of equipment even though similar processes are involved.

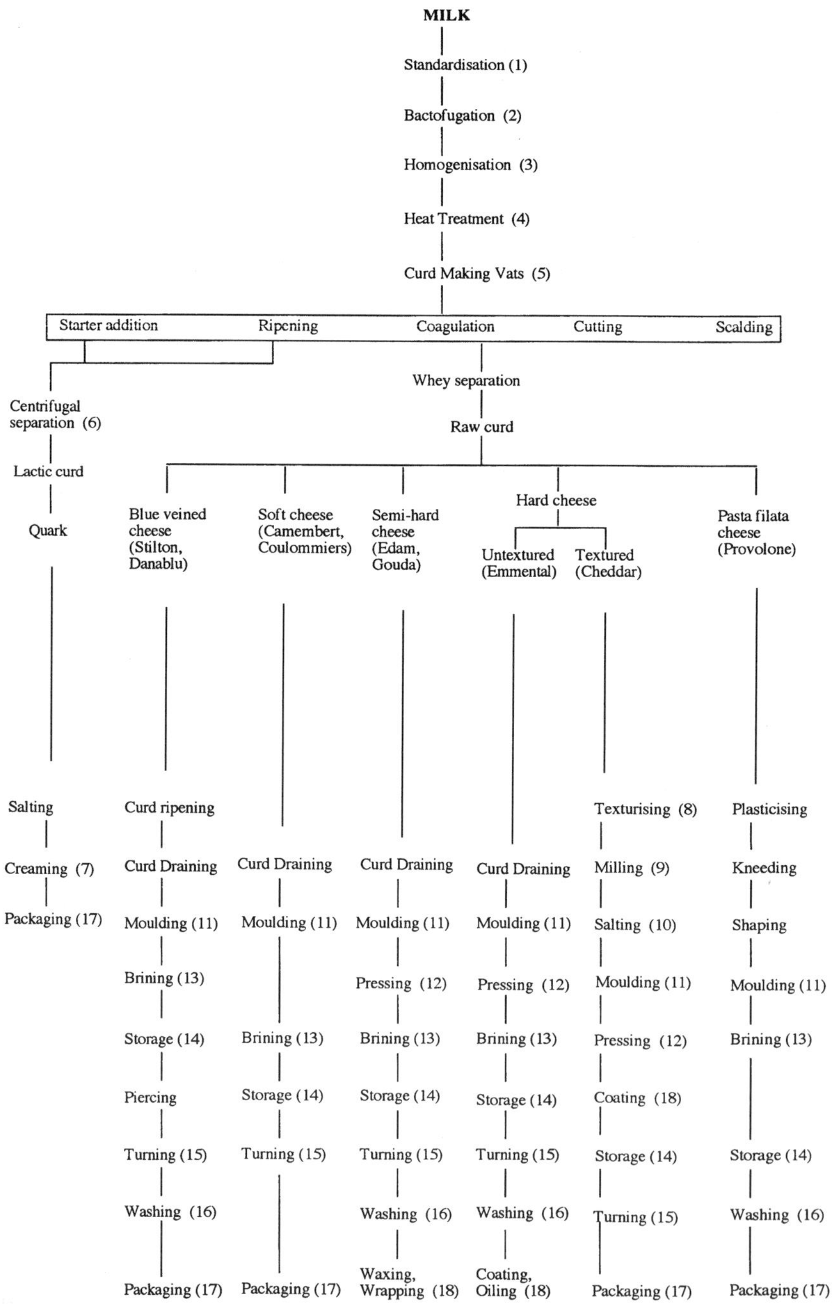

Figure 14.2 Schemes for the mechanization of cheesemaking.

The time lags in the growth of different bacteria in different growth media (milk or curds), together with variable enzyme activity in the curds would appear to make truly continuous semi-hard or hard cheese production more difficult than soft cheese production.

When different milk supplies follow one another in a continuous coagulation system, the differences in the rate of acid development upset the routine of production. Similarly, if a large volume of pre-ripened milk is coagulated in small batches in an attempt to make the process continuous, from that stage onwards there is a growing difference in maturity between the first and last curds. While the use of large silo tanks for bulking quantities of raw milk helps to smooth out the differences between supplies, tanks for pre-ripening milk should be of a size compatible with the rate at which coagulation takes place.

Except for soft and some smaller firmer cheeses, the trends in mechanization of cheesemaking are towards batchwise, continuously manipulated curds to ensure some time 'space' between the stages, by speeding up or slowing down transfer mechanisms. Thus, proper control can be exercised at the texturing, milling, salting and maturing stages for textured cheese. For non-textured cheese, e.g. Edam and Gouda, the batchwise continuous process works well since there is the opportunity to absorb the time lag between batches of curd, either in the vat or after hooping or between hooping and brining. The routine of production can be upset when curds are abnormal, i.e. slower or faster than normal in acid development. Whilst ideally mechanization should be capable of adjustment to allow for differences in the rate of development of the cheese curds to produce product of even quality, in reality the curd production process must be rigorously standardized.

Perusal of the literature registered at various National Patent Offices showed an increasing number of patents taken out in respect of cheese processing. While the acquisition of patent rights is of financial benefit to the patentee, or holder of the patent rights, patents may be a suffocating barrier to the development of new processes.

14.2 Homogenizers

Homogenizers used in the cheese industry are of two types. The first type is required to homogenize the milk fat, i.e. cream, into a state of dispersion so that it will mix with the milk serum. This type of machine (Figure 14.3) consists of a high pressure pump feeding into one or more close-seated valves, which by cavitation and sheer of the fat globules causes a reduction in their size until 85% of the globules are below 2 μm (from 2 to 0.02 μm) in size (Wilbey, 1992). In decreasing the fat globule size, the surface area and hence the susceptibility to lipolysis increases. In the manufacture of some

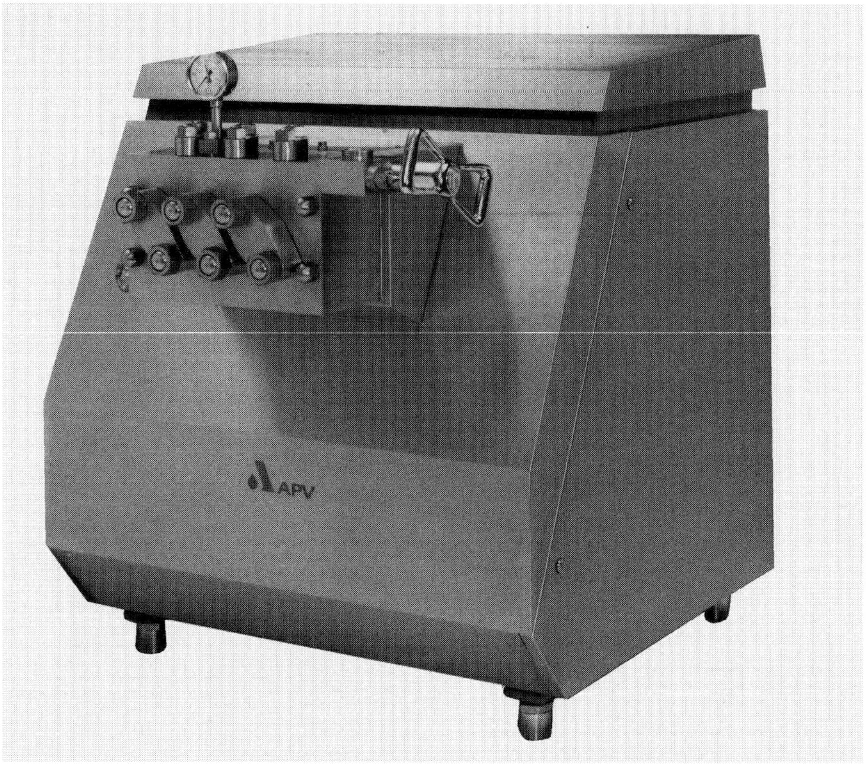

Figure 14.3 APV homogenizer for milk or cream. (Courtesy of APV Nordic Rannie AS)

varieties of blue-veined cheese where lipolysis is encouraged, it is general practice to homogenize the cream only and not the milk serum containing the casein.

The homogeneity of milk–cream–skim-milk mixtures can be achieved by the use of a high speed rotary mixer/emulsifier, which breaks up fat globule clumps rather than reducing the size of the fat globules appreciably. The rotary machines can take the place of the low pressure homogenizers in breaking down lumps of curd, as in lactic cheese, to produce a smooth-textured curd. Pressure-operated vibrating-reed homogenizers have been used, but the output is not usually enough for larger commercial practice.

Homogenization does not destroy bacteria in the milk, but by breaking up clumps of bacteria may induce more rapid growth after homogenization; hence the need for adequate heat treatment immediately after the process. Since the optimum temperature of homogenization is approximately 68°C, subsequent heat treatment is but an extension of the primary heating.

The clarifixator, a centrifugal machine similar to the cream separator, but used in place of an homogenizer for breaking up the milk fat in liquid milks, does not appear to have found use in the cheese industry.

14.3 Starter preparation equipment

Equipment for starter preparation must be able to be satisfactorily cleaned and sterilized, but perhaps of equal importance, the equipment must protect the starter from contamination by bacteriophage. The ingress of 'unsterile' air to the containers must be prevented at all times after the equipment has been sterilized, during inoculation and during fermentation. Wells of hypochlorite solution have been used to protect openings in the starter vessels against ingress of unfiltered air. Figure 14.1 shows one of the many installations. Filtered air is used in this type of installation.

Feeding the starter into the cheese milk by pump provides a possible source of contamination through the gland of the pump. Aseptic bulk starter tanks are often used. It is essential that temperature control can be exercised on the starter, not only to maintain even temperatures during incubation but to allow for cooling of the ripened starter to 5°C, ready for inoculation of vat milk.

The development of bulk starter manufacture systems has been paralleled by the development of direct vat set (DVS) starter systems where a powder or liquid concentrate may be added to the vat (Chapter 9).

14.4 Milk ripening tanks

The increase in cheese manufacture created a requirement for an increase in vat size or numbers of vats. The introduction of storage tanks as cheese milk ripening tanks filled this requirement without increasing the numbers of vats. However, the vats used simply as curd making and scalding vats had to be augmented by texturing drainers where such cheese as Cheddar was made, and by whey drainers for other cheeses like Edam and Gouda.

The introduction of ripening tanks also brought about the use of computerized rates of inoculation into the already warm milk being filled into the ripening tanks. This controlled starter injection also led to further automatic control of milk treatments and of milk flow. The ripening tanks needed to have large 15–23 cm milk outlets for rapid transfer of milk to the curd-making vats, and when empty the tanks needed adequate cleaning before being refilled. Thus, CIP systems became incorporated in the automatic control systems. The time lag in the growth rates of some starters can be allowed for because of the transfer from tank to curd-making vat on the first round of usage; but if the vats need to be filled a number of times in the day, then sometimes difficulties through overgrowth of starter may occur.

14.5 Soft cheese making equipment

Soft cheese manufacture occupies less time than hard cheese manufacture, which has encouraged development of batchwise continuous processes for curd making. The vats have tended to be of small capacity, 50–200 litres, arranged in series to serve simply as coagulation vats only.

In earlier systems the firmly set curds were coarsely cut and then poured directly into troughs which directed the curds into tall cheese moulds. The coagulation vats were fitted with cutters so arranged in a screw fashion that the flow of the curds into a succession of cheese moulds attached to wire grid trays could be metered.

The Alfa-Laval 'D' vat system used a succession of small D-shaped vats which had a discharge valve at the bottom of one end. A rail mounted above the vats carried agitation attachments which could be moved from vat to vat. A helical screw (Archimedian screw) shaped device was also suspended from the rail above the vat. This device metered the curds through the end gate into moulds placed below a distributing trough. The cheese in the moulds on the wire grid were handled through draining, brining or salting, inoculation and ripening rooms.

The Corblin system used extensively for Camembert cheese used long, horizontal, half-cylindrical vats for coagulation of the curd and partial whey drainage. Movable D-shaped partitions inserted in the vats divided them into smaller compartments, the curd from which could be filled into moulds attached to trays. Hansen (1968) described a soft cheese making system (Alfa-Laval) at Grieges in France using ten semi-cylindrical vats with overhead suspended curd cutters and stirrers. Moulds on an endless belt are filled directly from the vats.

Later systems (e.g. 'Alpma Fromat') used small coagulation vats which discharged the curd onto moving belts. The curds were partially drained of whey and then the curds were filled into troughs ending in vertical tubes in which the curds were compacted before being filled into cheese moulds on the usual wire grid trays. Thus, the wire grid tray with 20–100 cheese moulds became the handling unit for subsequent cheese processing, i.e. for *Penicillium* inoculation for Camembert cheese, or for brining or for ripening in the store.

Later, the Alpma coagulator equipment used a continuous flexible belt which was formed into a trough to hold milk or curd. The trough, divided into sections by half-circular slat divisions, gave the same coagulating facilities as a series of vats. The curds so produced could be cut, agitated, partially scalded and drained of whey by extension of the belt system. The curds subsequently could be filled into moulds and subjected to the usual treatments as indicated in Figure 14.2. The Alpma coagulator could be used for any of the softer cheese manufactures, including soft blue-veined cheese.

In spite of extensive mechanization for the coagulation of milk for soft cheese, the advent of the use of concentrated UF milk had a profound effect on soft cheese production, overtaking development of the Hutin-Stenne process for cold rennetting concentrated milk.

14.6 Curd-making vats (textured cheese)

The earlier square-ended, open-topped jacketed vats with swinging mechanical stirrers have been replaced in all but the smallest dairies by oval, cylindrical, enclosed vats complete with mechanical cutting and stirring devices. The swinging stirrer of the square-ended vats could be fitted with strip steel or wire curd knives and therefore was suitable for many types of curd from Cottage cheese (Figure 13.13) to Cheddar cheese. The whey was usually drained off the curd and then the curd dealt with manually. The square-ended vats were followed by round-ended vats with rotating agitation driven from an overhead gantry.

Demands for better process hygiene, CIP and reduced labour costs led to the production of totally enclosed vats for curd making and scalding, examples of the many types that were and are produced being:

(i) The former Alfa-Laval 'OST II' round vat with twin vertical mixer-cutters, still widely used.
(ii) Damrow 'Double O' vat.
(iii) Golden Vale 'W' cheese vat.
(iv) Wincanton enclosed vat, illustrated in Figure 14.4.
(v) New Zealand 'Vatmaster' vertical enclosed vat.
(vi) APV Curdmaster vat (Figure 14.5).

While the vertical enclosed vats represent a logical development of the open vats, overlapping developments have also led to the production of horizontal enclosed vat systems examples of which are:

(vii) Tetra-Tebel® OST-vat, illustrated in Figures 14.6 and 14.7.
(viii) Gadan Gasdjerg Maskinfabrik A/S Gadan OS 1000 cheese processing tank.

The curd-making tanks/vats are available in sizes from 2000 up to 20 000 litres capacity. The individual engineering firms have fitted their own vats with features demanded by their customers and so a standard vat is unusual. Most enclosed vats are fitted with CIP systems and some form of programmable logic control (PLC). Some problems were encountered with the earlier vats. Due to the construction of the external

Figure 14.4 18 000 litre Wincanton enclosed cheese vats including PLC with central data collection. (Courtesy of Wincanton Engineering Ltd)

Figure 14.5 APV CurdMaster vats with CheddarMaster belt system in background. (Courtesy of APV Nordic Cheese AS)

Figure 14.6 Sectional view of Tetra Tebel® OST IV vat. (Courtesy of Tetra Pak Tebel BV)

supporting steel work, hot spots on the walls of the vats sometimes caused burn-on of the curd, and thus needed manual cleaning. Furthermore, the hot spots on the sides/bottom of the vats caused hard, tough curds which clung to the curd knives when cutting. This caused crushing of the softer curds and the formation of excessive curd dust which was lost in the whey.

This problem of curd clinging to the knives was overcome in some vats by either reshaping the vat or reversing the rotation of the knives. The reversal of the knives washes off the clinging curds. In most instances the angle of the blades in the knife/agitator frame is designed so that the assembly acts as a cutting knife in one direction and as a mixer in the opposite direction.

Figure 14.7 Tetra Tebel® OST IV vat installation. (Courtesy of Tetra Pak Tebel BV)

The use of hot water circulation in the jacket of the vat, as against steam heating, overcame some of the locally heated hot spots. However, steam injection into the water jacket is the most usual form of heating.

The filling of milk into the vat and the emptying of curd from the vats of the enclosed type may be controlled by air/vacuum pressures in the vat. Some vats can be designed to deliver the whey/curd mixture 6 m above the vat. Where vats are filled from the top, frothing of the milk can be a problem unless the filling pipe is led below milk level. The frothing also causes fat loss and since the fat is liquid it also causes greasy surfaces which are more difficult to clean.

Since the vats are enclosed, automatic control in respect of the operations is a normal feature, i.e. for filling, adding starter, rennetting, cutting, stirring, scalding and emptying. Where the bottom of the vat is not made sloping the vat is provided with tipping gear, pneumatic or hydraulic, to tilt it for emptying. The speeds (2–15 rpm) and rotation of the agitators are controlled from an automatic control panel but it is desirable to have manual override of the operations for those occasions when difficulties arise. Since the functioning of the cheese-curd-making equipment is vital for the ultimate quality of the product, efficient control of the functions must be available to the cheesemaker.

Choice of a curd-making tank can only be made when all the details of construction and operation are considered in the context of the cheesemaking requirements. However, inspection of equipment under practical conditions is perhaps the best way to assess the capabilities of any newly developed equipment.

The practical cheesemaker, trained in manual methods, will have to readjust to the use of automatically controlled equipment. This process may take one or two seasons unless a competent practised cheesemaker is at hand.

There are problems with the maintenance and operation of the equipment which affect the efficiency of the process. Some of these points of concern are listed below.

(i) The nature of the steam supply and the water supply in respect of scale or rust formation.
(ii) The nature of the water used for rinsing, which may leave a dried-on deposit on the milk contact surfaces.
(iii) The positioning of spray balls in the CIP system.
(iv) The electrical/electronic control systems as affected by moisture laden atmospheres.
(v) The efficiency of the sampling and testing devices.
(vi) The efficiency of mixing is important, especially when rennet is added; otherwise a number of different areas of curd firmness will occur in the body of the vat.
(vii) The nature of the air supply for pneumatically operated valves; pressure controllers must be such that moisture cannot accumulate in the pipes.

One problem with tall curd-making vats has been that the hydrostatic head created by the height of the whey above the emptying pump causes a variable volume delivery as the vat empties. Thus, when texturing belts are used for de-wheying purposes the level of curd on the belt also varies unless a positive displacement pump is used for the transfer.

Some cheese processes in which water is sprayed, either hot or cold, into the curds, require a separate type of vat, fitted with water spray pipes along the length of the vat, which is usually of rectangular, square-ended shape. Washed curd cheese, semi-hard, some blue-veined cheeses, but particularly Cottage cheese, need this type of vat with water sprays. Stirring arrangements for these vats are various and include swinging cutter/mixer devices or overhead rotating paddles (e.g. the Damrow type).

The curds and whey are either run by gravity or pumped through to a whey strainer to separate curds and whey, usually the first stage as the curd goes onto the belt.

14.7 Curd-making vats (untextured cheese)

The vats already noted can be used for the manufacture of untextured cheese but some vats have been developed, usually in smaller capacities, as coagulation, cutting and scalding vats. These vats are used where the curd–whey mixture can be run directly into perforated moulds, which act as either conveyors or drainers, even though the cheese may be pressed subsequently. These cheeses include Emmental and Gruyère, Tilsiter, Havarti, Romadur, etc. Various types of curd-making vats have been developed by Broekmans *et al.* (1951), Schwartz and Schwartz (1957) and Lenz *et al.* (1957). Some of these vats are cylindrical with a row of glass portholes up the front. The agitators are driven from the centre of the base up a central tube. The top of the vat, domed in shape, is hinged so that it fits tightly to give a vacuum seal (for filling under vacuum). However, the designs varied from maker to maker (Figure 14.8).

The outlet of the vat is large, with a butterfly-type valve, so that the curd–whey mixture can be filled directly into perforated cheese moulds, either on a conveyor or on a movable drainer. For this reason the vats are elevated above the drainers so that the curd–whey mixture can run by gravity into the cheese moulds. The Steinecker mould filler (Lenz and Lenz, 1955) is part of such a conveyor system. When drained the cheeses are taken to presses to be pressed as required by the variety of cheese. In some instances, the cheeses are not pressed but repeatedly turned in their mould while on the drainer.

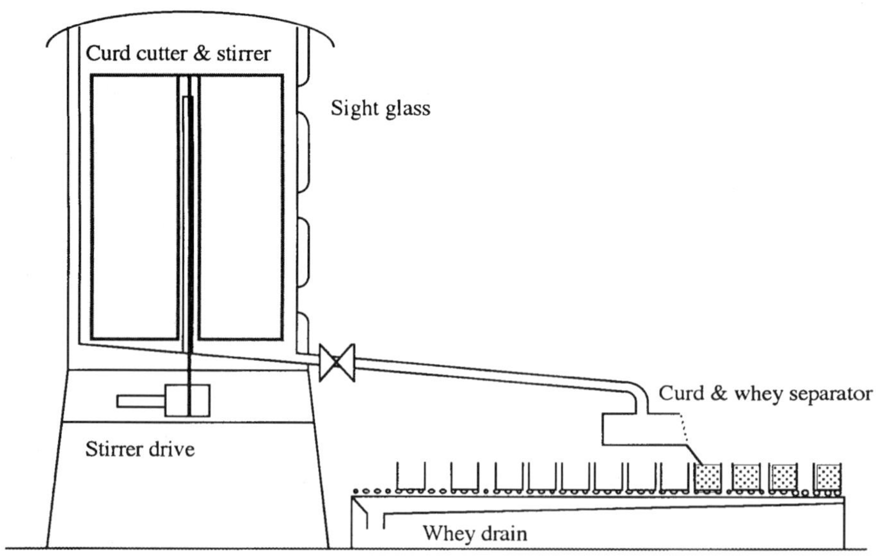

Figure 14.8 Diagram showing a cheesemaking vat with its outlet arranged for filling moulds directly from the vat. This arrangement is suitable for semi-hard cheeses which do not need pressing, but which are simply turned into the mould (e.g. Tilsiter and similar cheeses).

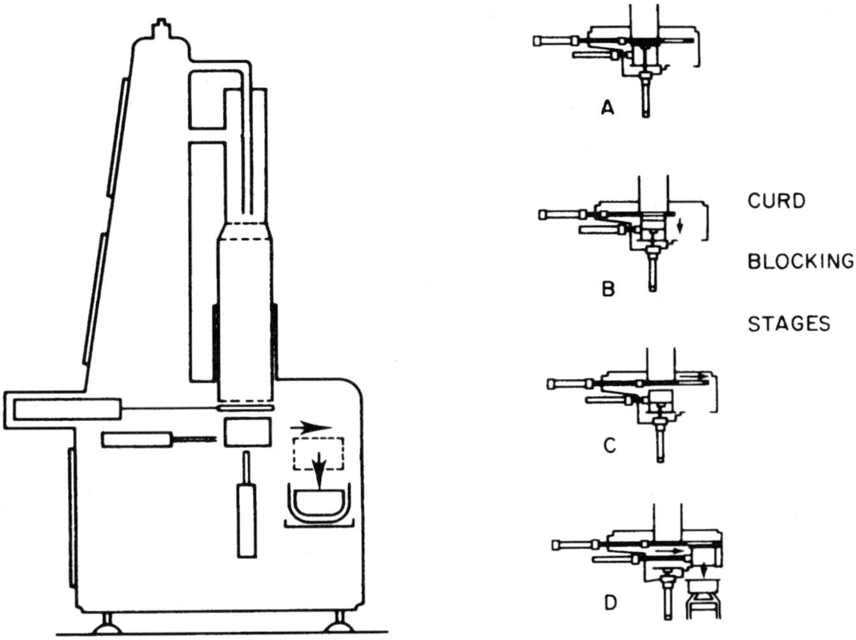

Figure 14.9 Sectional diagram of Tetra Tebel Casomatic® mould filling system for filling cheese moulds with the whey/curd mixture from the vats. (Courtesy of Tetra Pak Tebel BV)

Smaller, half-round, coagulating vats for soft cheese manufacture (Camembert cheese) are arranged to tip the curds into a multi-guide trough which guides the curd into cheese moulds. The vats can be arranged in series for semi-continuous curd-making.

Although it is not a vat system of curd coagulation, but a continuous curd-making system, the Hutin-Stenne (1964) system is worthy of mention at this stage. The system SH 13 is based upon the concentration of cheese milk to approximately 30% solids which is ripened at 15°C with starter, cooled to 2–3°C, and then rennetted while cold and held for 10 h, after the Berridge (1942) cold rennetting principle. The pre-rennetted milk is mixed under turbulent conditions with hot water to form a curd instantly. The curd–water–whey mixture is passed through tubes to agglomerate the curds into larger particles before the curds are dried off and moulded. The system SH 13 was incorporated into free-standing equipment with all the necessary ancillary equipment as in the APV 'Paracurd' machine (Brissenden, 1966). The curds produced by these methods tended to be very granular at first but later developed more normal textures as the protein broke down. Soft cheese of the Camembert type has been produced satisfactorily by this method. Some semi-hard cheese (Edam and Gouda) has also been made, but the method was not acceptable for harder varieties (Brissenden, 1966). A few

Figure 14.10 Buffer tanks with two Tetra Tebel Casomatic® MC mould fillers. The moulds filled may be varied, e.g. from large rectangular to small round. (Courtesy of Tetra Pak Tebel BV)

commercial enterprises in France, Belgium, Spain and Argentina were reported as using this system, but it was overtaken by development of ultrafiltration.

The continuous curd making NIZO system, developed at the Netherlands Institute for Dairy Research at Ede and its commercial version the NICOMA machine also used the two-stage rennetting system. Pasteurized milk cooled to 2°C was stored in a tank for up to 5 h with starter and the rennet additions. The milk was not agitated during this time. The milk was pumped through a heat exchanger at a carefully controlled temperature, 30°C, so that on leaving the exchanger the curd formed in a vertical coagulating chamber. A rotating knife at the top of the chamber pushed the curds into a drying rotary screen. It was possible to warm-water scald or cold-water wash the curds in this screen. The curds were discharged into a vertical tube with perforated walls. A sliding plate at the base of the column allowed the compacted curd to fall into cheese moulds. Although the system was devised for Edam and Gouda cheese, an important development of a similar system is the curd-forming tower and mould filler of the 'Casomatic' mould filling system used for other varieties of cheese (Figures 14.9, 14.10, 14.11).

Figure 14.11 View of the Tetra Tebel Casomatic® SC mould fillers in a semi-hard cheese line for traditional Gouda and Euro block cheese. (Courtesy of Tetra Pak Tebel BV)

Further continuous curd-making attempts to use the Berridge principle have not reached truly commercial status and are now overshadowed in the soft cheese field by the UF concentration processes.

14.8 Whey–curd separation equipment

Once the curd is made in the coagulating vats, whether intended for hard or soft curd cheese, the whey and curd must be separated from one another.

For soft cheese curds the Waldner Roiner system used a special vat which could be rotated so that more whey might be released from the curds. The whey was discharged separately and the curds transferred into long vertical cylindrical tubes. The curd compacted in the tubes and was guillotined in prearranged quantities into individual cheeses which then went forward for salting and ripening.

For the larger scale manufacture of semi-hard and hard cheese, the curds and whey from the cheesemaking vats (e.g. Steinecker type) were gravity fed or pumped through a rotary perforated drum, i.e. a whey–curd separator, so that the curds could be processed further or filled directly into cheese moulds. In small plants a vibratory strainer may be used (Figure 14.12).

Figure 14.12 Vibratory strainer for semi-hard cheese curd. (Courtesy of APV Nordic Cheese AS)

Cheese texturing equipment (section 14.9) usually incorporates its own whey-separation device. Since whey is often processed after drainage from the curd, but still contains small curd particles, the whey is usually passed over a vibrating screen, e.g. a Russell separator, which takes out the fine curd from the whey. Such curds would soon block up a centrifugal separator bowl and so the pre-screening device is useful. The curd fines in some cases are distributed amongst the new curds or are sent for manufacture of processed cheese. In large installations it may be economic to install a self-desludging whey clarifier; the decanted solids could then be returned to the appropriate section of the curd-handling system so that curd pH values match (hence avoiding problems with localized bleaching in the curd). While this approach helps yields, good process hygiene is essential as such recycling could easily introduce contaminants.

The table drainer has become a useful piece of equipment following the curd-maker vat, particularly for the smaller producers and where manual handling of the curd is preferred. Typically the drainer is a long open rectangular tank with a perforated longitudinal drain strip or movable screen belt as a base. One end of the tank has a liquid-tight but movable door (Figure 13.4). In use, the curd–whey mixture is run from the curd vat into the tank. The whey drains through the base screen, leaving the curd level and dry. Some tanks are fitted with plates to press the curd lightly. When dry the curd is moved through the opened end of the tank on its screen base and is cut into blocks by a guillotine attachment. The blocks of curd can be filled into cheese hoops for further pressing or salting.

The Tetra Tebel™ pre-press vat is typical of the whey drainage and curd pressing machines, but similar equipment is made by other specialist firms (Figure 14.13).

Figure 14.13 OPD pre-press system for semi-hard cheese. (Courtesy of APV Nordic Cheese AS)

14.9 Cheese texturing and draining equipment

Although many cheeses, e.g. Edam or Gouda, are close-textured with few gas holes and the Emmental group of cheeses develop their own texture by secondary fermentation, Cheddar needs a specialist operation to provide the so-called chicken breast texture in the curd before salting.

The traditional Cheddar cheesemaker achieved this striated chicken breast texture by repeated piling and squashing down of the curd to cause elongation of the curd particles in two directions during the fusion of the curds as the pH dropped. In order to remove the need for this laborious work, texturing machines were devised to follow on from curd production in the vats and include the following:

(i) The Bell Siro Cheesemaker 2, developed in Australia.
(ii) The CheddarMaster Systems, originally developed in New Zealand.
(iii) The Lacto-matic System, developed in Tasmania.
(iv) The Tebel-Crockatt combination, as used in Ireland but now virtually obsolete.
(v) The Alfomatic Cheesemaker, developed by Alfa-Laval, now Tetra Pak Tebel.
(vi) The Damrow matting conveyor and milling equipment, developed in the USA.
(vii) The Stoelting Cheddaring Vat, developed in the USA.
(viii) The Wincanton system.

This list is not exclusive.

Before considering these pieces of equipment there are two principles which are used to give the stretching or pulling and matting effect to the Cheddar curds. Drainage of whey from curd and compaction of the curd by the use of a tower was first demonstrated by Budd and Chapman (1961). The commercial towers are tall and hold up to 2 h production of curd. The height of the tower provides the pressure, and the change from round to the square lower cross-section provides for stretching of the curds. The CheddarMaster 2 and Lacto-matic machines use this principle.

The second group of apparatus uses flexible belts as conveyors (Figures 14.17–14.20). The curds passing through the machines are stretched and squeezed by the belts running at different speeds. The Damrow, Alfomatic and other machines use this principle. Both machines incorporate curd mills. The curds entering these machines are produced in normal cheese vats.

The original Stoelting cheddaring device used a cylindrical vat, with divisions capable of rotation at an angle. The curds were continually stretched and deformed during the rotation. Milling was external.

The Tebel-Crockatt system used the normal whey curd drainer until the curd was dry and then a rotating lawn-mower type of bladed rotor moved up and down the tank to cut and disturb the curd slices. Actual cheddaring was developed in drainers later. This apparatus could be used for Cheshire and other textured cheese. The system was popular in the UK in the 1960s but was superseded by belt systems.

Curd cheddaring systems using 'cheddaring boxes' such as the Tebel van de Ploeg and the Alfa-Laval 'Jongia' cheddaring box (Brookes, 1970) were also used for the production of cheddared curds. The cheddaring process takes 2–2.5 h and the curd was made to 'flow' by tilting the boxes through 90° every 15 min. The curds for cheddaring were produced in normal cheese vats and after cheddaring the curds were milled, salted and filled into moulds in the usual manner. Whey drainage from the boxes was allowed for in their design. It is important to remember that some cheesemaking machines were developed for cheesemaking from milk supplies different from, for instance, those in the UK. These systems have been superseded by belt systems.

'Off the peg' machines may need modification and some of the problems which have arisen include those due to speed of belts, the need to extend bearings, to avoid drips of whey onto the curd and modifications to improve the salting of the curds.

In some of the earlier CheddarMaster machines (Figure 14.14) cheddaring boxes were used but in later models these were replaced by a tower (Figure 14.15).

The CheddarMaster 2 is a complete curd handling, curd milling and salting unit. The speed of the belts can be varied from 10 to 30 min dwell time for the curd. The hold-up time for cheddaring in the tower may be from 1 to 2 h. Although the belts are slow moving, they require special attention to cleaning. The earlier machines were open but it was found advisable to fit covers over the machine to avoid contamination of the curd, control curd

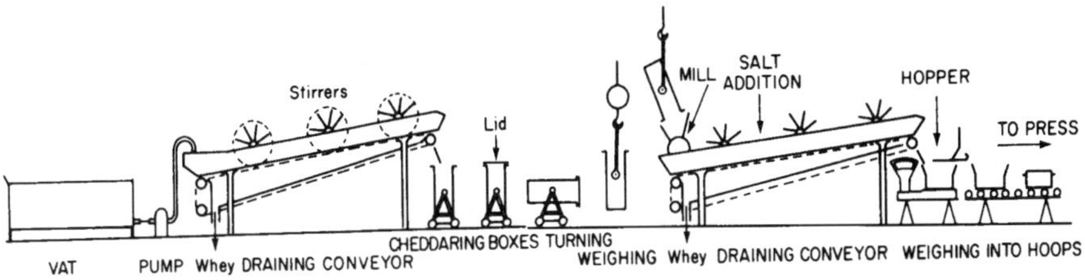

Figure 14.14 The original CheddarMaster system with cheddaring boxes.

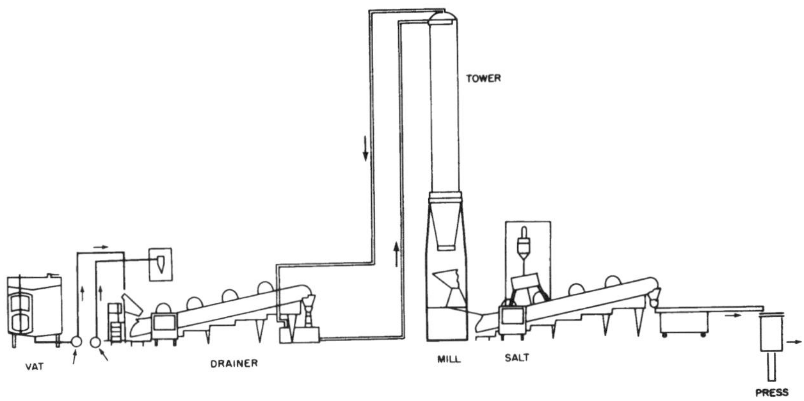

Figure 14.15 The CheddarMaster 2 curd draining, milling and salting machine with cheddaring tower.

temperatures on the belt and permit CIP. The CheddarMaster 2, having a large throughput, was sometimes designed to feed into large hoop ('Ton') presses capable of pressing 4500 kg of curd per hour. The latest Cheddar-Master 3 series offers the option of retaining the cheddaring tower (Figure 14.16) or using a belt-only system.

The belt-only systems can be illustrated by Figures 14.17 and 14.18. The Alfomatic machine was originally developed in Australia and used belts for de-wheying the curd, for curd fusing, cheddaring the curd and for conveying the curd to a mill. The system originally broke with tradition in that salting the curd was by strong brine before the curd was mellowed on another conveyor belt. Since brine salting adds water to the curd, a drier than usual curd must be obtained by the use of blades stirring the curd on a belt before the curd fusing belt.

The Alfomatic machine was intended to feed a curd Blockmaker moulding machine. The advantage of the Alfomatic machine is that rotary curd stirrers can be fixed along any of the belts, and thus the machine can be used for any of the textured curd cheese, i.e. Cheshire, Leicester or Derby. The use of brine in the Alfomatic machine is an innovation which ensures even salting

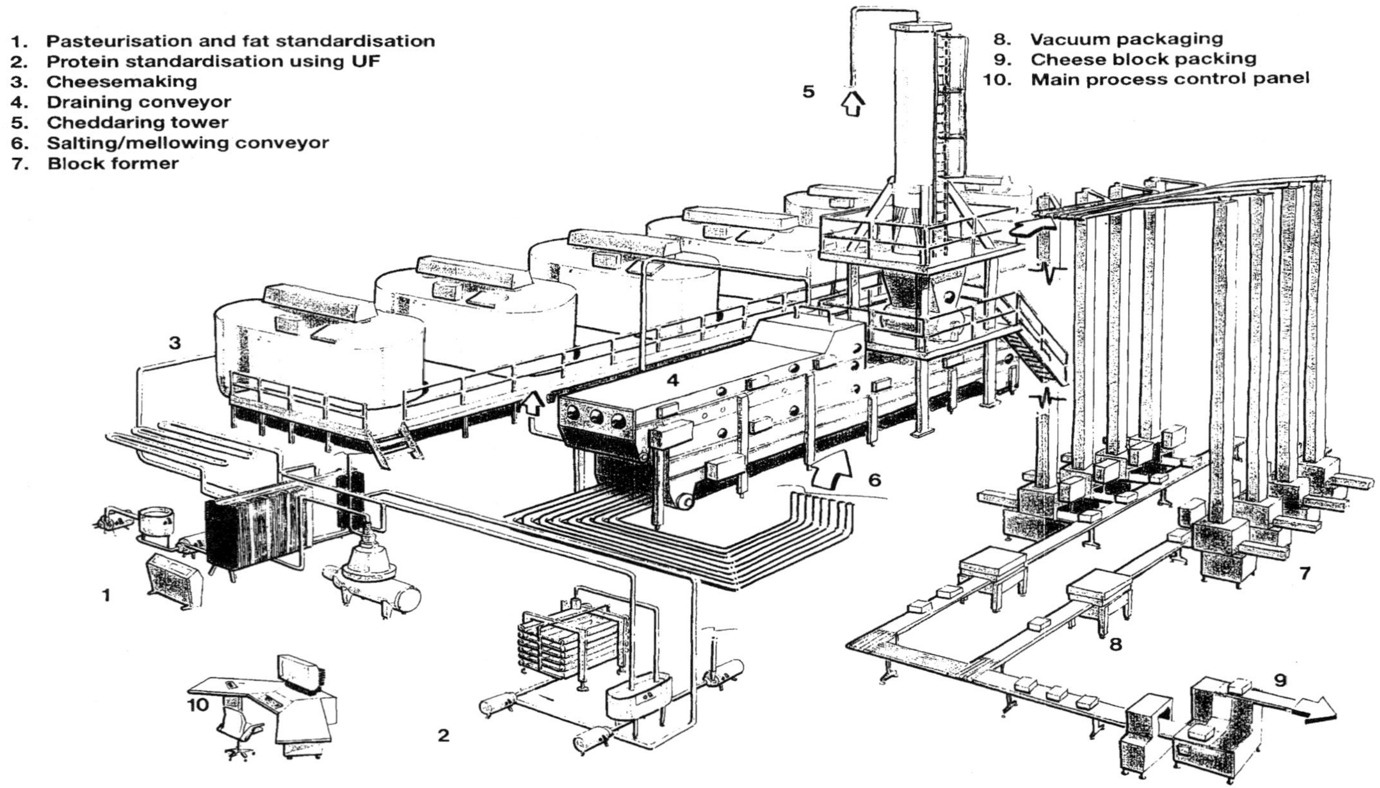

Figure 14.16 Large-scale Cheddar cheesemaking incorporating the CheddarMaster 3 system, in this example retaining the cheddaring tower option. (Courtesy of APV Nordic Cheese AS)

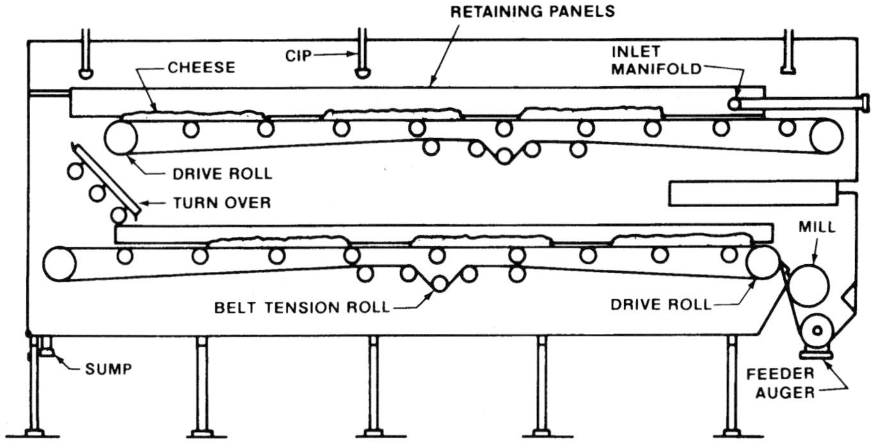

Figure 14.17 Diagram of the Damrow curd matting belts and curd mill for Cheddar cheese making. (Courtesy of Damrow Company, Wisconsin, USA.)

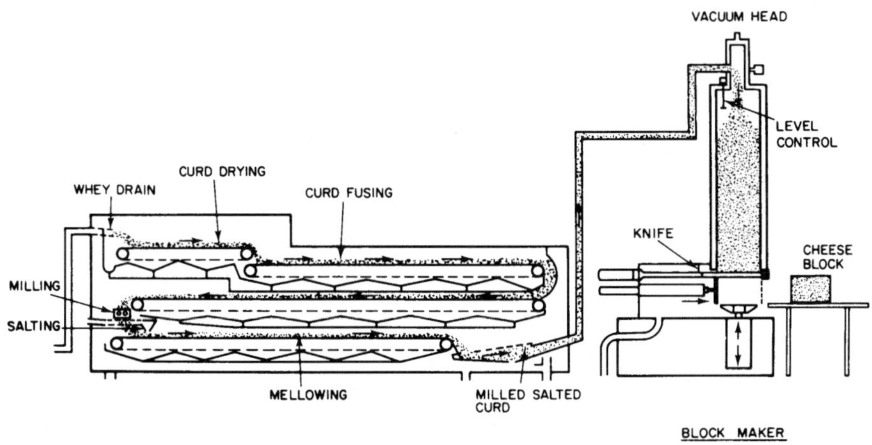

Figure 14.18 The Tetra Tebel® Alfomatic cheesemaker with a Block cheesemaker attached.

of the curd. The mellowing time for salt absorption is short, but if used in combination with the Block maker, this mellowing time is thereby increased.

Just as modifications of existing texturing machines can produce drier or non-matted curds, as referred to above for the Alfomatic system, so other modified machines have been produced for specific varieties of cheese.

The Bell Siro Colby cheesemaker system modified by Bell Bryant Pty. Ltd for the production of Colby and similar stirred curd cheese, utilized the Bell Siro Cheesemaker 3 (milling and salting machine), which was preceded by a special whey/curd separating screen and spiral-shaped rotating stirrers on the conveyor belt, before the dry curd fell into the milling section.

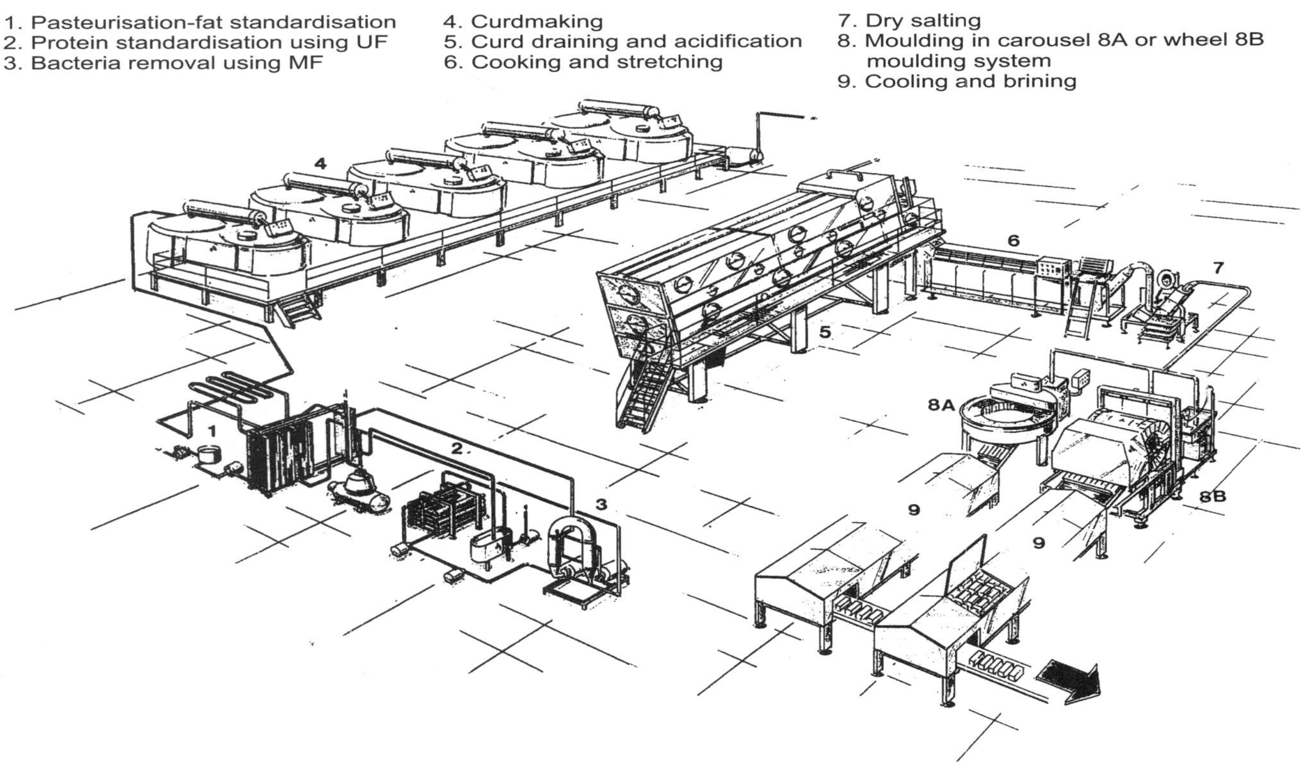

Figure 14.19 Large-scale process line for Pasta Filata cheese. (Courtesy of APV Nordic Cheese AS)

The CheddarMaster 2 machine was also capable of modification, in that extra stirring arms could be sited on the belts and even a third belt may be used for some non-Cheddar types of cheese. There were variations in the manner in which mechanisms could be arranged for the production of special types of cheese. For instance a Cheshire type of mill under the tower of a CheddarMaster 2 would enable a Double Gloucester cheese and perhaps even a Cheshire cheese to be made, although the texture was likely to be too close for some markets.

The belt system has proved very flexible and has been adapted to the manufacture of a very wide range of cheeses (Figure 14.19). The larger installations, handling 225 000 litres per day and above, tend towards the use of complete plants, mechanized and programmed for a specific variety of cheese. Flexibility for the manufacture of alternative varieties is being sought more often by manufacturers. In the choice of a plant, consideration of the mixing of curds should receive attention. Intensive mechanization when producing blocks of cheese tends to mix curds from different vats and of different maturity. This leads to problems in respect of quality. Mixed curds in highly coloured cheese may produce patchy discoloured cheese when ripe. The day-to-day manipulation of curds from consecutive vats should receive

Figure 14.20 Salting and mellowing module of a Wincanton 4650 kg/h continuous cheese machine. (Courtesy of Wincanton Engineering)

due attention during mechanized cheese plant operations. One of the biggest problems with large continuous curd-handling systems is maintaining constant pH conditions at the various stages of the process. Since the coagulation is a batch process, this requirement puts great emphasis on the consistency of production of the curd from the vats and this is the greatest challenge to the large-scale cheesemaker.

Some makers use a twin module approach in preference to a single module belt system (Figure 14.20), in which case a curd transfer mechanism is needed between the modules.

14.10 Cheese hooping or moulding machines

The material used for cheese moulds has changed from the traditional wood, which was normal up to the 1900s. In the first instance, iron and then tinned steel moulds were used for many years, but later aluminium or its alloys, then stainless steel and plastics, became popular for cheese moulds. Attempts to dispense with cloths as liners were aided by the introduction of finely perforated stainless sheet, as in the Perfora mould. However, the use of plastic of the right type and surface finish for the moulds allows cheese rind to mat without the cloth liners.

Moulds of various shapes, such as those made by Kadova (Rotterdam) in three-piece plastic are widely used for Edam, Gouda and similar cheeses (Figure 14.21), but where extra pressure is needed the plastic requires reinforcing. The Laude moulds are reinforced and can be used for Cheddar and similar hard pressed cheeses. Perforated stainless moulds are also suitable for many cheeses.

Figure 14.21 Perfora plastic mould module. (Courtesy of APV Nordic Cheese AS)

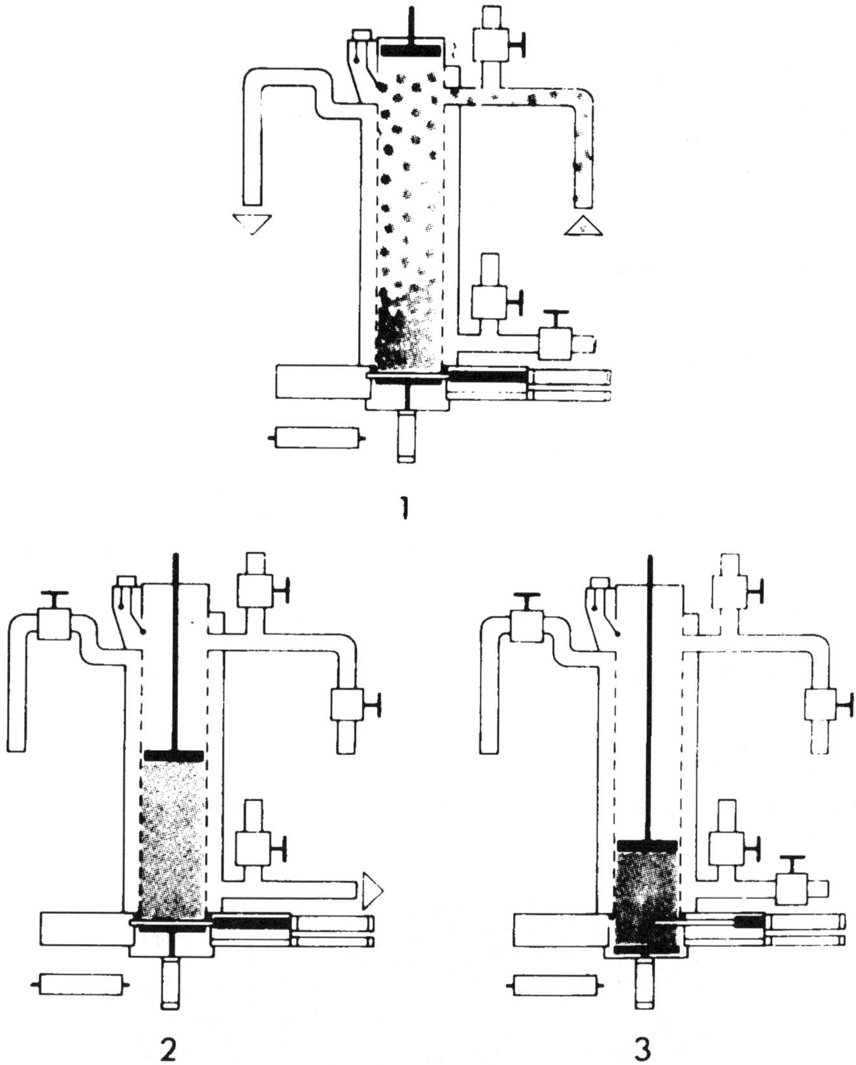

Figure 14.22 Diagram of operations for forming blocks in the Pasilac proportioning unit.

A number of the modern mechanized systems use devices in which the whey–curd mixture is fed into apparatus which releases the whey, and the curd is filled into the moulds, usually as blocks. One method of operation is illustrated in Figure 14.22.

Tebel curd drainers also produce blocks of curd ready to fit into the moulds. The Casomatic system (Figures 14.9–14.11) separates the whey and curd which is delivered into plastic moulds, but the Holvrieka system made by Gadan was slightly different in that the whey/curd mixture was accumulated in a tank before being pumped into the curd-blocking machine and thence to the moulds.

When cutting blocks of curd for consumer portions, the weight of the block is most important, rather than the height of the block. Vacuum may be used in some of the block makers to remove air from the curd even though the cheese requires pressing afterwards.

Block makers have been designed as a replacement for cheese moulds and presses. These devices take salted curd from the curd-maker machine into a tower approximately 6 m high and of 28×35.5 cm internal cross-section. The curd is under vacuum in the tower but is enclosed in a perforated liner so that whey can escape. A feed device at the top of the tower enables it to be kept full with curd. Examples of tower installations are shown in Figures 14.23–14.25.

Blocks of cheese are guillotined from the compacted curd and are ejected into a plastic 'barrier' bag on a packing table. The packs of green cheese are vacuum sealed and the plastic heat-shrunk onto the cheese. Since the curd is still warm and not 'set' the blocks tend to deform, particularly if stacked. Thus, the blocks are usually packed into plastic or wooden crates.

Where the shape and weight of the block is an important economic factor in wastage, e.g. with pre-pack operations, the block needs to be kept as close to a sharp-edged cuboid shape as possible. (The Uniblock system was

Figure 14.23 General view of a cheesemaking plant, showing Tetra Tebel® OST IV CH vats, Tetra Tebel Alfomatic 4B and Tetra Tebel Blockformers S. (Courtesy of Tetra Pak Tebel BV)

Figure 14.24 Wincanton Cheese Towers rated for 4 tonnes of Cheddar per hour at Wyke Farms, Somerset, UK. (Courtesy of Wincanton Engineering)

Figure 14.25 Cheese Tower installation at Milchewerke Oberfranken, Germany. (Courtesy of Wincanton Engineering)

designed to constrain the wrapped cheese within six plywood slats bound by steel tapes and achieved this objective at high labour cost.) Cooling of the blocks in a cold store before stacking is important from both a physical and a microbiological viewpoint.

14.11 Cheese presses

The majority of hard-pressed cheeses were pressed in either vertical or horizontal presses. The pressure was applied variously by spring, hydraulic or pneumatic heads. Enclosed vacuum presses have been used for some of the block moulds. Automated tunnel presses have become common for some cheese varieties where low pressures are sufficient. Figure 14.26 shows such a tunnel press in use. Some presses may use a combination of vacuum and pressure, e.g. 1.5 h under vacuum and 6 h under normal pressure, giving a total of 7.5 h press time.

Most presses have indicators of the pressure applied and some are programmed for automatic increases in pressure. With small to medium plants the pre-pressing and final pressing may be combined in the one machine (Figure 14.27) while in larger plants these functions must be separated.

Figure 14.26 SaniPress tunnel press – typically operating at 30–40 kPa. (Courtesy of APV Nordic Cheese AS)

Figure 14.27 CombiPress for pre and final pressing. (Courtesy of APV Nordic Cheese AS)

The large hoop press or 'Ton' press was a later development in cheese pressing for Cheddar cheese, so called because it held approximately 1016 kg (an imperial ton) of curd. The press chamber was of rectangular cross-

Figure 14.28 View of a semi-hard cheese line for loaf and block cheese. (Courtesy of Tetra Pak Tebel BV)

section, fitted with a piston from the bottom end for compression of the curd and for the extrusion of the pressed curd. The top of the press had a removable lid which could be replaced by cutters which produced four columns of blocks of cheese, each weighing 18 kg. The size of the block was determined by the height of the extruded curd.

The sides of the chamber with perforations were lined with plastic perforated sheets. The outer case was arranged for vacuum to be drawn. Vacuum at 85 kPa (approximately 0.15 bar absolute) was normally applied to the top, middle and bottom to produce an even vacuum and, therefore, an even curd. The process was normally program operated so that the vacuum was applied first and the pressure gradually applied afterwards. After the first 10 min a hydraulic feed pressure of 3.4 MPa was applied and after another 10 min 9.3–10.3 MPa, giving a final pressure of 0.48 MPa on the cheese. After a pressing time which varied from 7 to 20 h the pressed cheeses were extruded from the top of the press.

The original intention of this system was to produce a large block for maturation but it was found that such blocks matured very unevenly. Cutters were introduced to convert the large block into standard blocks. Since the press chamber was 56×71 cm in size, four 18 kg cheeses were cut off at each extrusion. Block size is important in avoiding waste when cut to consumer portions. After cutting, the cheese blocks were wrapped for

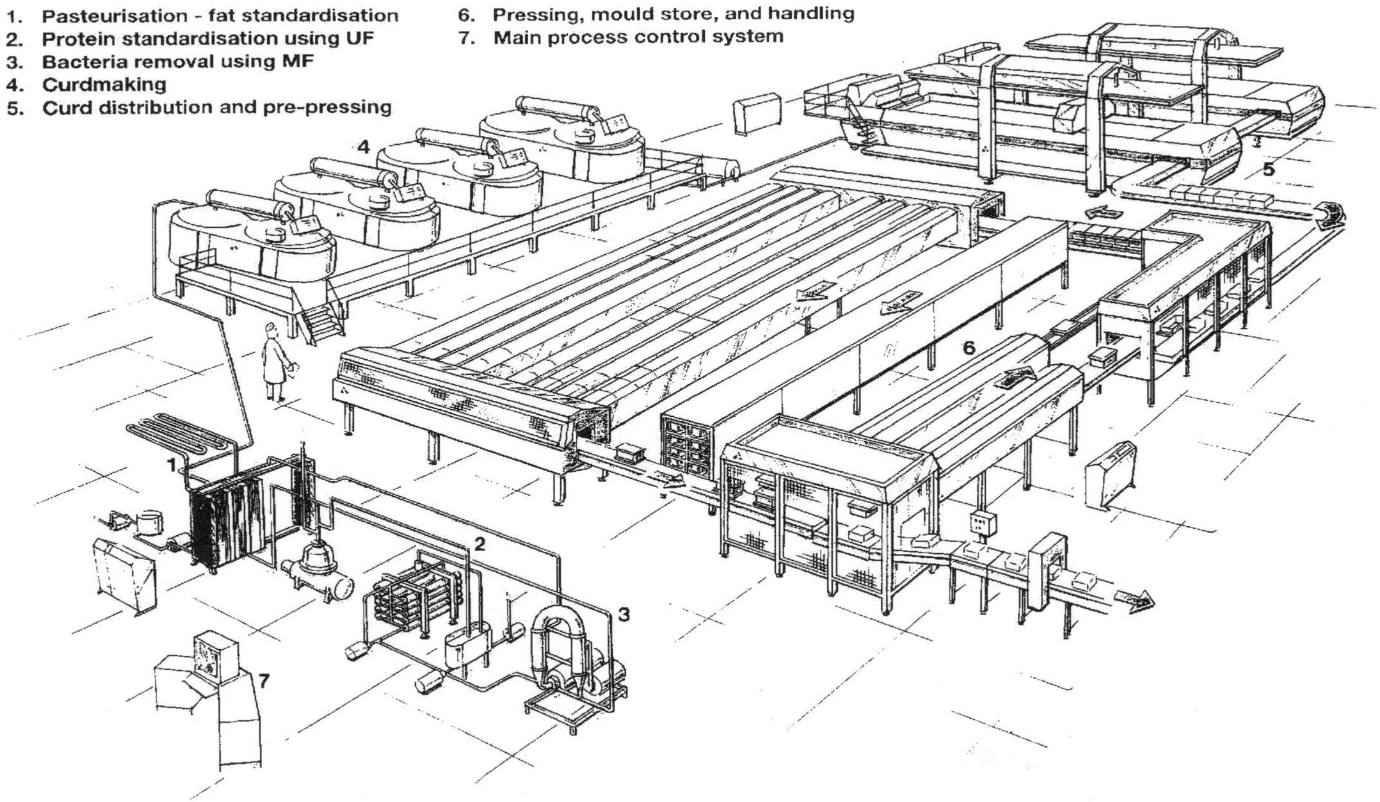

Figure 14.29 Process line for Gouda, etc. with pre-pressing system. (Courtesy of APV Nordic Cheese AS)

storage and ripening. Some cheesemakers used external cardboard boxes for protection of the wrappers, others used wood or plastic. Since the cheese was still warm, it was usual to cool it before stacking high in the store.

The first 2 or 3 h of pressing are the most effective time in press. The gradual cooling of the cheese to set it is also essential, since pressed warm cheese curd will expand if not set; and unfortunately this did not always happen when using the Ton presses due to lack of cooling. A further problem with the large hoop press was that it was less easy to follow individual vats through the process, and the system was overtaken by the development of cheese towers for large-scale producers, leaving a combination of smaller moulds (single or less than 20 kg) and towers (Figures 14.28 and 14.29) in the industry.

14.12 Cheese brining equipment

The older wooden, concrete and glazed ceramic brine tanks for cheese have given place to plastic tanks. Brine tanks may be above floor level or half-buried or even below floor level. Automatic filling and emptying conveyors for the cheese are normal in large installations. In some instances the cheese is loaded on racks which are lowered into sunken brine tanks.

Large tanks may have brine circulation within the tank since this speeds up the operation and avoids local variation in brine concentration. Temperature control is also required. Portable brine tanks have been

Figure 14.30 RackBrine system, showing the transfer of a rack of cheese from the loading unit to the brine vat. (Courtesy of APV Nordic Cheese AS)

Figure 14.31 TrayBrine system for water cooling and brining. (Courtesy of APV Nordic Cheese AS)

Figure 14.32 Membrane plant for microfiltration of cheese brine. (Courtesy of APV Nordic Membrane AS)

developed especially for the smaller cheesemaking units. Plastics are being used more and more for constructing these brine tanks, and for brining outside tanks using tray systems. Two examples of brining systems are shown in Figures 14.30 and 14.31.

Maintenance of the brine is critical to the safety of the process and the quality of the cheese. Brines need regular filtration to remove fines, plus treatment to remove any undesirable microflora. Pasteurization has been used for treating brine but the salt solution can be very corrosive in heat exchangers. Microfiltration may be used as an alternative method for removing microbial contaminants (Figure 14.32).

14.13 Cheese conveying and turning equipment

Where large numbers of cheeses are conveyed to and from storage, racks or grids on wheeled stillages are commonly used, e.g. for soft cheese like Camembert.

The larger cheeses, e.g. Edam or Gouda, need conveyors for transport of the cheese and automatic turning of cheese in the store. While many cheeses may be turned at one time, the larger cheeses, e.g. Emmental, are usually turned one at a time and this provides an opportunity for brine washing the moulds from the cheese rind.

Extensive mechanization of the cheese store has occurred in European and other large stores. However, block cheeses are usually handled on stillages since they require no turning.

14.14 Kneading equipment for Pasta Filata cheese

The traditional coagulating and scalding cheese vats can be used for the first stages in the production of Pasta Filata (stretched curd) cheese. The curd is made in the vats up to the point of whey separation. Sometimes the whey is retained for the secondary phase of heating but normally the curds are heated in hot water. After the secondary heating the curds are filled into a mixing or kneading machine similar to a baker's dough mixer. Cheese kneading machines for Pasta Filata cheeses are available, as discussed earlier. The outlet of such kneading machines may be linked to screw (Archimedian) extruders.

14.15 Miscellaneous equipment

Automatic control of the cheese process has now reached a state which is satisfactory, provided the correct information is fed into the control systems. Birkkjaer and Sigersted (1962) reported on a control system with synchronized controllers attached to each vat. Separate programs for each variety of cheese are necessary. The larger dairy equipment suppliers now specialize in control equipment with programmed operation for both processing and cleaning (CIP).

In addition to the major equipment for cheesemaking operations, other pieces of equipment are required for certain duties to replace manual labour. The piercing of blue cheese can now be automated. One machine has a number of vertical needles which are pressed into cheeses such as Gorgonzola or Roquefort. The Stilton cheeses require a machine with two rows of horizontal needles which pierce the cheese from side to side.

After the cheeses are mature, machines are required to wash the rind of cheese before packaging, except for Stilton cheese which is normally packed with a natural coat. Cheeses of some varieties need coating with wax or plastic resins, and machines are available for these duties.

14.16 Cheese packaging equipment

There are needs for two types of cheese packaging. The first is the wrapping of cheese for storage and ripening, the second is for consumer packs of from 15 g to 4.5 kg. Occasionally cheeses are consumer packed directly from the raw cheese blocks, but this practice is usually unsatisfactory except for short-life varieties.

The increase in output per hour of cheese and the cost of labour has had a great influence on the methods of packing as well as on the materials used. The effective cost of packing materials depends on the required storage life, and this in turn depends on the effectiveness of sealing, the moisture barrier effectiveness and the material's strength.

Vacuum packing has almost become an essential in order to obtain a close bond between film and rind, for both stored and consumer packs. Earlier systems used Pukka-film (DRG Flexible Packaging) for the 18 kg block cheese wrapping even though a hot press is needed to seal the film to the cheese rind (e.g. the Flowers press). This was continued by the smaller sized creameries for some time but most have moved to vacuum packing systems.

The system developed by W. G. Grace Ltd using Cryovac film as a shrinkable barrier is widely used. The film may be used in sheet or bag form according to the requirements.

There are many systems of packing used for both storage and consumer packing of cheese. Indeed, the cheese packaging industry has become a major sector in its own right in recent years, as discussed earlier. The two most common systems are vacuum packing and modified atmosphere packaging.

References

Berridge, N.I. (1942) *Nature*, **149**, 194.
Birkkjaer, H.E. and Sigersted, E. (1962) *Beretning Statens Forsogsmejeri*, **135**, 32.
Brissenden, C.H. (1966) *Journal of the Society of Dairy Technology*, **19**, 154.
Broekmans, T., Sell, W., Konnecicer, T. and Ahlborn, H. (1951) German Patent 820 346.
Brookes, H. (1970) *Dairy Industries*, **37**, 539.
Budd, R.T. and Chapman, H.R. (1961) *Journal of the Society of Dairy Technology*, **14**, 213.
Hansen, R. (1968) *Nordisk Mejeri-Tidsskrift*, **34**, 81.
Hutin-Stenne, G. (1964) *Tech. Lait*, **463**, 13, 15, 17, 18, 22, 23, 24, 27, 29.
Lenz, C. and Lenz, K. (1955) German Patent 927 718.
Lenz, C., Humplmair, N. and Zeiler, K. (1957) German Patent 827 375.
Schwartz, C. and Schwartz, M. (1957) US Patent 2 794 746.
Wilbey, R.A., (1992) *Journal of the Society of Dairy Technology*, **45**(2), 31.

15 Cheese maturation

If the cheesemaker has satisfied the instructions in the recipe, the end-product should be a raw curd formed into a cheese of a specified weight and size. The maturation of the curd into the final edible cheese now depends on four factors:

1. The temperature of the cheese store.
2. The humidity of the surrounding atmosphere; if the raw curd is hermetically packed, e.g. a block cheese or one coated in wax, the cheesemaker loses control of humidity and is left with only temperature to control the ripening stage.
3. The chemical composition of the curd, e.g. the fat content, as well as the level of amino acids, fatty acids and other by-products of enzymic action, all of which factors are determined by the previous cheesemaking operations.
4. The residual microflora of the curd, whether of starter origin, from the original milk or from the process plant and its environs. However, apart from making deliberate inoculations of penicillia for blue-veined or surface-ripened cheeses, or ensuring the growth of 'smear-coats' on cheeses like Port Salut, the cheesemaker can do little further to influence the microbiology of the curd.

Although the cheesemaker appears to have little control over the ripening process, it is important to appreciate the nature of the activities occurring during maturation, for these do have a dramatic effect on the flavour and aroma of the retail product. An indication of possible changes in the microflora is given in Figure 15.1.

15.1 Cheese curd

In hard cheeses, the raw curd, after pressing, is a rubbery and elastic mass, perhaps with gas or eyeholes, perhaps with slits or mechanical fractures and usually some free moisture. The curds will still be warm from the previous operations and, as the mass has not become 'set' or 'rigid', it may require support to keep the shape desired in the final cheese. The curd has a coarse structure, and individual particles of curd can still be identified, in spite of some 'matting' together that will have taken place during pressing.

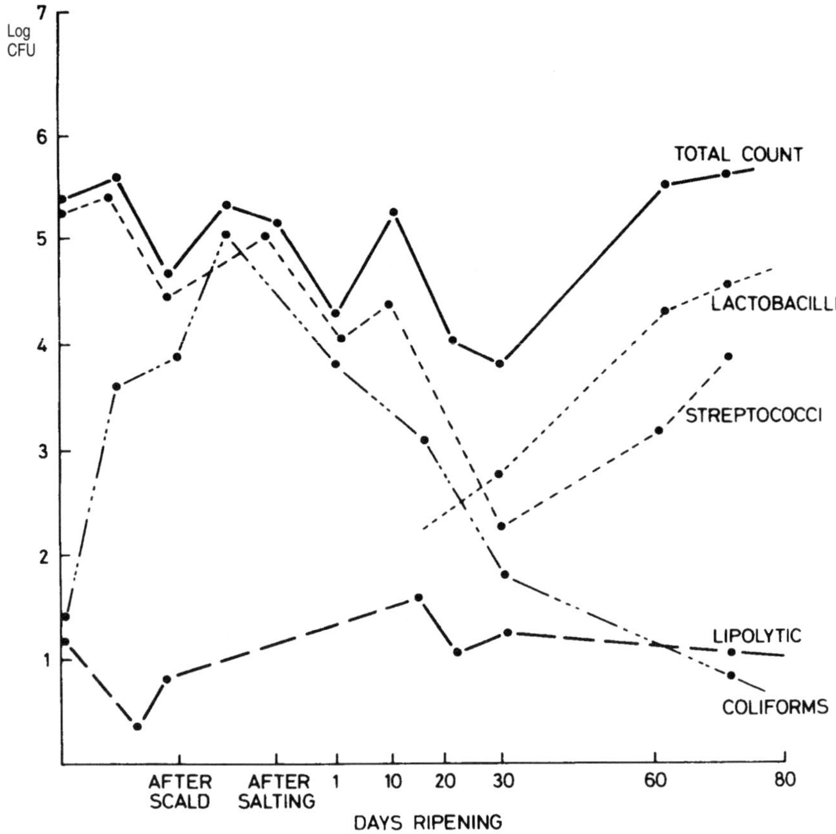

Figure 15.1 Graph illustrating the changes in the bacterial population during ripening of a cheese contaminated with coliform organisms.

In due course the curd cools into a less elastic, plasticized homogeneous mass, having lost its coarse structure, free moisture and particulate nature. There may be mechanical fractures or 'eyeholes' in the curds, and the body, flavour and aroma changes and develops new characteristics as the result of a variety of biochemical changes. These changes mainly involve the protein, fat and carbohydrates in the curd, but, even so, the proportion of fat and protein that is altered is relatively small in many of the hard-pressed varieties, and the lactose soon disappears. In fact, two-thirds of the protein remains intact in many cheeses, as does much of the fat, although some of the latter fraction is transformed into fatty acids and other lipid components which later provide flavour and aroma. However, in the soft cheese varieties like Camembert, the curd becomes, when fully ripe, soft and almost liquid. In these instances, the protein has been subject to intense proteolysis, and little of the native protein is left intact. Similarly, the milk fat is broken down by lipolytic enzymes to provide a share of the composite flavour characteristic of the cheese.

The enzymes associated with the ripening of cheese are produced by bacteria which have grown or are growing in the curd. In blue-veined, surface mould-ripened and smear-coated varieties, specific moulds and bacteria are responsible for secreting an additional complement of enzymes.

The bacterial populations in a cheese curd are continually changing both in total numbers and in species distribution. Species of lactobacilli often increase during the maturation of varieties like Cheddar, while propionic acid bacteria are major components of the microflora of Emmental. It would seem that the typical flavour of any one variety of cheese develops through the action of a succession of different species of bacteria and their enzymes and, in general, the more complex the microflora, the more strong smelling and highly flavoured the cheese is likely to be.

Increases or decreases in the growth of organisms, as a source of enzymes, are influenced by at least five factors:

(1) Moisture in the curd;
(2) Temperature(s) in the cheese store;
(3) pH of the curd;
(4) Inhibitory substances in the curd, i.e. antibiotics from the original milk or released during the metabolism of the microflora, or salts that may have been added during processing, e.g. sodium nitrate or sodium chloride;
(5) Oxidation–reduction potentials.

These factors are also involved in the activity of the enzymes produced by the bacteria in the curds, and the important substrates are carbohydrates, protein and lipids.

15.2 Lactose and related metabolites

Lactose and, in some cheeses, citrate are the main energy sources for the bacteria in cheese, and both are metabolized to lactic acid via pyruvic acid. Both citrate and pyruvic acid are important precursors of aroma and flavour compounds, e.g. citric acid → diacetyl, pyruvic acid → acetylmethyl carbinol → diacetyl, but, in general, lactose is the primary carbon source.

At salt-in-moisture levels below 5%, most starter bacteria, along with many adventitious contaminants like lactobacilli and pediococci, can metabolize the residual lactose in the curd to lactic acid, with the L(+) or D(−) forms dominating as dictated by the organisms. In some cheeses like Edam or Gouda, the level of lactose in the curd will be <0.1% but, in Cheddar cheese, the value may well exceed 1.0% (Farkye, 1993). This latter point means that acidification can continue during and after pressing, a point of vital importance both with respect to the inhibition of pathogens (Keceli and Robinson, 1997) and flavour production. For example, lactose can be metabolized to ethanol, butanol and acetone, as well as acids like acetic and formic, and all such compounds add to the complexity of the cheese flavour. The lactic acid itself acts as a substrate for further fermentation, so that, in Emmental cheese for example, the propionibacteria can metabolize lactic acid to propionic and acetic acids, while the CO_2 released at the same time serves to produce the essential 'eyeholes'. This reduction in lactic acid content also influences the nature of cheese, for it is one of the factors that leads to a rise in pH in cheeses like Brie and Camembert from around pH 5.0 at manufacture to 6.8–7.0 at the time of consumption. In many hard-pressed cheeses, the residual lactose appears to have a 'sparing' effect, so that the carbohydrate is metabolized preferentially to leave sufficient lactic acid to hold the pH within 0.1–0.2 units of the initial value; Cheddar cheese, for example, may be pH 5.0 at the end of pressing but only pH 5.3 some 12 months later.

15.3 Protein metabolism

As most of the whey proteins are lost during manufacture, the dominant proteins in cheese are the α- and β-caseins. It is these proteins, therefore, that are broken down by enzymes in any residual rennet, natural proteases in the milk and/or proteases and polypeptidases from starter or adventitious bacteria, and such activity is essential for the production of flavour and aroma in mature cheese. An indication of the trend during maturation is shown in Figure 15.2.

Some examples of the more important groups of these enzymes found in curds are given below, and it is important that each group may contain many individual enzymes which are quite specific in their activity:

Proteases acting on proteins to produce peptides
Peptidases acting on peptides to release amino acids
Aminases acting on fatty acids to produce amino acids
Decarboxylases acting on amino acids to produce amines
Transaminases acting on amino acids to produce keto acids
Deaminases acting on amino acids to produce keto acids
Decarboxylases acting on keto acids to produce aldehydes

Although it would be technically possible to identify many of the specific enzymes and their breakdown products, it is quite futile to imagine that a true mature cheese could be created in a 'test tube'. Even the identification of allegedly 'important' flavour compounds is of little practical value to the cheesemaker, particularly as most quality cheeses are produced quite simply by growing selected bacteria in the curds.

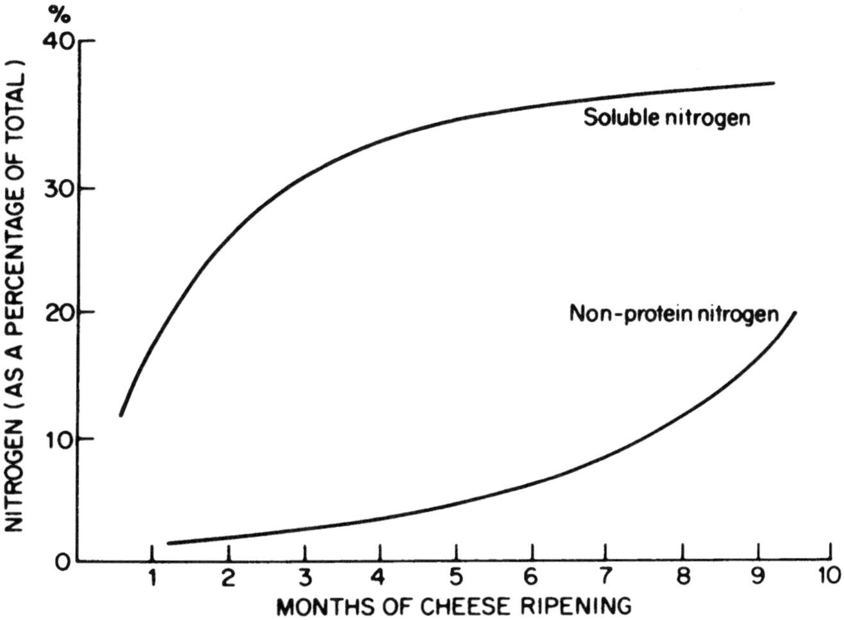

Figure 15.2 Graph showing the probable breakdown of protein in cheese during ripening as indicated by the production of soluble and non-protein nitrogen.

Nevertheless, the addition of certain enzymes has proved valuable. Most rennets, for example, contribute to the proteolysis that occurs during maturation, even though $<10\%$ may be retained in the final cheese (Farkye, 1993). Other methods of increasing flavour in cheese at an earlier age have involved the use of cheese slurries, in which the growth of micro-organisms has already provided the enzymes necessary to give the slurry a high degree of flavour and aroma. Such methods may be useful under conditions where cheese is urgently required, but not where cheese needs to be stored for a long time.

The list of groups of enzymes shows that amino acids are often the final products of their reactions, and many of these contribute to the general background flavour, although some with bitter flavours may have more impact than others. Free amino acids may, to some extent, indicate the degree of ripening in cheese, and it has been suggested that the pattern of amino acids could be employed to 'fingerprint' a variety. However, the quantities of amino acids found in any cheese can, in reality, only suggest a general trend.

One of the functions of the free acids is to form a 'pool' from which other components of flavour or aroma are formed, as illustrated in Figure 15.3. Many of the amines are formed by decarboxylases and, in addition to those highlighted, putrescene, cadaverine, histamine, taurine and tryptamine have been identified.

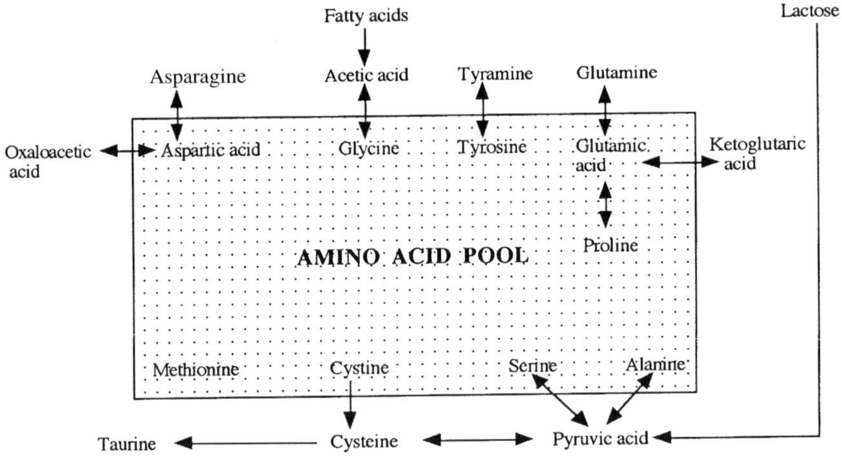

Figure 15.3 Diagrammatic representation of a possible amino acid pool developed during cheese ripening.

Although the 'aftertaste' may differ slightly, most amino acids have, depending on the individual taster, a quite distinctive first taste, which can be described as follows for the amino acids shown:

Bitter	– methionine, histidine, lysine, tryptophan, leucine, isoleucine, arginine, phenylalanine and tyramine.
Sweet	– serine, glycine, alanine, hydroxyproline, proline, aminobutyric acid, valine and threonine.
Broth-like	– aspartic acid and glutamic acid.
Little or none	– asparagine, glutamine and tyrosine.
Rubber-like	– cystine.

Peptides arising from the degradation of proteins exhibit flavours according to those amino acids which are terminal in the peptide chain. Thus, the peptides leucine–glycine–leucine–leucine–tyrosine, or alanine–tryptophan–phenylalanine–lysine–arginine, or phenylalanine–phenylalanine–valine–alanine–proline, are all bitter tasting peptides. Cleavage of the terminal acid by other enzymes removes the bitter element, and this happens when a cheese, which is bitter at one stage, loses its bitterness when fully mature. Interactions are also extremely important, and an oft-quoted example is the reaction of hydrogen sulphide produced from cystine with methionine to give methanethiol; all four components are capable of contributing to flavour.

The keto acids produce compounds leading to distinctive flavours or aromas, and are formed from lactose, fatty acids, proteins or amino acids. Various aromatic and aliphatic ketones are similar to the aldehydes in regard to their reactions, and both groups have intense flavour or aroma characteristics.

15.4 Lipids

The most important class of lipids in cheese is the triglycerides, upon which lipases act to give mono- and di-glycerides, and ultimately free fatty acids. However, many bacteria are unable to break down triglycerides so that, unless a system is present to release mono- and di-glycerides, the level of free fatty acids in the cheese remains low. In Cheddar cheese, for example, the levels of C_4–C_{18} fatty acids may be less that 0.1% at 3 months, and <0.2% even after 12 months (Law, 1984; Farkye and Fox, 1990); indeed, values above 0.2% can give rise to unacceptable, rancid flavours in hard-pressed cheeses. This latter figure equates to around 500 mg/kg of cheese and, given that the aroma or taste threshold for free fatty acids will vary from 0.3 to 100 mg/kg, depending upon the acid concerned, it is reasonable to propose that the release of free fatty acids is an important facet of maturation.

In blue-veined varieties like Stilton, the high degree of lipolysis is due to the activity of the so-called 'acid' lipase, which can give levels of free fatty acids three times higher than those found in Cheddar or similar cheeses. In some Italian varieties like Parmesan and Provolone Piccante, lipolysis is mainly due to esterases and lipases added in the form of pastes or powders.

Overall, Farkye (1993) summarized the involvement of lipids in flavour and aroma formation as:

- liberation of free fatty acids from mono-glycerides
- oxidation of free fatty acids to give β-keto acids
- decarboxylation of β-keto acids to ketones
- reduction of ketones to alcohols

and, of these, the free fatty acids, ketones and alcohols all contribute to the complex sensory properties of cheese or are further metabolized to relevant compounds.

It is important though that the rate and extent of all these reactions involving carbohydrates, proteins and lipids can, to some extent, be modified by the actions of the cheesemakers for, as mentioned earlier, a number of physico-chemical factors determine the path of maturation.

15.5 Curd moisture

The cheesemaker has, during the early curd-making operations, controlled the moisture content of the green cheese entering the maturation stage, while the humidity of the store helps to further control the moisture status in the cheese (unless covered by moisture-proof wrapping, i.e. block cheese).

The water in cheese is in a number of states: (a) bound in the structure of a component of curd, i.e. the protein; (b) loosely held by attractive forces in contact with the curd particles, e.g. hygroscopic moisture; (c) moisture which is free to move within the curd and which acts as a carrier for solutes. These moisture states in curd are in equilibrium, so that the picture changes as reactions in the curd release more bound water or require water molecules for a reaction. The same is true for the free moisture and, as the volume of free water controls the concentration of solutes, so those solutes can affect the metabolism of bacteria by exerting an osmotic effect on the cell constituents. Thus, the growth of micro-organisms is more prolific in high moisture curds than in low moisture curds, and the rate of maturation is faster in high moisture cheeses. Conversely, the evaporation of moisture from the curd rind leaves an accumulation of solutes at the rind, including calcium salts, a change that could be advantageous in curtailing microbial activity at the cheese surface.

15.6 Temperatures during maturation

The cheesemaker has control of another factor, temperature, during the manufacture and storage of a cheese, and, provided that other conditions are satisfied, temperature controls both growth of bacteria and the rate of biochemical reactions in the curd.

Temperature affects the generation times of bacteria according to the phase of growth. Thus, during the logarithmic phase, the generation time of some bacteria, e.g. *Escherichia coli*, can be as short as 20 min at 37°C, but this figure increases to 75 min at 25°C. Similar generation times have been shown by *Lactococcus lactis* sub-sp. *cremoris* and *L. lactis* sub-sp. *lactis*. A population of *Lactobacillus helveticus* takes approximately 75 min at 40°C to double in number, while a comparable increase for *Streptococcus thermophilus* takes 57 min at 40°C.

In general, low temperatures slow down both bacterial growth and any biochemical reactions occurring in the curd, while warmer temperatures speed up these activities. Indeed, the activity may be increased so much that the character of a cheese may be changed adversely, for it is relevant that not all reactions increase at the same rate, and the relative dominance of a particular pathway could give rise to off-flavours. Thus, a cheese matured slowly at low temperature (4°C) will not have the same flavour and aroma as a cheese held at 15°C, and hence the use of high temperatures must be carefully controlled. For example, in the early stages, Emmental cheese is held at low temperature in order for the lactic bacteria to grow and for some of the intermediate reactions in the curds to take place. Later, the store temperature is increased so that the propionic bacteria can grow and produce the flavour characteristic of the variety as well as the essential 'eyeholes'. If the low-temperature storage was to be maintained, a so-called 'blind' cheese, defective in flavour, would be produced, for along with the propionic acid generated during the fermentation, lactic, succinic and acetic acids are produced and add their contribution to the flavour of Emmental. In addition, alanine, aspartic acid, methionine, serine, arginine, tyrosine and cystine are decomposed by the propionic bacteria at the higher temperatures, and this action again influences the flavour of Emmental cheese.

With the blue-veined cheeses, the situation may be reversed, so that maturation at low temperatures is preceded by a short period in a warm store. This initial period allows the mould to spread throughout the curd while, later, a whole range of acids, such as lactic, malic, formic, acetic, isobutyric, caproic, caprylic, capric, pyruvic and ketoglutaric, are formed which contribute to flavour.

15.7 Acidity of the curds

As with temperature, the pH value of curds influences both the growth of micro-organisms as well as the rate of biochemical reactions, and these values can vary from pH 4.7 to 5.0 in acid cheeses, and from pH 4.9 to 7.0 in mould-ripened cheeses like Brie.

The early cheesemaking operations govern the rate of acid production up to the point at which salt is added, and this acidification, coupled with the loss of lactose, determines the lowest pH value that will be achieved by that particular cheese. Thereafter, the activity of the bacteria, and more especially yeasts and moulds, cause the curd components to break down into neutral or even alkaline components which cause the pH to rise. Where proteolytic activity is strong, the curd may reach pH 7.8 or even higher as the result of ammonia released under these conditions. The gradual change in pH value from stage to stage enables maturation to proceed in a somewhat stepwise fashion so that, at any one time, the flavour and aroma compounds reach a specific level of intensity. As the process proceeds, some of the first-formed flavour (or aroma) compounds may be metabolized, and the harsher or more strident sensory characteristics are replaced by more acceptable mellow flavours.

Buffering capacity is primarily a function of protein content and pH, as shown in Figure 15.4 for milk. Due to the high levels of protein, the buffering capacity of the curds is high in the early stages of maturation, but the gradual breakdown of the proteins reduces their impact. In the harder cheeses, like Emmental, Tilsiter and Svecia, there appears to be a greater degradation of α-casein than β-casein, whereas in soft cheese, e.g. Camembert, it is the β-casein that is broken down earlier; in blue-veined cheeses, the caseins seem to be degraded almost equally. In the acidic, sour milk cheeses, decarboxylation of glutamic acid, lysine, tyrosine, aspartic acid and arginine occurs early, while ornithine is transformed into putrescene and arginine to citrulline; valine is likely to decrease gradually.

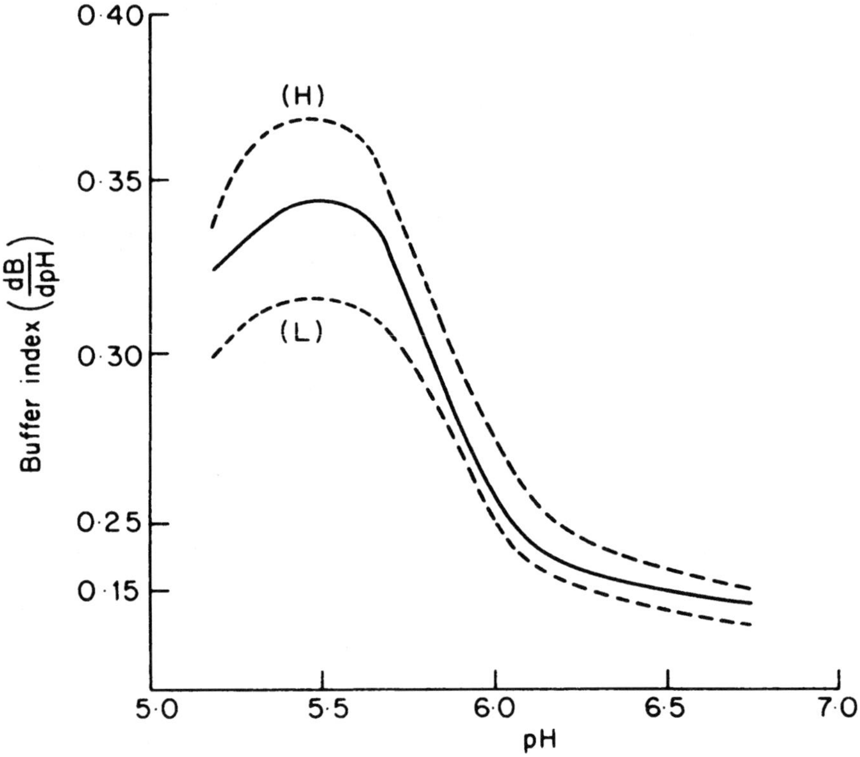

Figure 15.4 The buffer curve for an average cheese milk, showing a higher buffering power in the region between pH 5.2 and pH 6.0. The dotted lines indicate changes expected for high (H) and low (L) quality milks.

The activity of enzymes is also affected by pH, although time, concentration and solubility of reactants, and the oxygen content of the curd are also controlling influences. Thus, each enzyme has an optimum pH for maximum activity, and this level declines either side of the optimum. It may be important also that some enzymes are active within two regions of pH, i.e. at specific alkaline and acid values, even though the end-products may differ in each region of activity. In most hard cheeses, the enzymes relevant for maturation are active in the region pH 4.9–5.5, and these low pH values tend to stimulate hydrolysis of fats and the production of soluble nitrogenous substances, amino nitrogen and ammonia. In soft cheeses, the enzymes tend to be more active in the pH ranges from pH 5.3 to 6.0.

15.8 Inhibitory substances and oxygen

The careful selection and monitoring of milk supplies, control of the bacterial flora of milk by effective heat treatment, and the selection of appropriate starter cultures can assist the cheesemaker by reducing the risk of incorporating inhibitory substances into the cheese curd and/or overcoming the action of any that are present.

Antibiotics from udder treatments, nisin from lactococci, or inhibitory substances produced by other organisms growing in the cheese milk can cause an inhibition of the growth of starter bacteria and, sometimes, block enzymic reactions. In addition, the deliberate introduction of compounds like sodium nitrate and sodium chloride to milk or curds for the control of gas-producing or acid-producing bacteria, respectively, can influence the rate and direction of maturation, either directly or through the oxidation–reduction potential in the curds.

The oxidation–reduction potential in milk is brought about by the transfer of oxygen (oxidation) or of hydrogen (reduction) from one component to another. Some enzyme systems are able to oxidize and not reduce, while others can do both.

The major components of milk, protein, fat or lactose do not, as such, influence the oxidation–reduction potential, but ascorbates, lactates and riboflavin help to maintain stability. When bacteria grow they consume oxygen and thus change the oxidation–reduction potential. This change, which is an indication of biochemical activity, can be measured by colour changes in methylene blue, resazurin, or 2,3,5-triphenyltetrazolium chloride solutions if added to milk or curds. The change in potential when a reaction takes place indicates which direction the reaction is taking, i.e. an oxidation or a reduction. Thus, the oxidative deamination of alanine gives pyruvate and ammonia, while the reduction of pyruvate gives lactic acid.

Normally, *Lactococcus lactis* sub-sp. *lactis* is regarded as a lactic acid producer, but it also produces acetic acid after the lactose has been broken down to galactose and glucose; carbon dioxide may be also released during this fermentation. If the same species is grown under almost anaerobic conditions, acetic and formic acids are produced along with ethanol, diacetyl, acetone and 2,3-butanediol. The anaerobic bacteria require a negative oxidation–reduction potential for active growth.

The detectable respiration of cheese curd reaches a maximum in 6–9 days, and then declines rapidly. No decarboxylation reactions occur under these aerobic conditions but, thereafter, the redoxases catalyse the process of fermentation, leading to a complex mixture of substances formed by the decomposition of primary breakdown products.

15.9 Aroma in cheese

The aroma and the flavour of cheese are often not easily distinguished from each other, since each can have an influence on the other.

The aroma of cheese arises mainly from the volatile components released by the curd as ripening proceeds, and these components include esters, fatty acids, aldehydes, ketones, alcohols, amines, hydrogen sulphide and ammonia; the aldehydes, ketones and alcohols are probably the most important (Law, 1997). The deamination of methionine has been regarded as a major source of sulphydryl groups, but the degradation of the sulphur-containing amino acids (cystine, cysteine and methionine) releases mercaptans, acetic acid, formic acid, ammonia, carbon dioxide and hydrogen; ammonia reacts with acids to give salts that are slightly bitter. In addition, methionine may be transformed into methional, which has a bread-like taste and is regarded as a carrier of aroma, so giving a 'muted' character to the aroma.

Micrococci do not always show lipolytic activity when growing in milk, but lysis of the cells can release intracellular enzymes (lipases) which produce volatile fatty acids which contribute to aroma. Reactions in the citric acid cycle also provide precursors for important aroma compounds, i.e. acetolactic acid which forms acetylmethylcarbinol and diacetyl, while acetone, 2-undecanone, 2-tridecanone, formaldehyde, 3-methyl-tripropanol, 3-methyl-butanol and propionaldehyde are also found in the aroma fraction of cheeses at exceedingly low levels measured in parts per million. At higher levels, some of the gaseous sulphydryls can cause defects by bleaching the colour of annatto used in some varieties of cheese and, as the bleaching tends to be in patches, the product acquires a mottled appearance.

15.10 Flavour in cheese

Most bacteria involved in cheese ripening produce a number of compounds during fermentation, although they may be regarded as a major producer of one component, i.e. the sub-species of *Lactococcus lactis* are employed to produce lactic acid. Similarly, propionibacteria are known for the synthesis of propionic acid, but they can also liberate acetic acid and carbon dioxide from the metabolism of lactic acid and lactates.

Unwanted flavours or aromas may arise from the metabolism of adventitious bacteria like *Klebsiella aerogenes* and *Escherichia coli*, even though these coliforms can produce sufficient lactic acid to produce a curd. In small numbers, coliforms are regarded by some cheesemakers as useful for enhancing flavour since they deaminate glycine, alanine and glutamic acid, while even *Enterococcus faecalis* is sometimes added because it uses tyrosine decarboxylase to produce large quantities of tyramine. Bitter-tasting peptides can contribute to poor flavour as well, but these bitter peptides may be broken down during the later stages of maturation and lose their bitterness.

The keto acids, which are precursors of both ketones and aldehydes, are produced by enzymes secreted by bacteria and *Proteus vulgaris*, a frequent contaminant in milk, is known to secrete enzymes which deaminate at least

11 amino acids to their respective keto acids. The oxidation of short-chain fatty acids (C_4–C_{10}) also produces ketones through transformation of their keto acids, and large quantities of fatty acids are released through the action of lipases secreted by the moulds, *Penicillium glaucum*, *P. roqueforti*, *P. candidum* and *P. camemberti,* growing in blue-veined and white mould-ripened cheeses. The free fatty acids liberated by these lipases are oxidized to ketones, including acetone, propylmethylketone, amylmethyl-ketone, iso-butylmethyl-ketone, methylamyl-ketone, 2-butanone, 2-pentanone, 2-hexanone, 2-octanone and 2-nonanone; the lipase enzymes from the penicillia are active at 37°C and between pH 5 and 6.

The lipases arising from the bacteria in raw milk are mainly from the genera *Pseudomonas*, *Achromobacter*, *Alcaligenes* and *Serratia*, and while many of the bacteria may be destroyed during heat treatment, their enzymes are fairly heat resistant and remain active. The flavour of cheese is influenced by the short-chain fatty acids (C_4, C_6 and C_8) rather than the long-chain acids. However, both the long-chain fatty acids and the proteins in the curd form a background for the more specific flavour compounds produced during maturation.

In spite of the large number of flavour and aroma compounds derived from the degradation of both fat and protein, only small amounts of these latter components are utilized, and a considerable amount of fat and about two-thirds of the protein remain intact. In fact, the degradation of the protein and fat is mainly responsible for changes in the body and texture of the maturing cheese. Indeed, the progress of maturation can be followed in some cheeses by monitoring the extent of breakdown of the proteins into soluble and non-protein nitrogen.

15.11 Additional flavours and aromas

The groups of flavour or aroma agents which may be added to cheese include: (a) those which are added to impart aroma/flavour to the cheese, such as herbs or spices, and (b) those agents which are nutritional foods in their own right (e.g. ham, vegetables) and which are enclosed in a soft base of cheese, e.g. cottage and lactic curds.

Chopped herbs, fresh or dried, extracts of the herbs, and even entire seeds have been incorporated in curds at moulding and before the curds are pressed. Occasionally, the herbs may be mixed with minced, mature cheese curds which are re-pressed into shapes or containers (ceramic pots). In a few instances, the aroma/flavouring agent is pressed onto the surface of the cheese, e.g. the application of grape must to the coat of Tomme au Marc cheese, to provide additional flavour/aroma. An alternative is to allow the cheese to acquire aroma from the packaging material, as when Mont d'Or cheese is packed in resinous pine boxes.

Herbs which have been used in cheese include garlic, chives, sage, mint, melilot, tarragon, fennel, marjoram, basil, parsley and even lavender, but when herbs are incorporated in the curds, they can cause an unappetizing brown discoloration; a green colour may have to be added to mask the brown colour. Spices may be mixed with the curds or implanted on the rind, and some examples include caraway, clove, cumin, cinnamon, ginger, juniper, mace, nutmeg, aniseed, pimento, black pepper, mustard, poppy and celery. Extracts of spices, or even synthetic flavour/aroma agents have been used, but great care is needed to avoid the development of harsh flavours. Dry spices (grounds) are often mixed with salt to flavour cheese, as this provides an easy method of mixing the spice into the curd.

15.12 The use of wood smoke

It has been traditional to smoke certain varieties of cheese but, depending upon market demand, the number of cheeses of a particular variety that may be smoked will vary from time to time. Two methods of applying a smoked flavour are employed. In the traditional method, the cheeses are held in a smoke-charged atmosphere but at comparatively low temperatures, with oak or applewood shavings being allowed to smoulder and provide the smoke. For some low-volume cheeses, smoke is regarded as a preservative as well, but for the most part it is employed to add a distinctive flavour to a cheese. The second method is to employ liquid 'condensed smoke', with the liquid preparation being added either at milling or, like spices, along with the salt or immersing the cheese in a solution of smoke liquor.

Where smoking retains a preservative function, the effect of the process is to bring fat to the surface of the cheese, evaporate moisture and deposit smoke vapours containing phenolic substances; the latter have anti-microbial and anti-oxidant characteristics, and may impart an attractive colour to the cheese. The fat on the surface is also a deterrent to mould growth if the cheese is correctly stored.

15.13 Liqueurs, wines and beers

Culinary preparations for the dining table have often tended towards the use of liqueurs and wines in their preparation, so that it is not surprising that cheeses have been dosed with various beverages prior to their final maturation for consumption. Alcoholic beverages like beers, wines and liqueurs have been added to the raw cheese curd, or the whole cheeses have been immersed in the liquid. Many examples have a long history of usage, and include Ilminster beer cheese (UK), Beer cheese from Germany, Cider whey cheese, Formaggini di Montpellier (white wine – Italy), Cabrion (liqueurs –

France) and Red Windsor (port – UK) and even the traditional Stilton cheese does not always escape a dosage of port onto its surface. The mixing of flavours – beverages and cheese, vegetables and cheese, spices and cheese, meats (ham) and cheese – seems to break with tradition, but satisfies the taste of modern industrial societies for snack diets.

15.14 Cheese bases

The use of a bland cheese base, such as lactic or Cottage cheese, to act as a carrier for herbs, vegetables or chopped cooked meats is probably a spin-off from the use of processed cheese in the same manner. However, it is worth noting that processed cheese preparations are heat treated and thus safe from the public health standpoint, while in spite of the low pH (4.6–5.1) of the curds, the use of natural cheese curds could involve a public health risk unless the product is kept under correct refrigeration.

Typical mixtures use cheese as a base for the addition of lettuce, chives, onions, spinach, watercress, potatoes, carrots, coleslaw and chopped ham, but some traditional preparations of cheese and eggs have been used for many years, e.g. Eierkäse (Germany), Aggost (Finland), Osteaeggekage (Denmark) and the yolk of egg only in Dotterkäse (Germany).

15.15 Cheese slurries

As mentioned earlier, the urge to produce more flavour in cheese curds resulted in the production of liquid slurries with highly developed cheese flavours (Kristoffersen *et al.*, 1967) and, later, Singh and Kristoffersen (1971) extended the use of the slurry method to produce a system for the accelerated ripening of certain cheeses, e.g. Swiss cheese.

Cheese slurries for flavouring heat-processed cheeses, or freeze and spray-dried powders for possible use in bakery or other made-up foods have been developed, and the flavours can be modified to simulate Cheddar, strong Brick cheese, or even Romano depending upon the techniques employed (Kristoffersen *et al.*, 1967). The use of commercially available enzyme systems in cheese slurries was investigated by Kosikowski and Iwasaki (1975) and, although rancidity, burnt flavours and bitterness developed in some instances, neutral proteases and lipases gave rise to cheese flavours that lacked bitterness.

More novel uses of enzymes to produce flavourings imitating those of traditional cheeses have also been suggested. Jolly and Kosikowski (1975) used skim milk concentrated to 45–50% TS by ultrafiltration, which was then mixed with a preparation consisting of cream (40% fat) inoculated with lipase from *Aspergillus oryzae*, spores of *Penicillium roqueforti* and 0.1% of a commercial starter culture. Salt (2%) was added before the mixture was heated for 3 min at 77°C and finally chilled. The flavour and aroma of the product was similar to blue cheese. However, while such systems may find a niche in the ingredients industry, the mainstream cheesemakers will be constrained by tradition and/or regulations to follow a more prescribed route to achieve flavour.

15.16 Accelerated maturation of cheese

Cheesemakers have known for many years that the storage of cheese at elevated temperatures (18°C) accelerated the maturation process, but that unless the cheeses were moved into cool storage (5°C) as soon as the desired intensity of flavour had been achieved, the cheese would deteriorate quickly. For example, Freeman (1959) recommended that Cheddar cheese should be stored at 15°C for 4–6 weeks only, and that thereafter the temperature should be reduced to 5°C.

This urge to accelerate the maturation of cheese was motivated by the financial burden imposed on factories by the need to keep hard-pressed cheeses in-store for long periods, and accelerated maturation offered a means of shortening the time required before the cheese could be sold.

Maturation is achieved largely through the agency of enzymes (Law, 1980), and the incorporation of lipolytic enzymes (lipases) into cheese is relatively easy. However, because proteolytic enzymes attack the casein in milk and interfere with rennet activity, proteolytic enzymes have to be added to the curds and not to the milk. Salt is a convenient vehicle for incorporating enzymes into cheese curds before moulding (Law and Wigmore, 1982) but, even then, the choice of proteolytic enzymes is more difficult. For example, Law and Wigmore (1982) found that food grade proteinases used in cheese gave rise to bitter, soft, crumbly curds, but that a neutral bacterial proteinase (*Bacillus subtilis*) gave a cheese with an enhanced flavour when stored at 18°C. Using the same enzyme, Ramet *et al.* (1982a) made Saint Paulin cheese from recombined milk in which proteolysis was improved over normal cheese, and the same authors (Ramet *et al.*, 1982b) achieved similar effects with UF concentrated skim milk with added fat. The use of intracellular enzymes from starter streptococci (Law and Wigmore, 1983) could offer a more natural approach to accelerating cheese maturation but, even today, control over the process on a routine basis is proving difficult to achieve.

In some countries, there may also be legal barriers to the addition of cheese slurries or other materials to cheese curds, so that most cheesemakers will have to rely on control of the cheese room to achieve the desired flavour and aroma in their products. For this reason, a detailed consideration of the biochemistry of aroma/flavour development is not entirely relevant in the present text, the more so as the volume edited by Law (1997) is devoted, in large measure, to this aspect of cheesemaking.

References

Farkye, N.Y. (1993) Cheese maturation. In *Encyclopaedia of Food Science, Food Technology and Nutrition* (eds R. Macrae, R.K. Robinson and M. Sadler), Academic Press, London, pp. 810–817.

Farkye, N.Y. and Fox, P.F. (1990) *Trends in Food Science and Technology*, **1**, 37–40.

Freeman, T.R. (1959) Bulletin Kentucky Agricultural Experiment Station, No. 667.

Jolly, R.C. and Kosikowski, T. (1975) *Journal of Dairy Science*, **55**(9), 1272.

Keceli, T. and Robinson, R.K. (1997) *Dairy Industries International*, **62**(4), 29–33.

Kosikowski, F.V. and Iwasaki, T. (1975) *Journal of Dairy Science*, **58**(7), 963.

Kristoffersen, T., Micolajcik, E.M. and Gould, I.A. (1967) *Journal of Dairy Science*, **50**(3), 292.

Law, B.A. (1980) *Dairy Industries International*, **45**(5) 15–22.

Law, B.A. (1984) in *Microbiology and Biochemistry of Cheese and Fermented Milks* (eds F.L. Davies and B.A. Law), Elsevier Applied Science, London, pp. 209–229.

Law, B.A. (1997) *Microbiology and Biochemistry of Cheese and Fermented Milks*, 2nd edn (ed. B.A. Law), Chapman & Hall, London.

Law, B.A. and Wigmore, A.S. (1982) *Journal of Dairy Research*, **49**(1), 137.

Law, B.A. and Wigmore, A.S. (1983) *Journal of Dairy Research*, **50**(4), 519.

Ramet, J.P., El-Mayda, E. and Weber, F. (1982a) *XXI Int. Dairy Congress, Moscow*, Vol. 1, Book 1, 437.

Ramet, J.P., El-Mayda, E. and Weber, F. (1982b) *XXI Int. Dairy Congress, Moscow*, Vol. 1, Book 2, 546.

Singh, V.K. and Kristoffersen, T. (1971) *Journal of Dairy Science*, **54**(3), 349.

16 Cheese faults and cheese grading

16.1 Faults in hard-pressed cheeses

Cracks may develop in the rinds of cheeses that are not waxed or shrink-wrapped, and such openings may allow moulds to penetrate into the curds or cheese flies (*Piophila casei*) to lay eggs in the cracks so that the cheese becomes infested with maggots. Such cracks, which may be deep or exist as a network of small superficial lines, are caused by the curd shrinking during maturation, especially at low humidities (60% RH). The cause of such problems may be:

- the moisture or acidity of the curd is too high at pressing
- the curd is too cool (below 24°C) at pressing, so that the separate particles do not link up and cracks develop
- there is a lack of tat in the milk, especially in spring, with the result that the curd does not hold sufficient moisture and the hard curd may crack.

Sealing of rind with hot water helps, at least in Cheddar cheese, to prevent cracking of the rind, but the use of resins for coating the rind is not necessarily a cure.

Discoloured curd, whether white or coloured, is a defect caused by casual fermentations that are not intended, and these fermentations are usually accompanied by unpleasant flavours or aromas. The discoloration may exist in streaks or patches, and appears as a mottled surface when the cheese is cut open. The cause of the defect is usually contamination of the milk or curds, or uneven ripening of the curd before pressing. Thus, cold spots in the vat of curd can cause serious discoloration in the final cheese, and the mixing of curds from different vats during matting, milling or pressing may also affect the colour. Where curd fines retrieved from whey filtration are mixed with a more acid curd, a mottled colour or open curd may develop in well-aged cheese. Similarly, sectors of curd with excess salt do not mature properly, but give rise to harsh flavoured, hard, open-structure patches in the cheese. Parcels of cheese of uneven character may also arise in mechanical systems where curds of different maturities (pH) are mixed together. Sour, over-acidic cheeses may become hard, dry and crumbly with split rinds, and whey may leak from the cheese in the early stages. Sweet cheeses become, at first, soft, elastic and somewhat 'corky', but later become harder and more 'corky' with objectionable flavours; such cheeses may leak whey and become deformed under warm storage conditions.

To avoid problems of this type, the cheesemaker should monitor closely the acidity of the milk at rennetting, and the rate of acid development during curd formation; a lack of salt may make the defects worse in respect of sweet curds. Sometimes, defects are the result of inexperience and/or an inability to follow recipes, although poor quality milk or an inactive starter may confuse even the most competent cheesemaker; free whey left in the curd after pressing will ferment and may also cause defective aromas, flavours and colours.

While an open texture is desirable in some cheeses (Cheshire and blue-veined varieties), it is a defect in a close-textured cheese like Cheddar. This defect may arise in cheeses made from milk contaminated with gas-producing organisms, such as coliforms and yeasts, or it may result from the use of curds with acidities in excess of normal. Inadequate pressing of the curd, the production of a weak-bodied cheese with excess moisture or the use of starter that gives a very high rate of acid development are all features of the cheesemaking operation likely to give rise to an open texture in the form of slits or fractures.

Adherence to the rules of cheesemaking will usually correct these defects, but another fault of a similar type can arise during mould filling and pressing. This fault appears when a high initial pressure seals the rind of the cheese and so prevents whey from leaving the cheese. During maturation, the whey evaporates and the shrinking curd leaves open spaces. In addition, some of the whey remaining in the curd may have fermented, and the resultant gas forms holes in the curd and a so-called 'heaving' cheese results; unpleasant odours or flavours may also develop.

Cheeses like the Gruyère, Tilsiter and similar cheeses that have 'eyeholes' as the normal condition may also suffer in so far as the 'eyeholes' are irregular, some large, some small. For example, if the cheese has been cooled too much at pressing, or if the pressing was irregular, then the 'eyeholes' in the top half of the cheese may be small, while those in the bottom half may be large and irregular. A similar defect can arise if the turning of cheese fresh from the press or brine bath is neglected.

A defective body is perhaps one of the more common defects. A soft, soapy, pasty or sticky body is usually accompanied by low keeping quality, and the defect arises from lack of acidity in the curd, especially when the whey is removed. However, a too rapid rate of scalding may also lock whey in the curd particles by forming a hard skin around each particle, thus giving rise to a soft curd containing too much moisture (whey). In some cases, this entrapment causes excess acid production and an over-acid cheese or if the rate of acid development is slow, then the resultant curd is soft and pasty; a slow rate of acid production by a weak starter culture is likely to result in a weak pasty cheese.

Milk of better quality (in autumn) requires a slight adjustment to a standard recipe. The curds may be cut finer, and higher acidities and scald temperatures may need to be used. In the spring, when the milk may be of lower quality, lower acidities and scald temperatures will often cure emerging defects (see Cheshire recipe). A lack of fat in the milk or a loss of fat into the whey may produce a cheese which is rough, coarse, corky and is not acceptable to the palate. Such cheeses may also arise from a lack of acid production at some stage in the curd making (not necessarily at the end). Strict attention to storage temperatures, hygiene and correct curd manipulation, including pressing, are necessary to avoid many of these defects.

Feeding and other influences during milk production may also be responsible for milks which are not suitable for cheese made to a standard recipe. Cheese made from milk low in casein and/or from fodder lacking lime (CaO) will need special treatment and adjustment of the recipe. The acid produced is not taken up so quickly by the protein and the pH values decrease more quickly. Furthermore, the lack of fat that often accompanies these situations means that the curd does not hold so much moisture. Curd from these milks does not form a firm coagulum without added calcium, and the acidities need to be lower than the recipe suggests.

As mentioned earlier, taints in milk are often carried over into the cheese from weeds and other foods eaten by animals. Silage, sugar beet tops and brewer's grains are well known as being responsible for taints in cow's milk, e.g. butyric acid, fishy and malty taints, respectively. The surface growths of bacteria, yeasts and fungi on cheese rinds or coats can give rise to proteolytic activity in patches and softening of the curd at these points may lead to the formation of a putrid, bitter flavour in the cheese itself.

16.2 Faults in blue-veined cheese

The defects in blue-veined cheese are similar to the defects in hard-pressed cheeses as far as the curd is concerned (except that most blue-veined cheeses are not pressed and the curds remain open). Thus, the aim of the curd production stage in blue-veined cheese manufacture is to provide a medium in which moulds can grow, so that the mould can spread evenly over the surfaces of the curd particles. If the curd is in too large pieces at moulding (hooping), then large patches of white curd will be visible with no growth of penicillia.

Unfamiliar moulds, e.g. *Aspergillus niger*, and other coloured moulds may also contaminate the cheese curds; this problem mainly arises in cheeses manufactured under unhygienic conditions. *Penicillium roqueforti* and *P. glaucum* will not grow in high concentrations of carbon dioxide, so blue-veined cheeses are pierced to allow carbon dioxide to escape and oxygen to gain entrance. This essential procedure means that bacteria and moulds from the surface of the cheese may be carried into the curd, and their growth may form a discoloured line in the cheese; if the piercing rods are too large, a brownish growth of casual moulds may be clearly visible within the cheese.

According to the type of cheese and the hygiene of the stores, the coats of maturing blue-veined cheeses may acquire a number of different organisms. At first the cheese coat has the same organisms as the curd, but later other organisms begin to grow. *Penicillium*, *Oidium* and *Torula* species are components of the normal flora, but it is the growth of such organisms as *Proteus vulgaris*, *Streptococcus liquefaciens*, *Oidium lactis* and *Brevibacterium linens* which bring about softening of the curds near the coat and cause defective areas of maturation. The growth of coloured moulds, e.g. *Sporendonema casei*, a red mould, *Oidium auranlicum*, a yellow mould, or *Sporotrichm aureum*, a dark mould, do not appear to cause serious defects. Indeed, cheese buyers often look to the growth of some of these moulds as a sign of maturity and quality, especially in surface mould-ripened varieties (e.g. Livarot).

Excessive mould growth over the surface of a mould-ripened cheese which has a high moisture content allows both proteolysis and lipolysis of the curds to take place freely under the rind. If maturation occurs in a warm, humid atmosphere, then the coat may slip, leaving a sticky undercoat and the cheese will be unsaleable.

Since the coats of blue-veined cheeses tend to be rough, cheese flies (*Piophila casei*) lay eggs in the coats; the maggots then hatch out and live on the cheese curd; by contrast, the cheese mite (*Tyroglyphus siro*) infests dry-coated, mature cheeses.

16.3 Faults in surface mould-ripened cheese

Surface mould-ripened cheeses (e.g. Camembert) are soft curd cheeses, ripened from the outside inwards due to mould growth (*Penicillium camemberti*, *P. candidum*) on the outside allowing enzymes to penetrate into the curd. Proteolytic action turns the curds yellow and soft, and when the white central core has turned yellow the cheese is mature. The defects in these soft cheeses follow a similar pattern to other cheese as regards defects in the curd, e.g. taints, off-flavours and discoloration due to contaminants. In addition, this type of cheese can suffer from high acid development which produces a hard, shrunken, dry cheese, or low acidity which gives a sweet, soft, moist curd which may become proteolytic and liquefies in a short time.

In earlier years, Camembert cheeses were acceptable with blue or blue–grey mould growths, but recently the white mould *P. candidum* (sometimes called *P. caseolyticus* in older cheese books) has become normal for both Camembert and Brie-type cheeses.

Although the Camembert curd is close textured, automated techniques of manufacture, such as the Hutin-Stenne process, are reported to produce initially a more granular curd which, on maturation, loses the granular texture and is accepted as a normal Camembert cheese. The Alpma coagulators give an equally acceptable product (Shaw, 1993), and the quality of surface mould-ripened cheeses has improved considerably over the years.

16.4 Faults in smear-ripened cheese

The faults in the curd of smear-ripened cheeses are similar to those encountered in other types of curd, and include over-acid curds, sweet curds, lack of salt penetration and taints arising from fodder and contaminant bacteria. The growth of the dominant 'smear' organism, *Brevibacterium linens*, is controlled by the humidity of the maturation room and the temperature. Typically, Limburger cheese ripens at 15°C and 90–95% RH, but a milder flavoured cheese, Romadur, is ripened at 5–10°C so that smear development is less.

Adequate salting (dry or brine) and constant rubbing of the cheese rind provides the best conditions for satisfactory growth of the smear coat, but inadequate hygiene can lead to the growth of yeasts, moulds and other micro-organisms on the coat which may result in defective flavours. As *B. linens* is salt tolerant and survives well in the store rooms, it is not usually necessary to inoculate new cheeses from laboratory sources.

16.5 Aspects of quality

16.5.1 Chemical composition of cheese

Mention has already been made of the international meetings held to consider setting up Codes of Practice for milk and milk products. The first meeting in 1957 was under the auspices of the FAO, and Belgium, Denmark, Sweden and Switzerland took part. The WHO and the FAO have joined in giving combined reports since 1963 and in 1969 the Codex Alimentarius Commission issued their procedural manual (CX 8/7, 1969). Some 76 countries later accepted a Code of Principles issued by the Commission and in 1970 the Permanent Council of the International Convention of Stresa gave a list of 73 cheese names and recognized Roquefort cheese as French, and Sbrinz cheese as Swiss. The conditions for the use of cheese names outside the country of origin were also agreed.

Since that time most countries have issued regulations governing their own internal cheese industry and trade in cheese. The situation in the UK provides an example of the spread of subject matter and regulations in respect of cheese, and each country has similar widespread regulations which may need to be consulted from time to time with respect to international trade.

A list of the current legislative controls for the UK is given below, and these may be subject to further amendment from time to time.

England, Wales and Scotland
Food Safety Act 1990
Dairy Products (Hygiene) Regulations 1995
Dairy Products (Hygiene) (Scotland) Regulations 1995
Miscellaneous Food Additives Regulations 1995
Sweetners in Food Regulations 1995
Colours in Food Regulations 1995
Food Labelling Regulations 1996
Weights and Measures Act 1985
Weights and Measures (Quality Marking and Abbreviation of Units) Regulations 1987

Northern Ireland
Food Safety (NI) Order 1991
Dairy Products (Hygiene) Regulations (NI) 1995
Miscellaneous Food Additives Regulations (NI) 1996
Sweetners in Food Regulations (NI) 1996
Colours in Food Regulations (NI) 1996
Food Labelling Regulations (NI) 1996
Weights and Measures (NI) Order 1981
Weights and Measures (Quality Marking and Abbreviation of Units) Regulations (NI) 1991

One result of this legislation is that specific varieties of cheese have to meet certain criteria with respect to fat and moisture levels, and some familiar examples for hard cheeses are given in Table 3.2. In addition, some more general categories have emerged, and these have become particularly important with respect to labelling. For example, a consumer may expect to know that all manufacturers are employing terms like 'low fat' to mean the same values, and some typical examples are given in Table 16.1.

Obviously, some of the values are broad, and the final analysis for any given type of cheese will depend on the brand, but in other cases, the specifications must be adhered to if the varietal name of the cheese is to be employed (Table 3.2).

The methods to be employed for measuring the chemical parameters are also strictly controlled, and a country may adopt its own standard methods, for example those of the British Standards Institution or the American Public Health Association (Marshall, 1992), or it may opt for a procedure published by the International Dairy Federation. Some of these standard methods are regarded as reference methods as well, so that the Rose–Gottlieb method is retained as the 'bench mark' for milk fat, the Kjeldahl for crude protein and oven drying for moisture. In practice of course, the

Table 16.1 Classification scheme based on composition

Type of Cheese	Milk Fat (FDM %)	Milk fat (min. %)	Water (max. %)
Full fat hard cheese	> 48		48
Medium fat	10–48		48
Low fat	< 10		48
Full fat soft cheese	> 20		60
Medium fat	10–20		70
Low fat	2–10		80
Very low fat	< 2		80
Cream cheese		> 45	47–50
Double cream cheese		> 65	30–33
Full fat processed cheese	> 48		48
Medium fat	10–48		48
Low fat	< 10		48
Cheese spread	> 20		60

pressure for rapid results has led to the development of automated systems (Andersen *et al.*, 1993), which will enable a cheesemaker to check both the composition of the incoming milk and the composition of the final product. For many cheeses, the accurate measurement of salt content in the curd is vital as well, for while few regulations mention sodium chloride, its concentration is important both organoleptically and technically; for example, Cheddar will not mature correctly if the salt-in-moisture content is too high.

16.5.2 Microbiological standards

As mentioned elsewhere, it is assumed that cheesemakers will increasingly be under pressure to adopt Good Manufacturing Practice(s) and introduce a HACCP regime to avoid the introduction of pathogens into the retail chain. However, there appears to be a growing distinction between the large automated factories making hard-pressed cheeses like Cheddar or tonnage quantities of Brie or Camembert, and those plants that employ more traditional techniques. For example, an examination of the following list of disease outbreaks linked with the consumption of cheese reveals an interesting pattern:

Cheese	Organism	Country	Year
Camembert	*Salmonella bareilly*	USA	1953
St. Marcellin	*Shigella sonnei*	France	1964
Brie	*Escherichia coli* (enteropathogenic)	USA	1971
Mont d'Or	*Salmonella typhimurium*	Switzerland	1985
Queso Blanco	*Listeria monocytogenes*	USA	1985
Mozzarella	*Salmonella oranienberg*	USA	1989

Soft cheese	*Salmonella paratyphi*	France	1993
Soft cheese	*Brucella melitensis*	Malta	1995
Cheddar	*Staphylococcus aureus*	USA	1960
Cheddar	*Staphylococcus aureus*	UK	1968
Cheddar	*Salmonella heidelberg*	USA	1976
Cheddar	*Salmonella typhimurium*	Canada	1984

While the safety of hard-pressed cheeses appears to have improved over the last 10 years, soft cheeses remain a cause for concern.

The use of raw milk remains a major obstacle to improving the image of many cheeses (Keceli and Robinson, 1997), but even where pasteurized milk is available, some cheesemakers appear to make the mistake of considering that cheese is an intrinsically safe product. Some cheeses like Parmesan are never linked with food-borne diseases, but almost every other category has the potential to harbour pathogens like *Brucella*, *Salmonella*, *Listeria monocytogenes* and *Staphylococcus aureus*. Indeed, the values of pH, salt-in-moisture and available water of most cheeses are above the minimum requirements for growth of the common pathogens. The combination of sub-optimal factors may, of course, have an adverse impact, as may the gradual release of free fatty acids in cheeses undergoing maturation, but it is important that variations from a standard recipe can upset the anticipated stability. The survival of *Salmonella* in Cheddar cheese, for example, depends upon the acidity of the curd at pressing, and while the organism should not be present anyway, poor starter activity can turn a minor problem into a major public health incident (Jervis, 1988).

Clearly the role of the cheesemaker is to prevent contamination of the cheese with undesirable organisms by avoiding failures in the plant cleaning regime, to monitor pasteurization procedures, to employ a reliable starter culture and to eliminate risks of post-pasteurization contamination; the actual evaluation of the microbial quality of end-products may well be left to outside laboratories. Nevertheless, the cheesemaker should be aware of the procedures being applied, for it is essential that any samples are:

representative of an entire batch (one vat perhaps)
taken under aseptic conditions
stored, prior to examination, at a temperature that is appropriate
examined by approved procedures.

Equally important is the fact that any desired evaluation should be carried out on a routine basis and with a specific objective in mind. Thus, whilst a high coliform count in a fresh cheese may indicate low levels of hygiene, the reaction to the result should depend upon whether the cheese is to be consumed fresh or after storage for 12 months. As with the chemical analyses, standard methods of examination, as laid down by the British Standards Institution, the International Dairy Federation or the American Public Health Association, should be used, and some typical standards for a month-old hard cheese might be (Gavron and Luck, 1990):

Total coliforms	<10 cfu/g
Faecal coliforms	Absent in 1 g
Spore-formers	<10 cfu/g
Faecal enterococci	<100 cfu/g
Staphylococcus aureus	Absent in 1 g
Yeasts and moulds	<10 cfu/g
Other pathogens	Absent in 25 g

The generous figure for faecal enterococci is probably a reflection of the fact that *Enterococcus faecalis* is still regarded as a 'starter species' by some cheesemakers, but the specifications for pathogens may well become more rigorous. For example, for a soft cheese at the time of manufacture, it is proposed that *L. monocytogenes* should be 'absent in 25 g for five samples per batch', while for Cheddar cheese, the specification should be 'absent in 25 g for four out five samples per batch' (Keceli and Robinson, 1997). This easing of the requirements for hard-pressed cheese reflects the view that this pathogen will die out during maturation, whereas with a soft cheese, *Listeria* are quite capable of growing during chilled distribution. The level of spore formers in the milk can be a major problem with cheeses, particularly those requiring a secondary fermentation after moulding and pressing. An example of the effects of poor quality milk is given in Figure 16.1.

16.5.3 Cheese grading

Where cheese has been exported regularly, the grading of such cheese will have become an established practice. Thus, New Zealand, Australia, Canada, USA, South Africa, The Netherlands and Denmark have introduced grading schemes. (The New Zealand Dairy Industry Act 1894 made grading of cheese and butter for export compulsory, while the cheese transport agency on the Rhine introduced a voluntary grading scheme in 1923.) Cheese organizations in other countries formed their own quality schemes, but the Cheese Control Station at the Hague established in 1906 was one of the first of its kind.

The qualities of cheese to be marketed varied in composition, and some central organization was necessary, especially since the 20+, 30+, 40+, 45+ levels for FDM became accepted in practice (see above), as did the 'Volvet' standards. It should be noted also that standardization of milk for cheese has become a long-standing practice not only in EU countries, and the facilities to add protein to the milk is a critical aspect of cheesemaking in some dairy produce exporting countries. The ever-growing avalanche of EU regulations is but an extension of the earlier practices that were rarely

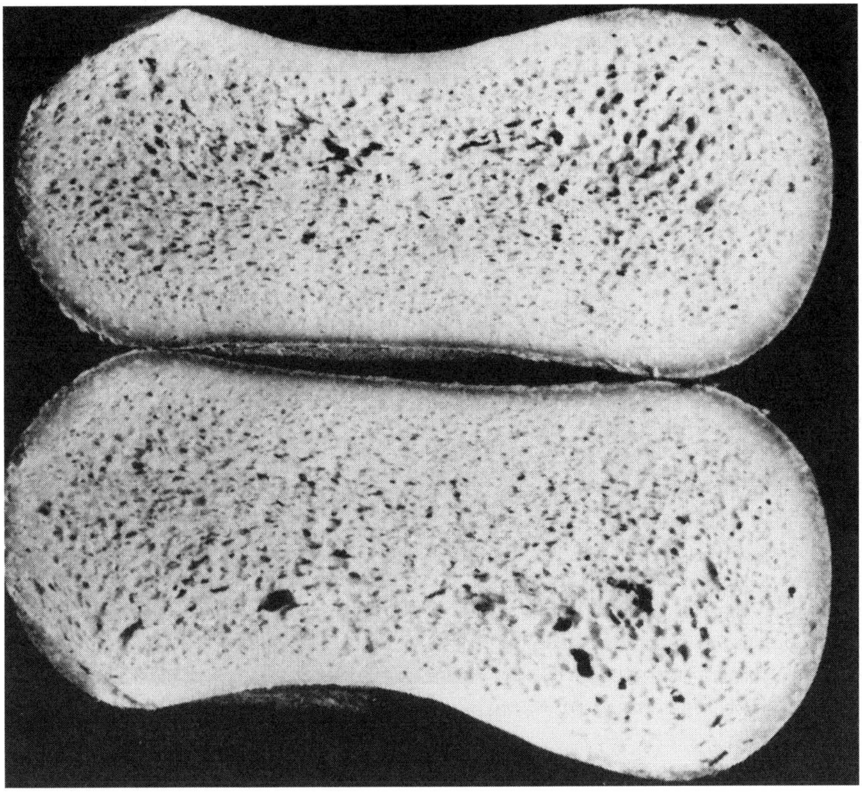

Figure 16.1 Blown cheese, made from highly contaminated milk.

monitored. Thus, anyone who has visited casual cheese fairs and auctions in Europe may well have been disappointed by the poor quality of much of the cheese, but while such cheese was only sold at local functions and markets, the trade was never in need of any form of control.

In the UK, the makers of the major varieties were more concerned, and the Cheshire Cheese Federation and the Cheddar Cheesemakers Federation were formed for the express purpose of marketing and grading cheese. A National Mark grading scheme was set up under the Agricultural Produce Grading and Marking Acts 1928–1931, and a National Mark Scheme evolved for the following cheeses:

Cheshire cheese, 1933
Stilton cheese, 1934
Cheddar and Caerphilly cheese, 1935
Derby, Leicester, Wensleydale and Lancashire cheese, 1937
Cream cheese, 1938

These schemes, which employed their own graders, were short lived, since the Second World War brought the schemes to an end in 1940. However, the Ministry of Food continued an inspection scheme in respect of edible quality, so as to build up stocks to make cheese rationing possible. The inspection scheme was not concerned with cheese varieties, but only with edibility and, to a lesser extent, handling properties. Thus, in order to cut cheese into small pieces for rations, the cheese could not be crumbly. Cheshire cheese had to be made more solid, and a somewhat 'Cheddary Cheshire' became the normal product. Indeed, this period of rationing, 1940–1954, almost destroyed the prevalent cheese varieties in the UK, but in 1954 the controls were lifted and a new grading scheme was inaugurated to safeguard cheese quality and the characteristics of the varieties.

Schemes were set up for: (a) England and Wales and Northern Ireland, (b) Scotland and (c) Farmhouse cheese; the wartime inspectors were brought in to run the schemes. The cheese industry had only the judgement of these inspectors to rely on for guidance on cheese quality, along with some assistance from the cheese factors and those marketing people who were knowledgeable in respect of market requirements. However, many of the cheese factors had gone out of business, so that some of the larger creameries had to market their own produce with only limited experience. A lack of younger graders induced research into mechanical and chemical methods of the assessment of cheese quality, but the research did not produce practical results due to the great number of factors that impinge on the quality of a product like cheese.

Meanwhile, public taste for cheese had undergone a marked change in favour of milder flavoured cheeses, since the imported cheese which the UK had relied on during wartime was more waxy in texture and milder in flavour than some of the traditional English varieties. Initially, the cheese industry took advantage of this trend, as it meant that young cheeses with more moisture could be marketed early, so avoiding the high cost of financing storage to maturity. However, the loss of cheese factors, who bought cheeses at an early age and matured them, made it necessary for the creameries to ripen and market their own produce, and this required heavy financial commitments. Later, increased milk production, which caused cheese stocks to rise and allowed longer times for maturation, again changed the situation. More recently, the increased cost of labour has led to the use of more mechanization to replace the heavy manual operations typical of the older cheese factories, and this move has again influenced the quality of the cheese being manufactured.

This short historical account shows how the quality of cheeses can be changed by economic and other outside influences, and hence it is not surprising that some adjustment to traditional grading schemes became necessary. Furthermore, there have been changes within the industry itself, so that the mechanization of the cheese process has, for example, made the use of the older method of grading cheese impractical, in that it is no longer possible in many factories for the grader to deal with cheeses vat by vat. The coating of cheese rinds also caused problems for the graders, for the introduction of waxing of cheese rinds in the 1930s, and the resin coating of smaller cheeses, caused more moisture to be held in the cheese. Softer-bodied cheeses and more instances of taints in mature cheese required the application of more rigid grading exercises. The use of refrigeration to overcome some of the problems caused by ripening cheese at 13–18°C (Figure 16.2) caused other problems at 4–7°C when, because of the shortage of labour to turn cheeses, 'soft crowns' developed. Soft crowns are caused by the migration of free moisture to the bases of the cheeses where soft patches of curd develop.

The film wrapping of cheese brought more problems for the cheesemaker in that the cheese had to be made with a lower moisture content at pressing. A traditionally made and bandaged cheese could lose from 8 to 15% of its weight during ripening, but with film wrapping, this loss was prevented and the characteristics of the cheese at grading time altered.

The large creameries, built up by amalgamation of smaller units, also caused grading difficulties. Thus, one of the older factories might have presented 1000 or so cheeses for grading, and even these numbers could tax the palate and nose of a grader, but when a factory presented 5000–6000 cheeses for grading, the problem became intense and only a few representative cheeses could be seen. Rigid boxing (wood or cardboard) of cheeses also limited the visual inspection of cheeses by the grader, and conflicts between graders and buyers became inevitable.

A further problem arose from the introduction of the large 'Ton' presses (block-formers), for a cheese taken from such a press had no true identity, and vat records were not true records of any particular cheese since curds from adjacent vats could be mixed in the press. The old grading system,

Figure 16.2 Three experimental cheeses showing the effect of storage temperature, from 7°C (left) to 13°C and 18°C (right).

whereby the grader assessed one cheese from each vat, was no longer applicable.

These changes only compounded the existing difficulties of the grading. For example, although cheeses made in each vat differ slightly in ripening characteristics, such as body and texture, even when made from silo bulked milk, these differences are accentuated during storage and maturation. In addition, the cheeses made at any one factory vary in character from those made at another, either due to the cheesemaker's intention or because of the variability of milk supply, starter quality or maturation conditions. All these differences caused strains to appear in the grading system and raise the question of what cheese grading is for – the factory or the consumer ?

In an attempt to unify procedures, suggestions have been made as to the time of grading, e.g. for Cheddar cheese, grading should be at 2–3 weeks for factory information, and again either at sale or at 5–6 months for retailer information; this approach has much to recommend it, in spite of the extra cost of 'double' grading.

Taylor, in his *Outlines of Agricultural Economics* (Bond, 1957), noted that 'grading as a marketing activity consists of classifying and sorting a commodity into groups, no two of which overlap, and in each of which commodity is of uniform kind, size and quality'. The grading of cheese is an act of estimating the quality, attributes or characteristics of a cheese in order to classify it into a group of similar quality. In order to satisfy this form of grading, there must be agreed standards, and this immediately raises the question as to how, when, by whom and what standards are to be set. Even this leaves aside the question of whether the cheese at any grading must meet the same standards during the remainder of its shelf-life. Gilles and Lawrence (1973) attempted to overcome this latter problem by correlating the compositional analysis of Cheddar cheese with anticipated quality, and figures for salt-in-moisture, for example, are able to provide an indication of the likely grade to be achieved later in the shelf-life.

Assuming that the price is right, public demand is usually for 'palatability' rather than flavour, aroma, body or texture, but, unfortunately, palatability covers more than the setting of 'grading points', for the palatability varies with the individual, his/her age and on whether his/her palate is trained or not. In general, the demand for highly flavoured cheeses is mainly from the older generations while the young prefer bland flavours, but, even then, travel abroad has 'trained' young palates to expect exotic flavours at home; palatability is, then, a changing standard.

For this reason, the consumer may not be the best arbiter of quality, the more so as local guidance is rarely available. Thus, the older type of retailer who knew about cheese could both 'average' customer requirements for his products and help with cheese selection, but the supermarket with a junior check-out cashier has replaced the knowledgeable retailer, and the link between producer and consumer has been broken. However, this crude generalization may no longer hold true for, more recently, the retail sector has become more aware of the need to provide service. The introduction of counters devoted to unpackaged cheeses in the major supermarket chains like Waitrose means that customers can once again seek expert advice and guidance as to what quality means with respect to any given variety.

16.5.4 Grading schemes

The cheese grading scheme in place in the UK between 1954 and 1974 gradually lost touch with both customers and with the new technology, and for many reasons needed to be replaced. However, the act of replacement highlighted the problems of grading, for in order to understand the objectives and principles of grading, a trained cheesemaker must be able to assess the cheese made by himself, as well as the cheese made by competitors. The cheese buyer must also have a high degree of discrimination between cheeses of differing quality, and standards set for grading must be understood by both the cheesemaker and the buyer.

An efficient grading scheme should be able to maintain a service which provides quality standards for production as well as marketing. Indeed 'quality', whatever it means, has become extremely important in consumer-led economies, and has raised the question as to whether a grading service should rely on human estimations of quality, or whether the service should provide regular analyses of chemical quality as well. Having provided such a service, should it be an advisory service to the manufacturer as it is in some countries, or should it be for the protection of the consumer?

The chemical compositions for various cheeses have been set out under agreements with both FAO/WHO and National Committees, but the establishment of sensory qualities have largely been left to individual countries; two typical examples are shown in Table 16.2.

The number of points awarded for each character of a cheese may vary from country to country, with some schemes awarding up to 5 points per character with 15 as a total maximum for all characters, while other schemes use up to 45 points per character to give 100 points as the maximum.

A 5-point system is used in some schemes as follows: 5, very good; 4, good; 3, satisfactory; 2, fairly satisfactory; 1, unsatisfactory; 0, defective. Under this system with four characters of the cheese to be assessed, 8 out of 20 points would be fairly satisfactory, i.e. the cheese would be marketable.

A scheme used in Scandinavia allocated the following maximum points:

Outer appearance	5
Colour and texture	5
Consistency of body	5
Flavour and aroma	5
Total	20

Table 16.2 Points systems for grading various types of cheese

Austria		Germany	
Hard-pressed cheese		*Hard-pressed cheese*	
Outer appearance	4	Outer appearance	5
Curd quality	8	Curd quality	4
Flavour and aroma	8	Aroma	3
		Flavour	8
Total	20	Total	20
Surface mould-ripened cheese		*Sour milk Quark*	
Outer appearance	8	Outer appearance	5
Curd quality	4	Texture	5
Flavour and aroma	8	Aroma	5
		Flavour	5
Total	20	pH	5
		Maturity	5
		Coliform count	5
		Yeasts	5
		Total	40

Schemes for Gouda, Edam and Samsoe awarded 6 points for outside appearance, 6 points for the combination of colour, aroma and consistency of the body and 8 points for flavour on a 20-point scale. The number ofpoints given for the outside appearance in these schemes denotes the importance of appearance, probably because it indicates what to expect inside.

A somewhat similar approach has been used in France:

	Gruyère	Emmental	Cantal
Form, shape and appearance	1.5	2.0	3
Rind defects	2.0	2.5	3
Texture and body	4.5	4.0	6
Flavour	7.0	6.0	8
Eyehole formation	5.0	5.5	–
Total	20.0	20.0	20

The points awarded on the 20-point scales differ from those given in 100-point scales, where more emphasis is given to certain characters of a cheese, such as flavour, or important varietal characters, such as texture and body.

A New Zealand scheme for Cheddar cheese and a similar grading system that had been operated by the National Association of Creamery Proprietors (NACEPE) until 1974 (since when grading has been continued along similar lines) are included in Table 16.3. As New Zealand is a major exporter of Cheddar cheese, much attention has been given to the possibility of predicting the likely grade of a mature Cheddar on the basis of its chemical

Table 16.3 Grading schemes that have been used in New Zealand and the UK

	New Zealand	NACEPE (UK)
Flavour (and aroma)	45	45
Body	20	40 (Body and Texture)
Texture	20	
Colour	10	5
Outer appearance	5	10
Total	100	100

Table 16.4 Suggested compositional ranges for Cheddar cheese grades (after Lawrence *et al.*, 1993)

	First grade	Second grade
FDM	52–56%	50–57%
MNFS	52–54%	50–56%
pH	5.1–5.3	5.0–5.4
S/M	4.7–5.7	4.0–6.0

composition at 1–3 months. In other words, there is little point in exporting a cheese which will be rejected or given a low grade 6 months later, and a system based on the use of compositional data for the segregation of atypical cheeses is given in Table 16.4.

The cheeses graded by the NACEPE system were placed into the following groups according to the points awarded:

Extra selected – awarded 93 points, but with not less than 41 points for flavour and aroma.
Selected – awarded 83–93 points with not less than 38 points for flavour and aroma.
Graded – awarded 70–83 points.
No grade – awarded less than 70 points, or for cheeses which do not have a legal fat content.

The NACEPE scheme for cheese grading and quality was based on the descriptions of UK cheese varieties given in Table 16.5. All varieties of cheese were to be free from gas holes and free from moisture. Selected and graded cheese could lose points on any of these qualities. The descriptions presuppose also that the grader is knowledgeable as far as the various parameters are concerned, and the scheme did allow for graders to meet and discuss the grading assessment of the different varieties.

The 'No grade' cheeses, which were not sold for human consumption, showed one or more of the following defects:

Table 16.5 NACEPE scheme for cheese grading (UK) – criteria for extra selected grade

Cheese	Graded at (weeks)	Flavour and aroma	Body	Texture	Colour	Finish
Cheddar	8	Superfine, buttery, no aftertaste	Firm not hard	Close, smooth, silky	Uniform	Superior, free from defects
Cheshire	2	Superfine, not acid	Firm not hard	Free from granules, silky	Uniform	Superior, free from defects
Derbyshire	4	Superfine, no aftertaste	Firm, not chalky	Fairly close, smooth	Uniform	Superior, free from defects
Lancashire	2	Superfine, no off flavour	Moderately firm	Slightly open, friable	Uniform	Upstanding, free from defects
Leicester	4	Superfine, buttery	Moderately firm	Slightly open, short texture	Uniform, bright colour	Regular shape

Flavour	– taints and off-flavours
Aroma	– unpleasant and strong
Body	– very acid and hard, or weak and pasty
Texture	– harsh, chalky, open or very pasty
Colour	– uneven, discoloured
Finish	– deformed, stained, damaged rind, mould growths, vermin damage (mites).

Excessive free moisture, gas holes, large splits, fat percentages of less than 48% in the dry matter and soft crowns were also causes for rejection. It is interesting though that the NACEPE scheme used only the top 30 points of a 100-point assessment, and was, therefore, only marginally better than the 20-point schemes. In many of the schemes, cheeses for export must acquire a minimum number of marks before the cheese can be exported. In one Danish scheme, cheese must carry 9–9.75 marks out of 15 before it can be exported.

The scheme in Denmark is interesting, since cheeses are graded by 12 people chosen from manufacturers, authorized wholesalers (exporters), a representative of the State Quality Control Board and an advisory officer to the creamery. The advisory officer is the link between the creamery and the grading and control service, while the exporters have links with the retailers. This reporting back to the manufacturer from the marketing officers is also the practice in other countries.

Whether schemes based on more objective systems would be possible is open to debate but, with respect to Cheddar cheese, Muir and Hunter (1992a,b) have explored the potential of techniques such as Principal Component Analysis. Similarly, Hort (1996) has been investigating methods for examining the physical properties of Cheddar cheese to see whether instrumental profiles might correlate with texture as perceived by a consumer. However, it is likely, at least for the foreseeable future, that grading on the basis of the 'human response' will remain an essential component of cheese manufacture and marketing.

16.5.5 Grading practice

The skill of grading has to be acquired by experience over a number of years, so that cheese made in various seasons can be assessed as well as cheese from different creameries. The following sequence of grading has been used by some cheese graders, although the procedures may vary.

1. The cheese must be identified by number, vat number, creamery and date of manufacture and since a cold cheese (4°C) has little aroma and the body is hard, it is advisable to 'temper' the cheese for 3–7 days according to size, at a temperature of 12–13°C. The ambient temperature is often too variable to be used.

2. The grader inspects the outside of the cheese in order to award points for appearance, care in finishing, cheese coat, bandage, film or coating. During the inspection, the grader will make mental notes about the cheese which subsequently will be confirmed or rejected. From experience, the grader will form an opinion, based on the convex top and sides of the cheese, as to whether the cheese is sweet – except in cheeses like Emmental which are naturally convex due to eyehole formation. This same evidence may also indicate contamination. A concave surface of the cheese indicates an acid, hard curd. Varietal differences accepted, the rind or coat of the cheese can also be informative; white patches can, for example, indicate acid cheese curds or, in some instances, warm storage. Rindless cheeses present problems for the grader, as a lot of points (10–20) are often available for finish; a different allocation of points must be made for such cheeses.
3. External pressure by the thumb is used to discover the hard, hollow or soft patches in the cheese. The rind may be intact, but mould growth during storage may have softened the curd just under the rind. In the case of cheeses like Emmental, the grader may tap the cheese with his/her thumb, fingers or handle of a cheese 'trier'; the tapping sound gives some indication of the internal structure in respect of 'eyeholes'. The information from these stages is stored mentally for future confirmation, although some graders award a proportion of points at this stage.
4. The grader then has to assess the inner quality of the cheese and uses either a spike or cheese 'trier'. The cheese trier or cheese iron is a semi-cylindrical steel tool which is used to take sample borings from the cheese. The pressure needed to insert either the spike or trier into the cheese gives some indication of the body texture and degree of ripeness of the cheese. An examination of the spike or trier immediately it is withdrawn from the cheese provides the grader with the first indications of aroma, and with a glimpse of the body and ripeness of any adhering curds. Points can be given for aroma at this stage, although the grader may allocate some extra or deduct some points later. The steel back of the trier gives an indication of breakdown (ripeness) of the cheese by the type of adhering cheese film. A greasy film indicates a more mature cheese than a rough rolled-up film, which suggests an immature cheese or one lacking fat. The core of cheese in the trier shows that the cheese has a good body if the trier is full, or a weak body if the core does not fill the blade of the trier.
5. Visual inspection of the core provides information on body texture and colour. The colour may be bright or dull, uniform or uneven, bleached or discoloured, and there may be fatty deposits between curd particles. Rarely, free moisture may also be seen. Eyeholes in the propionic types of cheese, gas holes in other types, mould growths, extraneous matter (grit, straw, seeds) are all noted, and points or lack of points noted down for future use.

6. The core of cheese is now handled between thumb and fingers. The feel gives confirmation of texture and body, and enables points to be awarded.
7. Working a piece of cheese to a paste in the hand warms up the curd to release more aroma. At the same time, the texture and body can be finally assessed with regard to ripeness and given points. One problem is that working the curds leaves a residue of aroma that may mask the aroma of the following samples, and even washing the hands between samples may not help. The flavour is, however, even more difficult to assess.

With past experience as a guide, the grader associates aromas with tastes, and so can give points for a combined flavour/aroma character. There are some flavours, i.e. salty flavours, which are non-volatile and so missed by this method, and to guard against this problem, the cheese must be tasted. However, the palate soon tires of flavours, so that it is not practical to taste many cheeses during a grading exercise; chewing an apple or piece of bread does cleanse to some extent, but it may take an hour for taste reception to fully return. Some graders find that chewing a small piece of cheese, spitting it out and then cleansing the mouth immediately is a practical procedure, but again it will only work for a small number of samples.

Some of the descriptions that have been applied by graders over the years to cheese characteristics are given in Table 16.6, and it is important that a cheesemaker should be aware of some potential faults so that they can be avoided.

There are economic problems facing both the cheesemaker and the creamery management which arise out of a grading system, and the following questions arise from interpretation of the individual points.

1. Is it more economic to include extra moisture in a cheese even though it may be down-graded – water is sold at the same price as fat or protein?
2. Is a full flavoured immature cheese to be graded and sold early?
3. Is a very mild flavoured cheese to be kept for better maturity?

Table 16.6 Terms applied to cheese describing certain characteristics

Taste:	Acid, alcoholic, burnt, bitter, cooked, flowery, buttery, bland, foul, fruity, mild, nutty, oily, off, resinous, rich, rancid, ripe, salty, sour, spicy, strong, superfine, tangy, sweet, weedy, metallic, creamy.
Aroma:	Burnt, foul, foetid, cowy, garlic, oniony, goaty, aromatic, spicy, floral, over-ripe, mouldy, pale, musty, clove, peardrop, fishy, etheral, ammoniacal, sweaty.
Body:	Hard, harsh, soft, brittle, elastic, flaky, springy, short, dry, pasty, leathery, gummy, wet, runny, rough, squeaky, close, chalky.
Texture:	Open, close, tight, gasssy, stringy, short, long, greasy, sandy, gummy, soggy, uneven, streaky, granular, friable.
Colour:	Natural, strong, weak, bright, dull, bleached, uneven, mottled, grey, discoloured, dirty, red, yellow, brownish.

4. In view of the price differentials between categories of cheese, should the cheesemaker aim for the highest quality cheese rather than the more economic?

It would seem that, having regard for the long-term, quality of produce is the better aim for the cheesemaker. In the UK, this premise is being followed by the farmhouse cheesemakers who combined into the 'Cheddar and Caerphilly Associations' and into the 'Cheshire and Lancashire Cheesemakers' to agree on standards for cheese which is largely hand-made, well finished and well matured, and commands a premium price. The use of milk from the cheesemaker's own farm was also included as a criterion for use of the term 'farmhouse', though nowadays the criteria have become somewhat flexible. Nevertheless, labels like 'Farmhouse Cheddar cheese' still imply that the product will be of excellent quality (Robinson, 1995), and cheesemakers will be well advised to protect the image that the name portrays.

Standards of quality for factory-produced cheese should take into account public taste and the need to handle competition from abroad. Some reviews on the subject of grading have been published, and while Alexander (1953) was concerned with grading Cheddar cheese, Keating and Young (1980) dealt with non-Cheddar cheese. Further relevant papers have been published by Rothe *et al.* (1978), McBride and Hall (1979) and Olson (1980).

References

Alexander, W.R. (1953) *Agricultural Progress*, **28**, 144.

Andersen, T., Brems, N., Borglum, M.M. *et al.* (1993) Modern Dairy Practice – 1: Chemical Analyses. In *Modern Dairy Technology*, Vol. 2, 2nd edn (ed. R.K. Robinson), Elsevier Applied Science, London, pp. 381–416.

Bond, K.W. (1957) *Agriculture*, London, **64**(7), 339.

Gavron, H. and Luck, H. (1990) In *Dairy Microbiology*, Vol. 2 (ed. R.K. Robinson), Chapman & Hall, London, pp. 345–392.

Gilles, J. and Lawrence, R.C. (1973) *New Zealand Journal of Dairy Science and Technology*, **8**, 141–145.

Hort, J. (1996) *Food Science and Technology Today*, **10**(2), 94.

Jervis, D.I. (1988) *Dairy Industries International*, **53**(3), 15.

Keating, D.C and Young, R.W. (1980) *Dairy Products*, **8**(2), 7.

Keceli, T. and Robinson, R.K. (1997) *Dairy Industries International*, **62**(4), 29–33.

Lawrence, R.C., Gilles, J. and Creamer, L.K. (1993) In *Cheese: Chemistry, Physics and Microbiology*, Vol. 2, 2nd edn (ed. P.F. Fox), Chapman & Hall, London, pp. 1–38.

Marshall, R.T. (1992) *Standard Methods for the Examination of Dairy Products*, American Public Health Association, Washington DC, USA.

McBride, R.L. and Hall, C. (1979) *Australian Journal of Dairy Technology*, **34**(2), 66.

Muir, D.D. and Hunter, E.A. (1992a) *Journal of the Society of Dairy Technology*, **45**, 23.

Muir, D.D. and Hunter, E.A. (1992b) *Food Quality and Preference*, **3**, 141.

Olson, N.F. (1980) *Dairy Field*, **103**(9), 75.

Robinson, R.K. (1995) *Colour Guide to Cheese and Fermented Milks*, Chapman & Hall, London.

Rothe, M., Engst, W. and Voigt, I. (1978) *Nahrung*, **22**(8), 695.

Shaw, M.B. (1993) in *Modern Dairy Technology*, Vol. 2 (ed. R.K. Robinson), Chapman & Hall, London, pp. 221–280.

17 Membrane filtration of milk and whey

17.1 Membranes

The term 'osmosis' is used to describe the spontaneous flow of water from a less concentrated to a more concentrated solution when the two solutions are separated by a suitable membrane.

In order to obtain potable water from a solution (e.g. sea water), the direction of the flow of pure water needs to be reversed, and this process has been popularly termed 'reverse osmosis'. Since this process requires the use of a suitable, i.e. semi-permeable, membrane, it has become a technique for the separation, concentration or fractionation of substances in fluid solutions employing membranes of different permeabilities. When the permeability is limited to that of the liquid solvent (e.g. water in a desalination process), the process has become known as 'reverse osmosis'. This is sometimes called 'hyperfiltration' as opposed to the term 'ultrafiltration' used for concentration of cheese milks, where the process becomes one of separation of the milk into two liquids using a more porous permeable membrane; the porosity of the membrane determines the composition of each resultant solution. Thus, the permeate is that part which passes through the membrane, and the retentate is that part which is retained by the membrane. The principle of the filtration process is shown diagrammatically in Figure 17.1.

The first major usage of reverse osmosis (RO) was in the desalination of sea or brackish waters to produce potable water, but the fluid solvent need

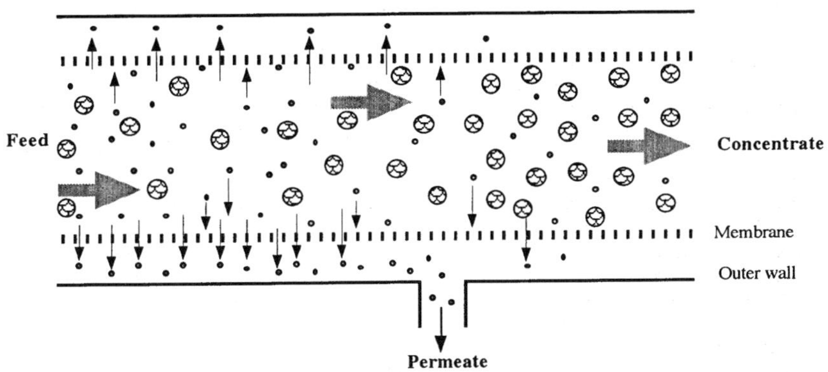

Figure 17.1 Diagram showing the principles of membrane filtration (e.g. reverse osmosis and ultrafiltration).

not necessarily be water; indeed the process is used in the paint and other industries for solvent recovery. A PCI plant for RO treatment of whey or skim milk is illustrated in Figure 17.2.

During the 1970s, the Dairy Industry was very hesitant in accepting the use of membrane filtration, partly for economic reasons and partly because the process parameters were not well known. As soon as the first practical

Figure 17.2 A PCI tubular RO plant for concentrating cheese whey or skimmed milk. (Courtesy of PCI Membrane Systems Ltd.)

results were achieved, especially in the processing of whey, the process as a whole began to expand more rapidly. The first membranes were made of cellulose acetate but synthetic polymers have now replaced the cellulose membranes on account of better resistance to heat and to detergent cleaning solutions.

17.1.1 Membranes for cheese milk and whey processing

The general concept of membrane structure is that of a barrier layer on a support. The barrier layer is thin, 250–500 Å thick, and is the separating semi-permeable surface. This layer is supported by a 0.05 mm thick microporous matrix of similar material, but more open in structure. In turn, these two layers are supported by an open-structured matrix of polyester or paper fibre 0.25 mm thick. Membranes made from polymer material for ultrafiltration may have pore sizes from 300 to 450 Å, and a detailed account of the current position has been given by Lewis (1997).

Maubois (1979) described a membrane incorporating zirconium oxide in a graphite base which enabled temperatures up to 400°C to be used, and Nielsen *et al.* (1972) gave some interesting data on the membrane retention of various substances on DDS (Danish Sugar Corporation) membranes (see Table 17.1).

The membrane material also affects the permeability to liquids, so that it is common to quote the molecular weight of substances retained. For RO membranes, the retention of sodium chloride is often used to indicate the retention capacity of the membrane (NaCl, MW = 58.4). Bennasor *et al.* (1982), using mineral membranes, suggested that the reduction of flux on the permeate from milk with added $CaCl_2$ is due to a loss of charge on the micelles which compact at the membrane boundary layer; calcium phosphate is also believed to play a role in modifying membrane performance (Daufin *et al.*, 1991). Pre-rennetted milk (with added $CaCl_2$), however, has larger aggregated micelles and this allows an increased flux of the permeate.

Whatever the choice of material, the membranes will be assembled into modules capable of treating a stated amount of liquid. The actual

Table 17.1 Retention of compounds by DDS membranes

Membrane	Operating pressure (MPa)	Molecular weight, 100% rejection	Ions rejected	
			Na (%)	Ca (%)
UF	2	35 000 Haemoglobin	0	0
UF	2	20 000 Chymosin	0	0
UF	2	6000 Pepsin	10	21
RO	5	1000 Peptides	65	78
RO	5	342 Sucrose	93	97
RO	8	342 Sucrose	99	100

membranes may exist as: (a) flat sheets, rectangular or circular in shape, and interleaved with specially formed rigid plates in which are formed passageways to convey the treatment liquids to and from the membranes, or (b) tubular membranes supported on rigid but perforated pipes that act as channels to convey the treatment fluids in and out of the module. The number of membranes per module depends on the desired throughput, and the installation also includes tanks for the liquids undergoing treatment and a pump or pumps of the positive flow type to provide the pressure necessary to cause a reversal of flow of liquid (water) through the membrane, i.e. a reversal of natural pressures. Piston pumps formerly developed for homogenization duties are often used, but lobe pumps may be used for lower UF pressure systems.

The successful operation of a membrane plant depends on the operating conditions and the cleaning routines. For these reasons, the manufacturers (suppliers) of membranes have produced detailed parameters for usage of their particular membranes. Thus, pressures, temperatures, pH, strength and composition of cleaning solutions – including enzymes (Coolbear *et al.*, 1992) – and the cleaning routines are usually set out in information leaflets.

A number of companies offer full-scale plants for the treatment of both milk and whey by UF, and the concentration factors are impressive. Thus, assuming that dried cheese whey contains some 10–12% protein, then concentration by a factor of 5 will give a powder with a 35% protein, a factor of 20 will raise the protein to 65%, while a combination of ultra- and diafiltration can give retentate which, after drying, has a protein content of 80% (Lewis, 1997). Figure 17.3 shows a spiral-wound plant for such protein

Figure 17.3 Spiral-wound UF membranes for whey protein recovery. (Courtesy of APV Nordic Membrane AS)

recovery. As the permeate from ultrafiltration contains high levels of lactose, efficient recovery and utilization of that sugar is essential to make membrane processing of whey economic.

17.2 Applications of membrane processing

In use, the membrane module can be considered the heart of the plant, and the liquid to be treated can be concentrated by several arrangements of modules.

(a) The liquid can be recycled through the module until the permeate is of the required concentration.
(b) The liquid can be recycled through the module until the permeate is of the required concentration, and then the retentate is bled away and replaced by an equal volume of raw liquid.
(c) The liquid can be processed continuously. This may require passage through several modules until the required concentration is reached.
(d) The liquid can be concentrated in UF modules and then further concentrated in RO modules.
(e) Effluents from UF modules may be treated in RO modules to provide pure process water.

Treatments (a), (b) and (c) may be used for the concentration of cheese milks or whey, while treatments (b) and (c) may be used for the pre-concentration of whey before electrodialysis to lower the mineral content (Houldsworth, 1980); deionized whey is required for infant feeding, for example. Treatment (d) can be used to provide whey concentrates for bakery or dessert manufacture or for spray-drying of the concentrate. Treatment (e) is being used to provide process water from the effluents derived from membrane filtration, as this approach both saves on the cost of mains water and also reduces the amount of effluent to be discharged.

Another process, ‘diafiltration’, has been developed for processes which require the further separation of the lower molecular weight substances (lactose and salts). The process utilizes the normal UF membranes, but water is added to the retentate which is then treated again to remove more of the lower molecular weight solids. The effect is to produce a partially deionized whey, for example, that can then be dried for special dietary purposes.

17.3 Application of UF to cheesemaking

There is a close parallel between the selective retention of casein micelles and milk fat within the curd granule and the selection that takes place at the semi-permeable UF membrane. Pretreatment of milk by UF will increase the

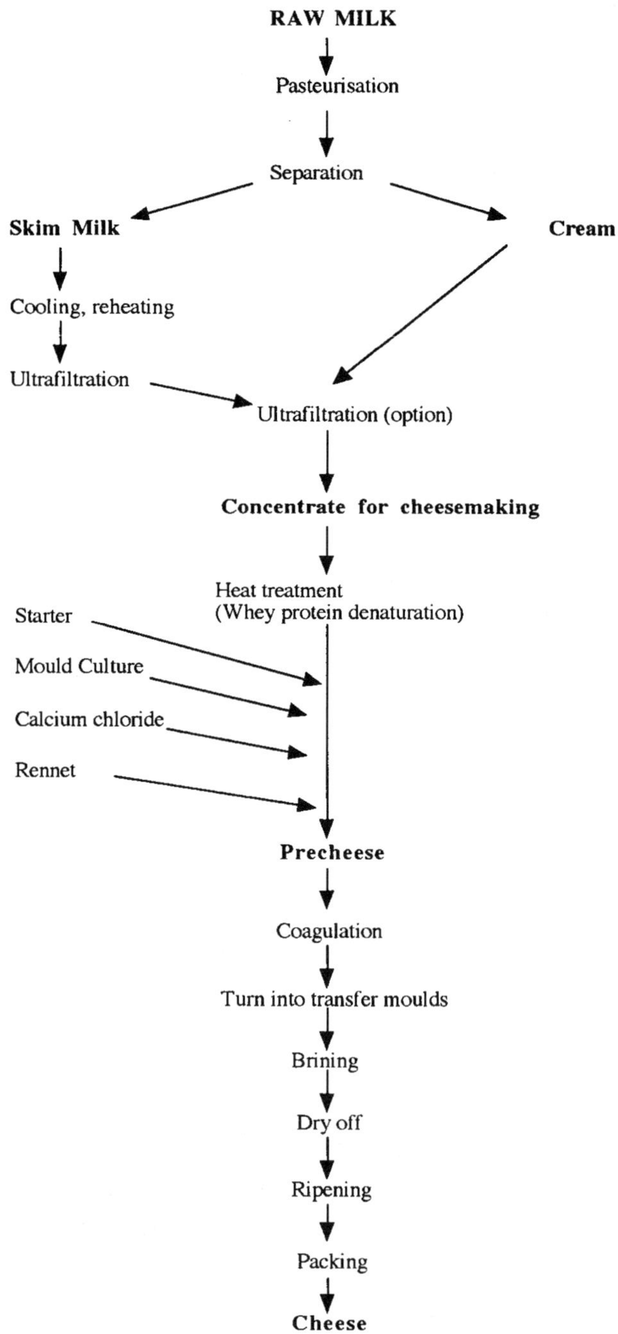

Figure 17.4 Flow diagram to illustrate the principles of the use of UF in cheesemaking.

levels of all milk proteins and the milk fat, though treatment is more easily carried out on skim milk than whole milk. The permeate will be a sweet solution of lactose and minerals plus traces of other low molecular weight compounds. Retention of whey proteins offers the potential for yield improvement and standardization to a lower protein/fat ratio. The process is viable for soft cheeses, and semi-hard cheeses like Hervé, Havarti and St. Paulin have also been made by using this process (Renner and El-Salam, 1991). Figure 17.4 gives an outline of the principles of the use of UF in the manufacture of cheese. The degree of concentration is not sufficient for hard cheese without drainage of whey, apart from any quality considerations.

Since the introduction of the Maubois, Mocquot and Vassal (MMV) process (Maubois *et al.*, 1971) for soft cheese making, the method has been used with extensive commercial success for small soft cheeses like Camembert. However, more practical investigations and basic research are required before the method is really viable for uses other than for soft cheeses. For example, Rousseaux *et al.* (1978) suggest that, for Comte cheese, the concentration by membrane techniques should not exceed 1:1.4; otherwise the flavour is affected, while at concentrations of 1:1.7 (i.e. a 5% protein level), the texture becomes fragile and the 'eyeholes' split. UF may also be used for the less rigorous role of milk standardization (Figure 17.5).

Mahaut and Maubois (1978) reported on the manufacture of blue cheese using Rhone–Poulenc membranes, and suggested that the cheesemaking process could be traditional if the protein content of the retentate was 10% or less, but if over 10% protein, then the recipe needed adjustment. Covacevich and Kosikowski (1978) developed a process for the production of cottage cheese using UF and diafiltration techniques and a starter culture

Figure 17.5 Protein standardization using a spiral-wound plant. (Courtesy of APV Nordic Membrane AS)

of *Lactobacillus delbrueckii* sub-sp. *bulgaricus*, while Veinoglou *et al.* (1978) and Georgakis (1978) applied ultrafiltration methods for Feta and Telemes cheese production. The former authors used PCI membranes and concentrated cow's milk, 1:4.5, and sheep's milk, 1:2.5, by recycling at 50°C; the concentration of milk for cheesemaking undoubtedly saves rennet.

Jacobsen (1978) gave some further information on the manufacture of Feta cheese with a dry matter content of 40%. The milk (3.5% fat) which was concentrated at 50°C under 650 kPa inlet pressure produced a retentate which was pasteurized at 65°C for 20 min. The retentate was then homogenized at 50°C and 5 MPa pressure. To the pasteurized retentate, 1% of starter and rennet (at 0.002% of milk) were added; both decolourize and lipase could be added at this stage. The rennetted pre-cheese was filled into plastic moulds for fermentation and setting at 27°C for 24 h, pH 4.85. The cheese was cut into 1-kg pieces and stored in salt brine, initially at 12–14°C and then at 6°C. Analysis of the fresh cheese gave 42% total solids, 18.5% fat and 23.5% solids-non-fat, but after salting, the dry matter increased to 45% due to loss of whey to the brine. Some cheeses were reported to exhibit defects, such as sliminess, grainy texture or yellow colour. However, these problems have been overcome, and Tamime and Kirkegaard (1991) have discussed in detail the commercial manufacture of Feta cheese from ultrafiltered milk.

Mottar and Delbeke (1978) reported on Hervé cheese production using ultrafiltration techniques. Skim milk was concentrated at 40°C under 200–280 kPa pressure to 87% of the volume, and 0.02% of hydrogen peroxide was added to avoid bacterial growth. Butter oil was then added to the retentate, and the mixture homogenized at 65°C and 20 MPa. The retentate, 19.4% protein, 22.4% dry matter plus 28% butter oil gave a cheese with 39.1% dry matter and 20.1% fat; the dry matter increased to 50% on ripening. The pH (5.0 before salting) increased to pH 5.4 at 6 weeks, and although the cheese at 2 weeks was bitter to taste, this fault disappeared as maturation proceeded.

In the production of Camembert cheese, Prokopek *et al.* (1978a) pasteurized skim milk and concentrated it by UF at 20°C to 16% protein. The fat was adjusted by adding 70% cream to give a retentate of 15% fat, 15% protein and 1.6% ash, and this was heated for 1 min at 80°C (or alternatively at 60–65°C for 30 min). The moulded, pre-cheese (rennetted at 32°C) was allowed to stand for 18 h at 16–18°C, after which it was salted and stored. Treatment of the retentate for 1 min at 80°C gave a smooth-textured cheese which ripened quickly, while treatment of the retentate at 62–65°C for 30 min gave a gritty, sandy texture which broke down on ripening. Treatment by homogenization at 3.5 MPa gave gluey, tough, soapy curds. Prokopek *et al.* (1978b) also showed that heating to 80°C for 1 min avoided the drainage of whey from the cheese, and that the addition of calcium chloride to the milk was not necessary; only 10–20% of the normal rennet addition was needed.

Maubois and Kosikowski (1978) described the production of Ricotta cheese from UF milk concentrated to 10% protein, but working on Mozzarella cheese by the UF method, the same authors found that the 19% protein retentate obtained at maximum concentration was insufficient to match the 22% protein in normal Mozzarella; they suggested that addition of milk powder might correct the deficiency. Spangler *et al.* (1990) demonstrated that Gouda could be made successfully from ultra-filtered milk, but highlighted also the need to modify any traditional cheesemaking process for use with concentrated milk since, for example, normal usage rates for rennet gave rise to cheeses with a very bitter flavour.

Early reviews by Maubois (1973) and Bungard *et al.* (1972) confirmed the success of both ultrafiltration and hyperfiltration on skim milk for the production of dairy products, including Havarti, Camembert and Cottage cheeses, as well as Quark and yoghurt. Although some success with the semi-hard cheeses was achieved, hard-pressed cheese production by the UF methods remained a problem area. Chapman *et al.* (1974) reported on the use of milk concentrated 1:2.0 by UF techniques for Cheddar and Cheshire cheeses, and for Cheddar cheese a *Lactococcus lactis* sub-sp. *cremoris* (NCDO 1986) starter (2%) was used at 30°C with a ripening time of 20 min; only 40% of the normal rennet quantity was required to produce a firm coagulum in 30 min. The curds and whey were stirred for 20 min before scalding to 35°C. The curds were milled and salted normally after cheddaring, and the time from starter addition to milling (5 h) was normal. The final pH of the Cheddar cheese was approximately 5.1, and while the cheese was of good quality, it lacked the strong flavour of traditional cheese.

Experimental Cheddar cheeses continued to be made (van Leeuwen, 1982; Sutherland and Jameson, 1982), but problems remained, such as the loss of calcium, phosphate and lactose into the retentate, and the adverse effect of excess whey protein on the texture of the cheese. Nevertheless, workers persisted with their trials, and while Green *et al.* (1981a) reported that curd formation in concentrated milk was faster than in normal milk, in spite of a reduced amount of rennet, Friis (1981) suggested that the concentration of milk could be used to increase vat capacity. Green *et al.* (1981a) also studied the slower rate of pH change in concentrated milks because of their higher buffering capacity, even though curd formation was faster and a lower level of rennet was used. The concentration of the milk did not affect starter growth, but the increased buffering action meant that the pH of the curds was higher than normal and fat was lost into the whey when the milk was concentrated more than $\times 2$. Green *et al.* (1981b) found that the curds from concentrated milks were coarser than normal. Reuter *et al.* (1981) showed the existence of a correlation between rennet action and protein concentration, in that the time of the first phase of rennet action decreased with protein content, while Storry and Ford (1982) reported that coagulum strength was correlated with the casein and divalent ion concentrations rather than the fat/casein ratio. Homogenization reduced the rate of syneresis by binding fat into the curd, but altered the structure of the coagulum.

Equally positive was the evidence that increased whey protein incorporation into the cheese served to increase yields (IDF, 1991), while financial savings on rennet and vat space could be achieved with milks concentrated by a factor of up to 2 (Kosikowski, 1988). Soft cheeses like Camembert are made successfully from concentrated milk (Shaw, 1993), as are semi-hard cheeses of the white-brined variety. Obviously, a semi-hard cheese made from the ultra-filtered milk, e.g. Danish Feta, has a very different texture from the traditional Feta cheese of Greece (Tamime and Kirkegaard, 1991), and the Danish cheese is usually referred to as 'cast-Feta' to indicate that it has been made from concentrated milk without pressing/free drainage of the whey.

Some of the problems associated with the use of concentrated milk for cheesemaking have been reviewed by Lelievre and Lawrence (1988), and Tamime (1993) has described in detail recent developments in the manufacture of hard-pressed cheeses. Although UF appears to have potential for utilization in soft cheese manufacture, the application of RO has not been exploited to the same extent (Grandison and Glover, 1994). Fouling of the RO membranes is one major problem (Kulozik and Kessler, 1988), but the retention of the lactose and minerals has an adverse effect on product quality.

References

Bennasor, M., Rouleau, D., Mayer, R. and Tarodo de la Fuente, B. (1982) *Journal of the Society of Dairy Technology*, **35**(2), 43.

Bungard, A.G., Olsen, O.J. and Madsen, R.F. (1972) *Dairy Industries International*, **37**, 539.

Chapman, H.R., Binnes, V., Glover, F.A. and Scudder, P.J. (1974) *Journal of the Society of Dairy Technology*, **27**, 151 .

Coolbear, T., Monk, C., Peek, K. *et al.* (1992) *Journal of Membrane Science*, **67**, 613.

Covacevich, H.R. and Kosikowski, F.V. (1978) *Journal of Dairy Science*, **61**, 529.

Daufin, G, Labbe, J.P., Quererais, A. and Michel, F. (1991) *Netherlands Milk and Dairy Journal*, **45**, 259.

Friis, T.L. (1981) *Dairy Science Abstracts*, **43**, 601.

Georgakis, G.A. (1978) *XX International Dairy Congress*, Vol. E, 800.

Grandison, A.S. and Glover, F.A. (1994) In *Modern Dairy Technology*, Vol. 1 (ed. R.K. Robinson), Chapman & Hall, London, pp. 273–312.

Green, M.L., Glover, F.A., Scurlock, E.M. *et al.* (1981a) *Journal of Dairy Research*, **48**(2), 333.

Green, M.L., Turvey, A. and Hobbs, D.G. (1981b) *Journal of Dairy Research*, **48**(2), 343.

Houldsworth, D.W. (1980) *Journal of the Society of Dairy Technology*, **33**(2), 45.

IDF (1991) *New Applications of Membrane Processes.* Special Issue No. 9201, International Dairy Federation, Brussels, Belgium.

Jacobsen, M.K. (1978) *XX International Dairy Congress*, Vol. E, 799.

Kosikowski, F.V. (1988) *Journal of Dairy Science*, **71**, 557.

Kulozik, U. and Kessler, H.G. (1988) *Journal of Food Science*, **53**, 1377.

Lelievre, J. and Lawrence, R.C. (1988) *Journal of Dairy Science*, **55**, 465.

Lewis, M.J. (1997) in *Separation Processes in the Food and Biotechnology Industries* (eds A.S. Grandison and M.J. Lewis), Woodhead Publishing Limited, Cambridge, England, pp. 97–140.

Mahaut, M. and Maubois, J.L. (1978) *XX International Dairy Congress*, Vol. E, 793.

Maubois, J.L. (1973) *Nordisk Mejeritidsskrift*, **39**(3) 57.

Maubois, J.L. (1979) *Technicien du Lait*, **12**, 934.

Maubois, J.L. and Kosikowski, F.V. (1978) *International Dairy Congress*, Vol. E, 791

Maubois, J.L., Mocquot, G. and Vassal, L. (1971) French Patent 2 052 121.

Mottar, J. and Delbeke, R. (1978) *XX International Dairy Congress*, E, 797.

Nielsen, I.K., Bungard, A.G. and Olsen, O.J. (1972) *Process Biochemistry*, **7**(9), 17.

Prokopek, D., Buckheim, W. and Knoop, W. (1978a) *XX International Dairy Congress*, Vol. E, 796.

Prokopek, D., Voss, E. and Thomasow, G. (1978b) *XX International Dairy Congress*, Vol. E, 795.

Renner, E. and Abd-El-Salam, M.H. (eds) (1991) *Applications of Ultrafiltration*, Chapman & Hall, London.

Reuter, H., Hissarich, D. and Prokopek, D. (1981) *Milchwissenschaft*, **36**(1), 13.

Rousseaux, P., Maubois, J.L. and Mahaut, M. (1978) *XX International Dairy Congress*, Vol. E, 794.

Shaw, M.B. (1993) in *Modern Dairy Technology*, Vol. 2 (ed. R.K. Robinson), Chapman & Hall, London, pp. 221–280.

Spangler, P.L., Jensen, L.A., Amundsen, C.H. *et al.* (1990) *Journal of Dairy Science*, **74**, 2809.

Storry, J.E. and Ford, G.D. (1982) *Journal of Dairy Research*, **49**(3), 469.

Sutherland, B.J. and Jameson, G.W. (1982) *XX International Dairy Congress*, Moscow, Vol. 1, Book 1, 450.

Tamime, A.Y. (1993) in *Modern Dairy Technology*, Vol. 2 (ed. R.K. Robinson), Chapman & Hall, London, pp. 49–220.

Tamime, A.Y. and Kirkegaard, J. (1991) in *Feta and Related Cheeses* (eds R.K. Robinson and A.Y. Tamime), Woodhead Publishing Limited, Cambridge, England, pp. 70–143.

Van Leeuwen, H.J. (1982) *XXI International Dairy Congress*, Moscow, Vol. 1, Book 1, 457.

Veinoglou, B., Voyatzoglou, E. and Antifankakis, E. (1978) *XX International Dairy Congress*, Vol. E, 798.

18 Cheese whey and its uses

Cheesemaking by traditional or modern methods inevitably produces a large quantity of whey (approximately 83% of the volume of milk used), and it is difficult to divorce whey disposal from cheese technology, because the disposal of whey is a major industrial consideration. In addition, whey contains valuable nutrients, i.e. whey proteins, lactose and minerals, and hence it should not be thrown away as waste, but should be used for stock and human nutrition.

18.1 Utilization and disposal of whey

An excellent review of the uses of whey has been published recently by Zadow (1994), Table 18.1 and Figure 18.1 are included merely to highlight a few of the ideas which are being examined.

Nevertheless, in spite of these options, cheesemakers have often regarded whey as a waste product to be dumped in the sea, rivers, quarries, mines and any convenient ground. Disposal on agricultural land as a fertilizer was at one time popular, but modern pollution standards designed to protect water courses from run-off have made casual disposal less attractive. This pressure, together with a greater appreciation of the inherent nutritional value of the

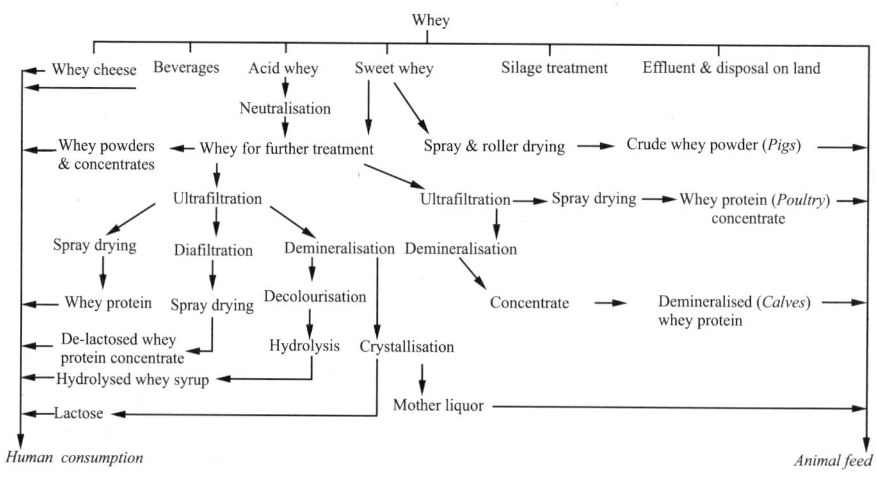

Figure 18.1 Examples of methods of whey processing and disposal.

Table 18.1 Some of the uses to which whey has been put

Pig and poultry feeds.
Drying for addition to bread to increase its nutritive value, or to infant, invalid and dietary foods for the same purpose.
Inclusion in carbonated beverages.
Fermentation to alcoholic drinks.
Precipitation of the whey proteins for addition to foods and elevation of nutritive value.
Pharmaceutical and cosmetic preparations.
Manufacture of potable-grade alcohol for liqueurs.
Manufacture of lactose and/or galactose/glucose syrups.
Production of Whey cheeses, e.g. Ziger, Urda, Ricotta.
Separation of riboflavin separation.
Manufacture of lactic or propionic acid for food purposes.
As a fermentation medium for the production of antibiotics, solvents like acetone and butanol, or biomass for human or animal feeding.

constituents of whey for both humans and livestock has initiated a complete change in attitudes towards whey and its handling (Zadow, 1992).

18.2 Quality of whey

The wheys from cheesemaking vary according to the type of cheese, and the content of protein, salts, fatty acids, lactose and lactic acid also vary. Wheys from soft cheeses or blue-veined cheeses tend to quite acid, i.e. 0.18–0.25% as lactic acid, while wheys from hard-pressed cheeses are likely to be even more acid, i.e. 0.22–0.4%. Whey taken off at pitching time for textured or from a non-textured cheese (i.e. Edam) is less acid than the whey that drains from cheese during the texturing process or from cheese during pressing. In addition, whey taken from the press is likely to have a high salt content, so that it cannot be used directly for stock feed or drying.

Consequently, the tendency is to segregate wheys for further treatment or disposal. For example, when extra sweet whey is required for the supplementation of dietary, invalid or baby foods, the whey is required to have a very low lactic acid content (i.e. acidity 0.12–0.13%). Such wheys are also suitable for obtaining high yields of lactose (Nielsen and Sprogo, 1997).

As mentioned earlier, concentration of the protein by ultrafiltration and diafiltration can give concentrates of over 75% protein (Morr, 1989), and some specialized products are available with over 90% protein (Horton, 1996).

The composition of a typical whey from a Cheddar cheese plant is shown in Table 18.2. The lactose content is fairly constant, but the precise level depends on the amount converted to lactic acid. By contrast, the protein and fat contents are largely dependent on how the coagulum has been treated; curd fines in the whey increase the protein figure, while rough curd handling increases the fat level in the whey. The mineral content of whey is fairly

Table 18.2 Typical composition (%) of cheese whey (Cheddar)

Water	93
Lactose	4.9
Crude protein	1.1
Fat	0.2
Lactic acid	0.2
Mineral Salts	0.6
Total solids	7.0

constant, but can be altered by the addition of chemicals to the milk, e.g. nitrates, calcium chloride or calcium hydroxide.

The screening and centrifugation of whey will alter both the protein and fat contents. Indeed, the separation of fat from whey for butter making is normal practice in many cheese plants; 3 kg of whey butter from 1000 litres of whey is a typical yield. The amount of casein in the whey from either centrifugation or screening is a valuable measure of the efficiency of the process, and is being used as such by managements. High levels of casein in whey indicate rough handling of the curd, inadequate curd formation through, perhaps, a lack of rennet or excess calcium binding into the curd, or an unsuitable milk supply.

18.3 Whey cheese

The urge to use whey more effectively has given some impetus to the extended use of whey or whey protein for human food. The manufacture of whey cheese is an old craft still practised in many areas. It was a farmhouse speciality in earlier days when goats and sheep provided both meat and milk, especially in the remoter mountainous regions like Norway; whey cheeses like Mysost, Gjetost and Niesost are still familiar.

Although all whey cheeses are made by similar methods, the choice of ingredients produces different varieties. For example in Norway the following cheeses are made from whey–milk combinations:

Gjetost	Goat's milk whey with added skim milk;
Flotemysost	Cow's milk whey with fresh milk or cream;
Blandet Gjetost	Goat's milk whey with cream added;
Gudbrandsdalost	Cow's milk whey with goat's milk or cream added;
Mager-Mysost	Cow's milk whey with some milk added;
Pultost Surost and Suprim	Sour whey with skim milk added.

Such cheeses are high-energy foods as indicated by a typical analysis of Gjetost:

Fat-in-dry-matter	33%
Dry matter	82%
Fat	28.5%
Protein	12.0%
Lactose	36.0%
Ash	4.5%

The cheese is sweet with a typical cooked flavour, and when goat's milk is used, there is a piquancy to the flavour.

Ricotta and similar cheeses of the Mediterranean basin are made from whey–milk or directly from whey (e.g. Ricottone). Production is on an industrial (Figure 18.2) as well as an artisanal scale.

Manual methods for making whey cheese, such as Mitzithra (Greece), require little more equipment than heated vats to cause separation of the protein curds from the whey, but more sophisticated methods use vacuum evaporation equipment for concentrating the whey mixture. Whey, milk, skim milk and cream are mixed according to a formula for each variety, and multistage vacuum evaporators may be used for the first concentration. The second concentration to 80% solids produces the final viscosity of the cheese and induces the desirable Maillard reactions (protein–sugar interactions) for extra flavour. The final cooling, mixing and packing units are designed to encourage crystallization of the sugars to prevent 'sandiness' in the product. Control of temperature is essential for satisfactory evaporation, 'browning' and cooling, and instrumentation is a necessary part of the equipment (Skaltveit, 1958).

The common faults found in whey cheese are listed below.

(1) Unstable fat – exuded during stirring and handling.
(2) Spots of light colour – inclusion of particles of casein.
(3) Dark brown colour – excessive heating.
(4) Short texture – due to hard milk fat, soured milk or an uneven mixture of fat, protein and water.
(5) Hard texture – too high a dry matter content.
(6) Soft texture – too low a dry matter content or too soft fats.
(7) Sandy texture – quick cooling that produces large crystals of lactose.
(8) Burnt flavour – high temperature heating.
(9) Insipid flavour – unsatisfactory browning of the liquid cheese.
(10) Sour flavour – soured/too acid milk used.
(11) Mouldy flavours – mould growth in the cheese or on the surface.

Sauge (1960) wrote about the equipment required for making whey cheese, and Holland (1980) described both the equipment for, and the manufacture of, various varieties of whey cheese. A discussion on the manufacture of Gudbrandsdalost whey cheese by Bakkene (1980) updates previous references. Zerfiridis and Manolkidis (1978) reported on the pro-

Figure 18.2 Vat and drainer for cooking, acidification and draining of Ricotta curd. (Courtesy of APV Nordic Cheese AS)

duction of whey cheese from Telemes whey, and Anifantakis (1991) gives details of all the major whey cheeses manufactured in Greece. Ilardi (1980) reported on the production of Ricotta cheese from cheese whey, and a more recent account of Ricotta manufacture has been provided by Shaw (1993).

Employing whey in a different manner, whey powder has been proposed as an additive to cheese milks to increase the yield of cheese (Brown and Ernstrom, 1982). However, Banks and Muir (1985) found that whey proteins tended to weaken the body of cheese, and could also produce off-flavours in mature hard-pressed cheeses; as mentioned earlier, similar problems have been encountered using ultra-filtered milks for the production of Cheddar cheese. Roller-dried whey is not suitable for incorporating into cheese curds.

18.4 Whey treatments

Whey treatments demand a pasteurization process to destroy micro-organisms (especially lactic bacteria) prior to subsequent usage. Neutralization of the whey with soda or lime produces lactate which, unless removed, causes difficulties when used in calf feeds. Neutralization and subsequent demineralization is, therefore, essential to bring the properties of the whey nearer to those of sweet whey values.

Heat treatment may be required also to remove any residual rennet because, during cheesemaking, a proportion of the rennet is lost with the whey. If this enzyme is not removed prior to spray-drying, the resultant powder retains sufficient enzyme activity for a suspension of the powder in water to gel spontaneously on heating; the powder would be unsuitable, therefore, for many food uses (Harper, 1992). To avoid this problem,

Table 18.3 Effect of different heat treatments on the milk clotting activity[1] of some commercial rennets

Coagulant	Clotting time (min)			
	No heat	55°C/5 min	55°C/10 min	62°C/30 min
A	3.5	> 120	–	–
B	4.0	> 12	–	–
C	3.6	19.8	42.6	> 300
D	3.5	3.7	5.3	16.3

[1] As determined by adding 1 ml of diluted enzyme to 10 ml of reconstituted skim-milk powder (10% total solids +0.01M $CaCl_2$) at pH 6.5 and incubating at 30°C (Berridge, 1952).

pasteurization of the whey during subsequent processing is essential, as is knowledge of the heat stability of the commercial rennet being used in a given cheese plant. Thus, as shown in Table 18.3, the heat lability of commercial preparations does vary, and the end-user of a whey powder needs to know if contaminant enzymes are likely to be a problem.

18.5 Waste disposal

In view of the regulations regarding the disposal of waste waters into water courses and rivers, the disposal of normal effluent from a cheese plant (average BOD 80–200) is capital expensive. Since whey has a BOD of 4000–4800, it is clear that, apart from the loss of valuable whey proteins, the disposal of whey through a treatment plant would add greatly to the cost of cheese production. Royal (1974) has given a resumé on the prevention of pollution by dairy wastes, and Wheatland (1974) has described the treatment of waste waters from dairy factories. Fisher (1968) has written a review on dairy wastes. More recent general reviews include Forster and Wase (1987) and Dazell (1994) while specific information on equipment and processes is given by Dégrement (1991). Donkin (1997) reviewed the problems of sludge settling in aerobic biological systems treating dairy processing waste waters.

Within the EU, waste treatment facilities will need to meet the Directive 91/271/EEC 'Concerning urban waste water treatment'.

References

Anifantakis, E.M. (1991) *Greek Cheeses*. National Dairy Committee of Greece, Athens.

Bakkene, G. (1980) *Meieriteknikk*, **5**, 25.

Banks, J.M. and Muir, D.D. (1985) *Journal of the Society of Dairy Technology*, **38**(1), 27.

Berridge, N.J. (1952) *Analyst*, **77**, 57.

Brown, R.J. and Ernstrom, C.A. (1982) *Journal of Dairy Science*, **65**, 2391.

Dazell, J.M. (ed.) (1994) *Food Industry and the Environment: Practical Issues and Cost Implications*, Blackie, London.
Dégrement (ed.) (1991) *Water Treatment Handbook*, Vols I and II, Lavoisier Publishing, Paris.
Donkin, M.J. (1997) *International Journal of Dairy Technology*, **50**(2), 67–72.
Fisher, A.N. (1968) *Dairy Science Abstracts*, **30**(11), 567.
Forster, C.F, and Wase, D.A.J. (eds) (1987) *Environmental Biotechnology*, Ellis Horwood, Chichester.
Harper, W.J. (1992) *New Applications of Membrane Processes*, IDF Doc. No. 9201, International Dairy Federation, Brussels.
Holland, K. (1980) *Meieriteknikk*, **7**, 41.
Horton, B. (1996) *Dairy Industries International*, **61**(5), 43.
Ilardi, S. (1980) *Rivista della Soc. Italiana di Scienza dell Alimentazione*, **9**(6), 441.
Morr, C.V. (1989) in *Developments in Dairy Chemistry*, Vol. 4 (ed. P.F. Fox), Chapman & Hall, London, pp. 245–284.
Nielsen, A. and Sprogo, J. (1997) *Dairy Industries International*, **62**(5), 21.
Royal, L. (1974) *Journal of the Society of Dairy Technology*, **27**(2), 66.
Sauge, P. (1960) *Meieriposten*, **49**, 429.
Shaw, M.B. (1993) in *Modern Dairy Technology*, Vol. 2 (ed. R.K. Robinson), Chapman & Hall, London, pp. 221–280.
Skaltveit, T. (1958) *Nordisk Mejeritidsskrift*, **24**, 133.
Wheatland, A.B. (1974) *Journal of the Society of Dairy Technology*, **27**(2), 71.
Zadow, J.G. (1992) *Whey and Lactose Processing*. Chapman & Hall, London.
Zadow, J.G. (1994) in *Modern Dairy Technology*, Vol. 1 (ed. R.K. Robinson), Chapman & Hall, London, pp. 313–374.
Zerfiridis, G.K. and Manolkidis, K.S. (1978) *Milchwissenschaft*, **33**(10), 617.

19 Selected cheese recipes

As already stated, it is prudent for a cheesemaker attempting the manufacture of unfamiliar varieties of cheese to first gain experience with that variety. Although a recipe is a guide to the practice of cheesemaking there are often sound reasons, sometimes local, for certain deviations from written guides. These reasons may not apply elsewhere.

The following recipes were selected as being representative of farm, dairy or creamery practice. An endeavour has been made to include a range of recipes for the more familiar types of cheese as well as recipes which give variations from these curds by reason of processing temperatures, curd manipulation, brining or ripening conditions. Some recipes use traditional methods where heat treatment of the milk was not practised; if heat treatment is not used then stringent control of raw milk quality, process and environmental hygiene throughout the cheesemaking and maturation are essential in minimizing risk, as discussed in earlier chapters.

List of the cheese recipes

1. Appenzeller	Switzerland
2. Aragackij Syr	Armenia
3. Asiago	Italy
4. Awshari	Iraq
5. Baker's Cheese Curd	USA
6. Beaufort	France
7. Beli Sir u Kriskama	Serbia
8. Bjalo (Belo Salamureno Sirene)	Bulgaria
9. Blue Vinney (Dorset Blue)	England
10. Bresse Bleu	France
11. Brie	France
12. Brinza	Israel
13. Bukovina	Ukraine
14. Burduf	Romania
15. Butterkäse	Germany
16. Caerphilly	Wales
17. Caledonian	Scotland
18. Cambridge	England
19. Camembert	France

20. Cantal	France
21. Carrick	Scotland
22. Cascaval Dobrogen	Romania
23. Cheese base for the addition of solid flavourings	–
24. Cheddar	England
25. Cheshire	England
26. Colwich	UK
27. Cotswold	England
28. Cottage Cheese	USA and Europe
29. Coulommiers	France
30. Cream Cheese	–
31. Danablu	Denmark
32. Derbyshire	England
33. Domiati	Egypt
34. Double Cream Cheese	–
35. Double Gloucester	England
36. Dunlop	Scotland
37. Edam	Netherlands
38. Edelpilzkäse	Germany
39. Emmental	Switzerland
40. Estepe	Brazil, Argentina
41. Feta	Greece
42. Feta from UF concentrated milk	–
43. Fromage Frais (Triple cream Suisse, Demi-Sel, Neufchâtel)	France
44. Gamelost	Norway
45. Gorgonzola	Italy
46. Gouda	Netherlands
47. Gruyère	France/ Switzerland
48. Halloumi	Cyprus
49. Herregaardsost	Sweden
50. Hervé	Belgium
51. Italico	Italy
52. Kashkaval (Kaschkawal)	Bulgaria, Hungary
53. Kefalotyri	Greece
54. Kesong Puti	Philippines
55. Kingston	England
56. Lactic Curd Cheese	UK
57. Lancashire	England
58. Leicester(shire)	England
59. Limburger	–
60. Liplovska Bryndza	Slovakia

61. Livarot	France
62. Manchego	Spain
63. Mascarpone	Italy
64. Monterey	USA
65. Mozzarella	Italy
66. Parenica	Slovakia
67. Parmesan	Italy
68. Peakland	England
69. Pecorino Romano	Italy
70. Pelardon	France
71. Petit Suisse	France
72. Pont L'Evêque	France
73. Provolone	Italy
74. Quark	Germany
75. Reading Yellow	England
76. Ricotta	Italy
77. Roquefort	France
78. Saint-Maure de Touraine	France
79. St. Paulin	France
80. Sage Derby (or Derbyshire)	England
81. Samsoe	Denmark
82. Sbrinz	Switzerland
83. Single Gloucester	England
84. Steppenkäse	Austria, Hungary, Germany
85. Stilton	England
86. Surti	India
87. Svecia	Sweden
88. Telemea	Romania
89. Tilsiter	Germany
90. Trappist	Bosnia
91. Vacherin	France
92. White Pickled Cheese	Eastern Europe
93. Wensleydale	England
94. Whey Curd Cheese	England
95. Whiteknights	England
96. Yorkshire Cheese Curds	England

1. Appenzeller

Swiss cheese made in two qualities, quarter fat and full fat. Flat, round cheese shape. Various sizes: 20–30 cm diameter by 10–15 cm high, weight 6–8 kg.

Rind	Firm with yellow to brownish smear coat.
Curd	Ivory white to yellowish colour.
Texture	Soft, full, fatty but sliceable. Gas holes few but round (0.5–10 mm) maize corn size.
Flavour and aroma	Clean (not tainted) mild, aromatic. Quarter fat cheese flavour is stronger than the full fat cheese due to the longer ripening period.
Milk	Cow's milk, fresh 3.3–3.6% fat (morning and cooled evening milk mixed).
Heat treatment	Mainly none. Cooled to 30–31°C.
Starter	Usual to use small quantity of starter (0.05–0.1%). Sometimes add propionic culture along with lactic starter.
Rennet	Rennet extract, 20–25 ml per 100 litres milk but sufficient to give firm curd at 30°C in 30 min.
Cutting	Cut slowly to hazelnut size. Stir to float the curd.
Scalding	Scald to 42–46°C for full fat curds, or 36–38°C for quarter fat curds; when the curd is firm take of the whey.
Hooping	Fill the hoops with curd and press for at least 60 min.
Pressing	(Traditional presses are stone weights).
Salting	After pressing to shape the cheese is left for 12 h in moist atmosphere and then immersed in brine bath 18–20°B (18–20% salt), 10–14°C for 1–3 days. Dry off for 10 days. During this time the smear coat forms.
Maturation	Coating with the 'Sulz' follows. ('Sulz' is a liquid mixture of wine, yeasts, salt and spices (seasoning).)
Final maturation	Store cheese for 4 weeks. Then in cheese factory's store for 4–6 months for full fat cheese, 6–7 months for quarter fat cheese.
Faults	Blown curds, deformed cheese shape. Whitish greasy appearance developing later. Curds spongy, short texture, crumbly, soft, weak. Large Emmental-like gas holes. Secondary fermentations, bleached curd, mite infestation. Sour, bitter, rancid flavours.

2. Aragackij Syr

Typical smear coated semi-hard Armenian cheese. Approximate sizes 21–23 cm diameter by 8–10 cm high and weight 3–5 kg. Also in block shapes.

Rind	Thin, smooth with yellowish smear coat.
Curd	Bright yellow colour.
Texture	Pliable with round to oval irregular gas holes.

Flavour and aroma	Clean, lightly ammoniacal, but typical of the type of milk (sheep's or goat's).
Milk	Sheep's milk but mixture of sheep's and goat's milk may be used. Cow's milk may also be used in the mixture.
Heat treatment	Milk used raw but may be heat treated, 70–74°C. Cool to 30–34°C.
Additive	Heat treated milk only, 0.02% calcium chloride (added in solution).
Starter	Heat treated milk only, 0.5–1% as necessary.
Rennet	Temperature 30–34°C, 30–35 ml rennet per 100 litres, sufficient to give firm curd in 30–35 min.
Cutting	Cut to size of maize (6–8 mm). Stir to float the curds.
Scalding	Scald to 40–42°C taking 10–20 min.
Drainage	Ladle curd into hoops and leave to drain for 4–6 h.
Pressing	Turn curd into cloth and return to hoop. Press for 1–2 h with light pressure, 2 kg per cheese. Allow cheese to dry off.
Salting	When the cheese is dry, salt the curd for 2 days or immerse in brine (18–20% salt) at 12°C for 4–5 days, to give 2–3.5% salt in cheese.
Maturation	Ripen first at 15–16°C and at 90% RH for smear coat to form. Finally store at 10–12°C and 85–87% RH. Mature in 2 months.

3. Asiago

There are two types: (i) Asiago da Allievo: FDM 36%. Size: 32–45 cm diameter by 8–12 cm high, weight 8–14 kg. (ii) Asiago Grasso de Monte: FDM 48–52%. Size 38–42 cm diameter × 11–13 cm high, weight 11–13 kg. Some cheeses, Vecchio di Montagna and also Pecorino d'Asiago, are sheep's milk cheeses.

Asiago da Allievo

Rind	Yellowish, smooth, regular and not too thick.
Curd	White in young cheese, straw colour in older cheese.
Eyeholes	None to medium sized, regular distribution of eyeholes in older cheese.
Milk	Cow's milk used raw. Evening milk, acidity 0.2–0.21% in the morning. Partly skimmed, then morning milk added.
Heat treatment	None.

Starter	1% of starter (whey culture 0.82–1.00% acidity).
Rennet	Liquid extract or powder. Temperature 30–36°C. Quantity of rennet, 20–28 ml per 100 litres to enable coagulation in 20–35 min.
Cutting	Cut curds to wheat grain size.
Scalding	Heat to 38–40°C in 10–20 min with stirring, then adjust to 42–46°C. Stir until curd is firm.
Moulding	Use coarse cheesecloth in hoops and fill curd into hoop.
Pressing	Press using 4–5 kg per cheese for 1–2 h. Take out of moulds and dry off. Turn frequently.
Salting	Dry salt or immerse in salt bath.
Maturation	Wipe with salt water. Stack cheese one on another, stack four deep in summer, eight deep in winter. For consumer cheeses mature for 4–6 months at 10–12°C, 85–90% RH. Grating cheeses for 12–18 months. In mountain country, cheeses may take 2 years to mature, hence flavour is strong and piquant.

4. Awshari

Kurdish name given to a hard, ball-shaped cheese. Made in mountainous regions, mainly from sheep's milk. Sizes variable, methods of manufacture variable; flavour strong, piquant, and the curds become so hard they have to be soaked at times in order to use them.

Typical method of manufacture (mainly Spring and early Summer)

Milk	Sheep's milk (but may have admixture of goat's milk).
Heat treatment	None.
Starter	None (but sour whey may be added).
Colour	None.
Rennet[1]	Extract, 50–60 ml per 10 litres of milk. Temperature 35°C. Curd firm enough to cut in 30–80 min (variable due to rennet strength).
Cutting	Curd cut with perforated spoonlike blade.

[1] Rennet: Lamb vells are cut to strips and soaked in 5% salt solution (boiled water is used for solution). The vells may have been dried and mixed with sugar, alum, black pepper and cloves. Hence the rennet extract is spicy. Water melon seeds and orange peel may also be used in the rennet extract which takes 3–4 days to prepare.

Stirring	Stir to float curd and add 1% of coarse salt.
Scalding	Knead the curd in the whey then scald to 45°C taking 15–20 min.
Drainage	Transfer the curds to long, open-weave bags and knead the bags by hand until the curd has formed a ball-shaped mass in the bottom of the bag.
Pressing	Press and twist the bag to assist drainage.
Salting	Press the curd into ball shape and rub with salt for 4 days.
Maturation	Press balls of curd into animal skin prepared with salt. The leg openings of the skin provide drainage holes. In 8–10 weeks transfer the hard cheese balls into another skin and fill the skin spaces with Jaqi[2]. Maturation takes 2–6 months or longer. Hard cheese balls need soaking in water for 2 days to 1 week before use.

Variable characteristics of the Awshari cheese

Body	Variable from very hard and brittle, to soft and sliceable, elastic curd.
Texture	Rock-like with a few mechanical holes leading to cracked curds, also whey pockets and gas holes.
Flavour	Sharp, peppery, rancid to cheesy with piquant flavours and aroma. Sharp, rancid, salty or pungent aroma.

5. Baker's cheese curd

Used by the bakery industry for cheesecake or other fillings.

Milk	Skim milk separated from good quality cow's milk
Heat treatment	Pasteurize at 72°C for 15 s and cool to 31°C.
Starter	Culture as for butter manufacture (*Lactococcus lactis* sub-sp. *lactis* and *cremoris. L. lactis* sub-sp. *lactis* biovar *diacetylactis*, plus *Leuconostoc mesenteroides* sub-sp. *cremoris*). Add 5% of starter to skim milk.
Rennet	Temperature 31°C add 0.4–0.5 ml rennet per 100 litres milk. Leave to set for 5 h until the pH is 4.5 (or 0.55% acidity). Keep warm.

[2] Jaqi is a dried fermented spiced milk product made from whey proteins and milk curds which have been heated to 90°C to precipitate the proteins. The proteins are mixed with wild garlic, onions, dog rose petals, salt and other local weeds. The Jaqi, mixed with the already spiced curds, adds other piquant flavours to the cheese.

Drainage	Scoop the curd into calico bags. Pile the filled bags in a vat for drainage. Re-pile every 15 min. When the curd is drained collect the curd at the bottom of the bags and then pile the bags with a layer of ice between. (Note: the curd should remain moist but not wet for the baker to use effectively.)
Storage	Store below 8°C. The pH should not be lower than 4.4.

6. Beaufort

One of the more recent cheeses of the Gruyère family of cheeses made in the Western Alpine regions. Flat cylindrical shape in various sizes, 30–70 cm diameter by 10–16 cm high, weight 12–70 kg. Farmhouse or small dairy production but the cheeses are collected when new by local cheese stores for further ripening. Individual cheesemakers use temperatures for processing to suit their own methods. Mainly local sales.

Rind	Even, smooth, yellowish colour, no cracks or cloth marks.
Curd	Yellowish colour, fatty, waxy.
Texture	Close, greasy, sometimes small gas holes but slit openings are accepted.
Flavour and aroma	Mild, aromatic, clean taste, not bitter, slightly salty sometimes.
Milk	Cow's milk of good quality (3.7–4.0% fat).
Heat treatment	None in the mountain dairies. Animal heat is retained for starter growth.
Starter	20–50 ml of culture per 100 litres milk at 38°C. Culture includes *Streptococcus thermophilus*, *Lactobacillus helveticus*, *Lactobacillus lactis*. Skim milk for the culture is normally ripened to 20°SH (0.45% TA) to 40°SH (0.9% TA).
Rennet	18–30 ml rennet extract per 100 litres milk. Powdered rennet may be used (1.5–3 g). Temperature 25–32°C. (Note: Rennet quantity, acidity, and temperature combine to give a firm curd in 35 min.)
Cutting	Curd cut to size of hemp seed taking 15 min. Stir to float curd and then settle for 15 min.
Scalding	Scald slowly to 40–45°C and then heat to 52–54°C. Stir for further 40 min.
Hooping	Lift curd from vat (conical) with cloth on steel frame and place in wooden circular hoop (as for Emmental).

Pressing	Press for 3 h at 5 kg per kg cheese. Turn and press for 21 h at 10–15 kg per kg cheese. Acidity of whey from first pressing 13–27°SH (0.29–0.60% TA). Cool the cheese after pressing for 24 h at 10–15°C.
Salting	Immerse in brine bath (20–30% salt) at 10–15°C for 24 h. Alternative method – dry salt over a period of 10 days, twice a day for the first days, turning regularly.
Maturation	Mature at 12–15°C for first 3 months. Then at 10–12°C for 6 or more months (usually in cheese factory store). The cheese rinds are scraped to keep them clean, rather than the more usual brine wash.

7. Beli Sir u Kriskama

Typical white pickled cheese (Serbian White Cheese). Various shapes and sizes, typically 10 × 10 × 12 cm, weight 1–1.25 kg.

Rind	Moist, greasy.
Curd	Porcelain white.
Texture	Tender but firm, no gas holes usually but there may be a few small holes.
Flavour and aroma	Salty with typical aroma due to type of milk. Flavour slightly sour but clean.
Milk	Sheep's milk or cow's milk or mixed (sheep's milk 5–7% fat; cow's milk 3.1–3.3% fat).
Heat treatment	Farmstead dairy, none. Creamery HTST at 74–75°C for 15–20 s for cow's milk. Cool to 30–32°C.
Additives	Calcium chloride 0.03% in heated milk only.
Starter	0.3% of a mixed culture (*Lactococcus lactis*, *Lactobacillus casei*) in heat-treated milk, 0.1% in raw milk.
Rennet	Add at 25–30 ml per 100 litres. Sufficient to give firm curd in 70–90 min. Usually 30°C for raw sheep's milk, 32°C for cow's milk.
Cutting	Curd is cut horizontally and vertically with a knife to 60 × 60 × 20 cm pieces without crushing the curd, to release whey. After 15–20 min the curd is recut to smaller sizes then pressed for 1–4 h with a cloth over a frame laid on the curd. This curd is cut into 10-cm cubes.
Salting	Cubes are dry salted or immersed in brine (20–24% salt) for 10–20 h at 12–15°C.

Maturation	The salted curds are matured in parchment-lined casks or tinplate cans filled with 10–12% coarse salt pickling solution at 12–15°C. The curd is ripe in 4–6 weeks.
Storage	Longer term storage is at 10°C.
Faults	Curd may dissolve in the brine if too strong. Gas holes in the curd break it up into small pieces.

8. Bjalo (Belo Salamureno Sirene)

Typical of the mid-European, pickled cheese curds. Square shaped cheese. Various sizes 10 × 10 × 8 cm, weight 1 kg.

Rind	None to very thin, white in colour.
Curd	White, fairly soft.
Texture	Semi-hard, close, smooth, no gas holes.
Flavour and aroma	Slightly salty and acid. Flavour typical of this class of cheese (sheep's milk).
Milk	Sheep's milk, 6.5–9% fat.
Heat treatment	Low temperature, long hold heating at 68°C for 10 min. Cool to 30–32°C, normally 31°C.
Additive	30 ml of a 50% calcium chloride solution per 100 litres milk.
Starter culture	(*Lactococcus lactis*, *Lactobacillus casei* mixture). Add 0.1–0.3% starter.
Rennet	20°C. Curd should be set in 1–1.5 h.
Cutting	Cut the curd into 2–3-mm cubes.
Drainage	Put cloth over a long frame (90 × 20 cm); each frame holds the equivalent of 200 litres milk. Ladle curd onto the cloth. Cut the curd into strips. Tie the cloth over the curd to give some pressure. Leave to drain. Cover with wooden cover to keep warm.
Pressing	Apply 10–12 kg pressure to curd for 4–5 h. Room temperature 14–16°C.
Salting	Cut the curd into 12 × 12 cm pieces and salt in saturated coarse salt solution for 15–20 h, or dry salt for 24 h.
Packing	Pack salted cheese curd in metal or plastic drums up to 120 litres and also in smaller sizes. Fill up with brine (10–12% salt). Ripe in 2–4 weeks.
Storage	Store at 4–8°C for 10 months if necessary.
Faults	Contamination by micrococci, yeasts and moulds. Rusty brown spots or strips. Surface colour becoming dark grey colour. Rose red colour due to medication of cows (phenolthiazine).

Stringy curd due to *Bacillus viscosum*.
Inferior rennet.
Lack of acidity – antibiotics in milk.
Sharp and fermented flavour due to long storage.
Crumbly and/or hardened curd due to over-salting.
Uneven structure of curd due to high magnesium content of coarse salt.

9. Blue Vinney (Dorset Blue)

Blue-veined cheese, 30.5 cm diameter × 12.7 cm high, weight 6.3–8.1 kg. Rough brownish coat, long ripening period.

Characteristics	Firm open curd with irregular blue–green veins. Piquant, peppery, mild-to-strong flavour.
Milk	90 litres of low fat, traditionally hand skimmed, milk per cheese (i.e. 1–1.5% fat, separated from raw milk in shallow leads or pans – set for 12 h).
Heat treatment	Rarely pasteurized (which often leads to hard dry cheese).
Starter	None in raw milks but 11 ml per 100 litres in colder weather or milks chilled overnight.
Rennet	17 ml per 100 litres milk. Temperature 24–27°C. Rennet well diluted. (Raw vell extract – extra quantity may be needed.) Setting time 80–90 min. Acidity 0.35%.
Cutting	Both ways with vertical knife, only one way with horizontal knife. 13-mm blades. Leave for whey to rise (10–20 min).
Stirring	Stir to float curd in whey, then leave until whey acidity is 0.35% (same as before rennetting).
Whey off	Drain whey and allow curd to compact. Then cut into 15 × 15 cm blocks and pile double. Turn blocks to release whey every 20–30 min until acidity is 0.89–0.9%. Break by hand or coarse mill (break early, 0.65% acid, if curd is very wet).
Salting	Salt the broken up curds, 250 g per 100 litres milk.
Moulding	Wooden moulds (for warmth) with tin followers traditionally used (tinned steel with followers is the modern usage).
Pressing	Light pressure only for 4 h. Then turn and rub cheese with salt. Return to press at 70 kPa (500 kg per cheese) overnight. Wipe with cloth and remove to store, 95% RH.

Maturation	Temperature 10–16°C. Turn daily. Softer curds may mature in 8–12 weeks. Firmer curds require 6–9 months.

10. Bresse Bleu

Small cheese packed in carton box 125–500 g weight, 6–10 cm diameter × 4–7.5 cm high. Moist, full flavoured and resembling Gorgonzola. Formerly made on the 2-day curd system.

Milk	Fresh milk (3.8–4.0% fat) for 55% FDM.
Heat treatment	Farmhouse cheese – none. Otherwise either heated to 68–70°C for 16 s or 60–61°C for 30 min. Cool to 29–32°C.
Starter	According to weather and acidity of milk add 2–3% starter for heat treated and low acid milks, 1–2% starter for high acid milks. (Starters with predominance of *Lactococcus lactis* sub-sp. *cremoris* preferred.) Ripen to 0.18% acidity.
Rennet	20–27 ml per 100 litres to give very firm curd in 50–75 min.
Cutting	Cut into 2.5-cm cubes with wire frame knife. Leave for whey to form in the cut curds.
Ladling	Set out cloths (linen or fine weave) in bowls or basins on drainer tables. Ladle curd into cloths (even quantity in each cloth).
Drainage	Drainage in cloths assisted by tying cloth around curd for 60–90 min.
Moulding	Steel cloth lined hoops on mat are filled with curd from the opened up cloths. *Penicillium* powder is sprinkled on the curd in layers as they are placed in the hoop. Some makers prefer to add *Penicillium* cultures before renneting. Pack the first and last layer to give smooth outer coat. Turn the cheese frequently (6 times day one, 4 times day two). Temperature 17–20°C.
Salting	Some cheesemakers add a small amount of salt to the milk. Cheeses are dry salted by rubbing salt on the rind, but also by washing in 8% brine at 15°C for 8 successive days. The cheese should absorb 3.5% of salt.
Maturation	The cheeses are matured in humid atmosphere, 95% RH and 17–20°C for 1–2 weeks. The cheese is pierced 10–12 times and then transferred to a ripening cellar at 3–5°C and 92–95% RH. The cheeses are turned daily and should mature in 4–8 weeks. Some cheesemakers wrap the cheese in foil if the mould growth is too rapid.

11. Brie

Soft cheese made in several varieties:

Brie de Coulommiers: 14–16 cm diameter × 3 cm high, weight 0.45 kg.
Brie de Montereau (Melun): 22–24 cm diameter × 3 cm, weight 2.5 kg.
Brie de Meaux: 36–40 cm diameter × 2 5 cm, weight 3 kg. Also Pâté de Brie: 0.25 kg.

Varieties differ in external appearance and finish. Mainly the coat is close textured, whitish and shiny, sometimes with red–brown smear patches in older cheese.

Curd	Creamy white to golden yellow inside.
Texture	Close, smooth, without gas holes but sometimes crumbly.
Flavour and aroma	Prevailing flavour is mild, aromatic, slightly acid, but may become stronger, piquant in older cheese.
Milk	3.2–4.0% fat but may be standardized to 3.2–3.5% fat (to give firmer cheese). Acidity not to exceed 0.3%.
Heat treatment	Flash or HTST treatment 72–74°C for 40 s, cool to 30°C.
Colour	Sometimes slight colour used – 'Orlean' vegetable colour.
Starter	According to acidity of milk (none to 0.25%), 0.75% mixed culture starter. Note that a *Penicillium candidum* culture may be added at this stage.
Rennet	12–18 ml of rennet per 100 litres milk to give a firm curd in 2–3 h. Temperature 30–31°C.
Cutting	The coagulum is cut with wire hoops into 3 × 3 cm layers, or by a wire strung frame. The curd may also be ladled with a long-handled ladle with 2-mm holes for whey drainage.
Hooping	The two-piece hoops are set on cane, rush or wire grid mats and the curd is poured or ladled into the hoops. Temperature 20°C.
Drainage	After 18–24 h the top hoop is removed and the cheese turned over onto a new mat. This turning is repeated two to three times.
Salting	After 12 h the cheese is dry salted and a single strip hoop is placed round the cheese to keep it in shape. Coarse salt is used for the first salting (both sides) and fine salt is used for the second dry saltings with 12 h between.
Drying off	The cheeses are dried for 6–7 days at 12–14°C.
Storage	Store for 4 weeks at 10–14°C and 80–85% RH, when the cheese develops a white felt mat of *P. candidum* mould.

	Cheeses are then transferred to lower temperature 8–10°C until ripe in 4–5 months. Sometimes a patch or coat of *Brevibacterium linens* develops in some varieties.
Packaging	The cheeses are packed in greaseproof paper or parchment and placed in cardboard or woodchip circular boxes. Some cheese may be cut into portions and wrapped in foil or laminate before packing in boxes.
Faults	Chalky body results from over-acid milk and milk short of fat. Excess storage temperature may cause curd to liquefy. Red–brown smear growth causes high flavour and ammoniacal aroma.

Note: Fully mechanized systems for manufacturing Brie are now available – see earlier text.

12. Brinza

Brine pickled cheese variety made in Israel. Square or rectangular shape. Approximately 8 × 8 × 8 cm high, weight 300 g.

Rind	None.
Curd	White with many irregularly distributed small holes.
Texture	Elastic, soft, with increasing crumbly texture.
Flavour and aroma	Flavour according to milk used, i.e. sharp, acid or salty. Typical of pickled cheese. Piquant when mature.
Milk	Sheep's milk but may have 25% cow's or goat's milk added.
Heat treatment	Heat treatment required but varies from dairy to dairy (HTST or low temperature, long hold).
Starter	Not normally used.
Rennet	Temperature 28–29°C. Rennet extract powder or tablets may be used to give a firm curd in 1–1.75 h.
Ladling	Wooden table with sides 10 cm high, approximately 1 m square is lined with open drainage net and covered with cheesecloth. The curd is ladled with a special shovel type ladle to fill the table mould.
Pressing	The cloth is folded over the partly drained curd and the curd is pressed (15 kg per kg curd) for 2–3 h.
Salting	The curds are dry salted for 2–3 days, then the firmed curd is cut into blocks (8–10 cm) and immersed in a brine bath (9% salt) for 4–5 days.

Maturation	For immediate consumption the cheese is dried off. For longer storage time the cheeses are packed in layers in canisters and the spaces filled with 14% brine. The containers are stored in a cold room until required.

13. Bukovina

Partially washed curd cheese. Tall cylindrical cheese 10–12 cm diameter × 10–45 cm high, 4–6 kg weight. Also a brick-shaped cheese 25–30 × 14–15 × 10–12 cm, weight 5–6 kg.

Milk	Fresh, low acidity 0.17%, standardized to give cheese with 45% FDM.
Heat treatment	70–72°C for 15–20 s, cool to 28–30°C.
Additives	15–30 g calcium chloride per 100 kg milk. 10–30 g sodium or potassium nitrate per 100 kg milk.
Starter	Traditional culture with *Lactococcus lactis* sub-sp. *cremoris* and *L. lactis* biovar. *diacetylactis*. Use 0.75–1.5% culture added at 28°C.
Rennet	Dried or powdered rennet 2–3 g per 100 kg milk. Temperature: 28–30°C. Coagulation completed in 30–35 min.
Cutting	Coagulum is cut into maize-size pieces. Stir to float curds for even cutting.
Whey off (1)	Stir curds for 20–30 min till whey acidity is 0.1–0.13%. Take off 30–40% of the whey.
Scalding	Add sufficient pasteurized water at 50–60°C (about 15%) to raise the temperature to 32–35°C. Stir for 15–40 min until curds are firm.
Whey off (2)	Take off whey and cut curds to fit hoops. Turn curds into cloths and replace in hoops.
Pressing	Press curds in hoops (moisture content 44–48%, pH 5.4–5.6).
Salting	Immerse cheese in 20% brine at 10–12°C for 2–3 days, to give 1.5–2.2% salt.
Storage	Store cheese to dry off then mature at 12–14°C and 86–88% RH for 30 days, pH 5.3–5.45. Wax the dry cheese, store at 10–12°C for 2–3 months until mature (final cheese moisture 40–44%).

14. Burduf

Cheese made in high mountain huts from sheep's milk and matured in skin bags. One of the older established cheese types. Weight of curd in bags 2–8 kg or 30–60 kg.

Curd	White to yellowish colour.
Texture	Crumbly, Quark-like, spreadable, not hard.
Aroma and flavour	Strong flavour, slightly salty, piquant to sharp, typical of sheep's milk.
Milk	Sheep's milk, 7–8.5% fat. Milk well filtered.
Heat treatment	None.
Starter	None.
Rennet	Temperature 35–40°C. Rennet quantity according to source of rennet (25–30 ml rennet extract per 100 litres) but sufficient to produce firm curd in 35 min (lamb vell rennet may be used).
Cutting	Cut to size of wheat grains (3–4 mm), leave to firm up.
Whey off	Repeatedly knead the curd to express the whey.
Ripening	Ripen the curds for 6–10 days at 8–12°C in loaf-shaped pieces.
Salting	Break the curd pieces, add 2.5–3.5% salt and mince through mill or mincing machine to a fine paste.
Storage	Put the homogeneous mass of curd in sewn-up skin bags (sheep or goat). Press the curd to exclude air. Sew up the skin. Store for 10–15 days at 12–16°C. Tighten the skin after beating the air/gas out of the curd and store at 6–10°C. (Note: empty spaces in the skin allow mould to grow and the leather bags should be pressed to keep air out.)

15. Butterkäse

Semi-hard cheese made firmer today than formerly. Flat round shape, 19 cm diameter × 5–6 cm high, 1.5 kg weight. Sausage form 25 cm long × 8–9 cm diameter, 2 kg weight. Made mainly in two qualities, 45 and 50% FDM (but 60% FDM is also made).

Outside appearance	Yellow–brown to reddish, elastic rind.
Curd	Yellowish.
Body	Half-hard, equally ripe throughout.
Texture	Soft, pliable, not pasty or runny.
Flavour and aroma	Mild and slightly acid.
Milk	Cow's milk, fat 3.0–3.2% and 3.5–3.8%.
Heat treatment	HTST pasteurization 70–74°C.
Starter	1.5–2%, ripened at 30°C.
Rennet	Temperature 40–43°C, 40 ml of rennet extract per 100 litres milk to give firm curd in 14–18 min.
Cutting	Curd is cut to cherry-size pieces (2–3 cm) with harp or quirl. Stir to float the curd.

Washing	Take off 50% of whey and add 10% water, stir for 45 min.
Moulding	Scoop curd into moulds when firm but not solid. Turn three to five times in the first 2 h. Keep the room temperature at 38–40°C.
Salting	Salt in salt bath, temperature 23–24°C for 16–18 h, pH before salting 5.0–5.1 (brine strength 16%).
Storage	Dry off the cheese in store, 1 month at 5–6°C and 90% RH. Turn four to six times. Cheese coat may be coloured with annatto to yellow colour. Pack in film or foil and enclose in cartons. Store at not more than 6°C.

Note: The Austrian Butterkäse is softer and nearer in texture to the Italian Bel Paese but is sold in sausage or roll form.

16. Caerphilly

Normal cheeses are made from approximately 40 litres of milk. Weight 3–4 kg. Flat round shape, 25 cm diameter × 6 cm high.

Rind	Thin. Slightly mouldy coat.
Curd	White.
Texture	Close with open spaces.
Flavour and aroma	Mild flavour, pleasantly acid.
Milk	Mixed, 3.2–3.5% fat. Heat treated to 72°C for 15 s. Cool to 32°C.
Starter	Amount: 0.25% mixed cultures (in winter 1–1.25%).
Ripening	Variable, 10–90 min – normally 45 min at 32°C.
Rennet	Temperature 32°C. Rennet, 25–30 ml (diluted with 200 ml water) per 100 litres to set in 40–45 min. Acidity 0.19–0.21%.
Cutting	Cut both ways with vertical knife, 6-mm blades. Allow whey to rise then cut with horizontal knife. Acidity 0.13–0.14%. Curd firm.
Scalding	Scald to 33–33.5°C in 20–30 min. Stir until dry, i.e. no pockets of whey inside curd, or until imprint of hand remains on pressed curd (approximately 45–50 min). Acidity 0.15–0.165% (winter 0.18–0.185%).
Drawing whey	Acidity 0.2–0.21%. Curd scooped to conical mass at corners of vat and also along vat sides.
Texturing	Curd cones cut with long knife and pieces of curd piled along vat sides. Curd broken up or disturbed until acidity is 0.22–0.24%.

Milling	Some cheesemakers prefer to cut or break curd, others mill in coarse mill. Acidity 0.28 (summer)–0.40% (winter).
Salting	1% salt is mixed in the curd (42 g per 4 kg cheese). (Note that a small quantity of salt is also used to rub onto the cheese while it is being turned in the press.)
Moulding	Mould lined with fine calico with extra steel collar. Plastic moulds are now used. Curd packed evenly.
Pressing	Light pressure only at first, 24 kPa (120 kg per cheese). In 10–20 min turn and rub cheese with salt. (Note: only half normal salt is used in the curd.) Turn several times with increasing pressure. Finally leave overnight in press at 100 kPa (500 kg per cheese).
Brine bath	Cheese removed from mould and cloth and placed in brine bath (strength 18% w/v, made up to strength each day and discarded when the brine becomes cloudy) at 15–16°C for 12–24 h, to give 2% salt in the curd.
Storage	The cheeses are dried off from the brine bath and before sale, rubbed with rye flour/lime mixture to coat and keep dry. Temperature 10–13°C, mature in 10–14 days.

17. Caledonian

Bleu d'Auvergne type cheese. Cheese developed by the West of Scotland Agricultural College. Blue-veined cheese from lowland area of Scotland.

Milk	Cow's milk heated to 72°C for 15 s, cooled to 32°C.
Starter	2.5–3% of mixed culture starter (3 litres per 100 litres of milk).
Mould culture	Concentrated liquid commercial culture of *Penicillium roqueforti*, 10–25 ml per 100 litres of milk according to directions on culture.
Decolourant	Chlorophyll mixture (commercial) may be added to whiten the curd, 10 ml per 100 litres.
Rennet	Temperature 32°C, 30 ml per 100 litres well diluted. Set for 90 min before cutting.
Cutting	Cut into 6-mm cubes. Stir to float curd then settle curd in whey for 20–25 min.
Whey off	Allow whey to drain out of the curd in 35–40 min. Make channel in centre of curd to assist drainage.
Moulding	Ladle curd into the moulds (hoops). Turn six to seven times in the next 5 h and leave to drain overnight (no pressing).

Salting	Wash in 20% brine at 15–18°C for each of 3 days. Then dry salt on the fourth day.
Maturation	Transfer to temperature of 10–12°C and humidity of 92% RH, leave for 2 days and dry salt again. On the eighth day after making, prick cheese for aeration on the flat sides. Return to shelf on the round side. Turn each day. On the fourteenth day after making, prick again and foil wrap.
Storage	Store at 2–5°C for approximately 3 months until mature.

18. Cambridge

Cambridge slice or York cheese. Traditional English cheese, unsalted, soft, with coloured layer in the centre. Sometimes with coloured whipped cream on top. Size: 20 × 10 × 4 cm, weight 0.45 kg.

Milk	Fresh, good hygienic quality milk (3.5% fat).
Heat treatment	None, but sometimes 63°C for 30 min. Cool to 30°C.
Starter	Not more than 0.1% mixed culture starter.
Rennet	Add 30 ml of rennet extract per 100 litres. Then divide the milk (30% and 70% by volume) after mixing the rennet into the milk.
Colour	Add quickly 1.5–3.0 ml of annatto per 30 litres to the smaller volume of milk and stir in well. Leave both white and coloured milks to set.
Moulding	Two-piece moulds on a wooden base. Place straws (more recently straw mats) in the bottom of the mould – warm with boiling water. Ladle the curd carefully into the warmed moulds with white layer, then coloured layer, then white layer on the top.
Drainage	Leave to drain for 36–96 h in room at 18–21°C. Drainage is slower at low temperature.
Packing	Remove cheese from moulds while still on the straw base. Pack in wooden trays in open greaseproof packing. (Whipped cream topping on the top of the cheese is usually placed just before consumption.)

19. Camembert

Soft, flat, cylindrical cheese with white mould surface growth. Sizes variable – typically 10.5–12 cm diameter × 2.5–3.5 cm high, weight 240–325 g. Smaller cheese 8–9 cm diameter × 2.5–3.5 cm high, weight 135–145 g. Occasionally 80-g cheeses are made. Composition varies according to quality, e.g. 40, 45,

50 or 55% FDM. 'Veritable Camembert' must have 45% FDM, the 'Camembert de Normandie' having an AOC. Characteristics are:

Rind	Thin, white with white mould growth of *Penicillium camemberti*. Very ripe cheese may have growth of *Brevibacterium linens* on the edge.
Curd	White to yellowish. Curd becomes semi-transparent with central strip remaining white.
Texture	Soft, plastic but young cheese may be crumbly. When ripe, elastic and fatty, sometimes semi-liquid.
Flavour and aroma	Delicate aromatic mushroom flavour. Slightly alcoholic with ammoniacal aroma when fully ripe. Becomes a strong flavour when over-ripe.
Grading	Shape, 1 point; rind and mould growth, 3 points; body, texture and colour, 6 points; flavour and aroma, 10 points.

Farmhouse manufacture

On average 100 litres of good quality milk will give 48 cheeses.

Milk	3.8% fat milk direct from the cow at body heat. Sometimes evening and morning milks are mixed. Occasionally heat treated. (Thermization at 63°C for 30 min.) Acidity 0.16% (7.2°SH).
Starter	In winter and in heat-treated milk 0.1–0.25% starter added at 30°C. Mesophilic cultures used, e.g. *Lactococcus lactis*, sub-sp. *lactis* and *cremoris*, *L. lactis* sub-sp. *lactis* biovar. *diacetylactis*, *Leuconostoc mesenteroides* sub-sp. *cremoris*.
Colour	None up to 3 ml per 100 litres.
Rennet	Quantity depends on strength. 10–20 ml of extract or 2.2–2.7 g rennet powder per 100 litres milk; to give coagulation time of 15 min in summer, 18 min in winter, but firm enough to ladle in 1–2.5 h. Temperature 30–36°C according to fat content and time of year.
Moulding	Ladle the curd in slices evenly into cylindrical moulds (11.5 cm diameter × 11.5 cm high) with perforated sides, placed on cane straw, hessian or wire mesh mats on a rigid frame or board. Sterilize before use.
Drainage	Leave for whey drainage. Turn the cheese when firm enough, in its mould at 6–7 h. Turn at intervals three to four times (the last time out of the mould). Temperature 18–20°C.

Salting	Rub both sides of the cheese with salt during turning (6–8 g salt per cheese). Dry off at 18°C and 70–80% RH for 2 days.
Inoculation	A suspension of mould spores is sprayed on to the cheese. (Alternatively 0.5% of a liquid culture of the moulds is added to the milk before rennetting.)
Maturation	Store at 12°C for 10–12 days and when satisfactory mould growth has occurred pack the cheese and store at 4–5°C until sold.
Packaging	Whole cheeses are packed in greaseproof paper in a cardboard or chipwood box. Cut portions, 1/6 or 1/8 of whole cheese, are individually wrapped in metal foil film or paper and packed in a box.

Factory manufacture

Size of cheese according to trade requirements. FDM according to requirements. Depending on the type of mechanization used, quantity of milk may be in batches of 1000 or 2000 litres, but is often only 100 litres.

Milk	Fat adjusted to 2.3% for cheese with 40% FDM, 2.7% for 45%, 3.2% for 50% or 3.6% for 55% FDM. Acidity 0.16 to 0.24% as lactic acid (7–11°SH).
Heat treatment	Pasteurization since 1967, HTST 68–71°C for 120 s; also LTLT at 60–63.5°C for 30 min. (Thermization 63°C for 15 s used for stored milk.)
Starter	Mesophilic cultures (i.e. *Lactococcus lactis* sub-spp. *lactis* and *cremoris*, *Leuconostoc mesenteroides* sub-sp. *cremoris* mixtures). 0.1–0.25% for milk ripened before rennetting.
Colour	None to 2–3 ml per 100 litres milk.
Additive	Since 1953, calcium chloride may be used to assist coagulation. Quantity: up to 0.02% (5–20 g per 100 litres).
Spore culture	0.5% of a liquid culture of *Penicillium camemberti.*
Rennet	Depending on strength of rennet and time required for complete coagulation, 10–15 ml rennet per 100 litres of milk up to 22–30 ml rennet for coagulation in 1 h. Acidity 0.20 to 0.24%. Temperature 23–34°C according to time of year and fat content of milk.
Cutting	The coagulum is cut (in 1 h 10 min to 1 h 30 min) sufficiently to enable curd to be poured into moulds. Alternatively the curd may be cut after 45–55 min into cubes of 1–2 cm and the whey allowed to rise for 1–2 h. The whey is released before pouring the curd into moulds.

Moulding	The curd is poured into moulds made of stainless steel, aluminium, nylon or hard PVC, which are single piece moulds fixed on to a plate or wire grid. The moulds vary in size from 8 to 12 cm diameter and 11 to 15 cm height.
Turning	The cheeses are turned at approximately 1, 2, 6 and 12 h, in a temperature of 18–20°C.
Salting	The firm cheeses are immersed in salt brine (18–20% salt) at a temperature of 16–18°C for a time according to the size of cheese, i.e. 125 g cheese for 80–100 min, 320 g cheese 120–150 min. Cheese with higher fat content may be retained longer in the salt bath. The average salt content will be approximately 1.8%.
Inoculation	The cheeses, when dried off for 2 days at 18°C, are sprayed with the mould culture if the milk had not been inoculated.
Maturation	The cheeses are dried off for 2 days at 18°C and 70–80% RH. The cheeses are then matured for 10–12 days at 12–13°C and 90–95% RH for the mould to grow on the rind. Ripe in 21–35 days.
Packaging	The whole cheeses are packed in greaseproof or waxed paper and boxed in cardboard or chipwood. Cut portions of the cheese (1/8, 1/6, 1/2) are covered with aluminium foil, plastic film or waxed paper and replaced in cardboard or chipwood boxes. The packaged cheeses are held at 4°C until required.

Use of UF concentrated milk for Camembert cheese

Milk	Standardized to give required composition (typically protein to fat ratio, 1:1.45).
Heat treatment	HTST, 68–72°C for 16–20 s. Cool to 30–35°C.
First ripening	Add 0.8% starter and ripen to pH 6.0–6.2.
Ultrafiltration	Process at 30–52°C to give a solids content of 30–34% (fat 14.5%).
Second ripening	Ripen at 35°C to pH 5.0–5.2.
Rennet	Add rennet extract, 10–25 ml per 100 litres concentrate. Temperature 35°C. Stir rennet in thoroughly. (Note: since the milk solids are concentrated, a smaller total quantity of rennet is used.)
Moulding	Pour rennetted milk into moulds and leave to set (1–1.5 h) and then to mature for 12 h. Moulds are set on plates (no drainage).
Salting	Immerse cheese in brine (18–20% salt) for 15–30 min after turning cheese onto wire grids. An alternative

	method is to add salt to the milk before rennetting to produce 1.5–1.65% salt in the curd.
Spraying	Spray the drained cheese, after salting, with mould spores (*Penicillium camemberti*)
Maturation	Mature at 7–13°C for 4–6 weeks or according to requirements for trade. Pack as for other Camembert cheese.

20. Cantal

Hard-pressed cheese made in the Central Massif area of France on farms (with more than 50 cows) and in creameries. Tall cylindrical shape 30–50 cm diameter × 40 cm high, weight 30–45 kg. Smaller cheese 'Cantalets' are also made, weight 8–10 kg.

Rind	Grey–white with orange–red spots; darkening in colour as the cheese matures, becoming reddish and then grey.
Curd	Light yellow to yellow colour.
Texture	Firm, pliable, becoming more brittle as the curd ripens. No gas holes but curd fissures.
Flavour and aroma	Slightly bitter, piquant but sharp tending to soapiness.
Milk	Cow's milk used fresh with natural heat (3.1% fat in creameries).
Heat treatment	Farmhouse none. Creamery HTST 72°C for 16 s (but other temperatures have been used).
Starter	Farmhouse, none. Creamery, 1.0% active mixed starter culture.
Rennet	Temperature 30–33°C. 20–30 ml rennet extract per 100 litres milk according to time of year and milk quality. (Note: Creamery practice is to use 0.016% calcium chloride with heat-treated milk.) Curd should be firm in 1 h (factory, 35–40 min).
Cutting	Cut with harp to hazelnut size (1 cm). Stir to firm curd for 10–20 min then pitch curd.
Whey off	Press curd in vat with racks or plates. Take off the whey. Cut the curds, turn over, pile and press five to ten times. Develop acidity in the curd for 24 h then cut and mill. Cool the curd.
Salting	Salt the curds with 1.8–2.5% salt according to the curd moisture. Fill into the hoops ready for pressing.
Pressing	Turn the cheese in the press three to five times. Pressure finally 40 kg per kg of cheese, average 1.5 T per cheese.

Ripening	Ripen for 3 months at 8–10°C. Turn regularly and keep rind of cheese clean.
Faults	Rancid strong flavour due to *Clostridium butyricum*. Cracked rinds due to cold air draughts. Soapy flavour due to lipases produced by bacteria. Retention of whey in cold atmosphere during production. Weak wet curds due to lack of acidity. Hard marble-like curd with little flavour.

21. Carrick

Recent type of cheese made in Scotland – high moisture (50%) and low fat, similar to Port Salut cheese.

Milk	Pasteurized cow's milk, or heated to 68°C for 15 s. Cooled to 31°C.
Starter	Mixed culture starter, 0.5% per 100 litres milk. Ripen 20–30 min.
Colour	7 ml standard annatto per 100 litres milk.
Rennet	30 ml standard rennet per 100 litres milk at 31°C to produce curd to cut in 40–50 min.
Cutting	Cut with 16 mm knife to cubes the size of beans. Allow the whey to rise between cuts.
Scald (washing)	Stir for 5–10 min after cutting, then syphon off the whey down to the level of the curd, approximately 40% of whey. Replace the whey with equal volume of water at 33–36°C in order to raise the temperature to 31°C again. Stir for 50–60 min. The whey acidity (now reduced by water addition) 0.05–0.06% acid.
Texturing	Push the curd into the end of the vat to form a 9–10 cm layer. Then weight with a stainless plate for 10 min or until the curd can be lifted in cut sections into the moulds.
Moulding	Perforated 5-cm hoops (e.g. Perfora) are filled with the unbroken curd.
Pressing	Press for 1 h at 90 kPa, or 18 kg on each cheese). Take cheese from mould, dip in water at 48°C for 30 s, then re-press.
Brining	Brine next day for 2 h at 12–14°C. Allow to drain – dip in brine/annatto bath. Dry for 3–5 days, then pack in film.
Maturation	Cure for 4–6 weeks.

22. Cascaval Dobrogen

Southern Romanian cheese of the Kaschkawal group. Typical of the high scald kneaded cheeses made from sheep's milk. Flat, round cheese approximately 30 cm diameter × 10 cm, weight 6–8 kg. Semi-hard cheese.

Rind	Smooth, dry, yellowish red colour.
Curd	Ivory white colour, moderately firm.
Texture	Close with occasional slits rather than gas holes.
Flavour and aroma	Piquant, sharp to aromatic, typical of a sheep's milk cheese, sometimes salty.
Milk	Sheep's milk (6–7% fat).
Heat treatment	None.
Starter	None.
Rennet	Temperature 32–38°C. Rennet quantity according to the sources of the rennet but sufficient to give firm curd in 45 min (20–25 ml rennet extract per 100 litres milk).
Cutting	Usually cut with harp to 5–6-cm pieces. Leave to firm and then in 10–15 min cut again to pea-sized cubes (8–10 mm).
Scalding	The curd is scalded to 40–42°C and then stirred for 15 min longer before settling down.
Pressing	Remove half the whey and then press the curd in the vat. Then press in cloth (6–10 kg per kg curd) to partially cheddar the curd. Curd must reach pH 5.2 (1.2–1.35% acid) in 2–10 h before final scald. Leave in 5-kg blocks to ripen.
Scald and kneading	Test a piece of curd in hot water at 72°C. Test should show long strings of bright shiny appearance. Curd is cut into thin slices (5 mm). Immerse approximately 10 kg of the slices in hot water at 90°C and stir for 1–3 min with wooden slat to a homogeneous plastic mass.
Salting	Knead 2–2.5% coarse salt into the plastic curd, then form into cylindrical metal moulds to cool. Turn five or six times and next morning remove the mould.
Maturation	Store at 16–18°C and 80–85% RH for 1–2 days. Then rub salt into the rind for 2 days.
Storage	Store at 4–6°C. Protect the rind with wax or resin coating against mould growth.

23. Cheese base for the addition of solid flavourings

Low fat cream cheese base.

Cream	12–15% fat by direct separation or by dilution of higher fat cream.
Homogenization	Homogenize 7 MPa at 58°C.
Heat treatment	63°C for 30 min or HTST at 72°C for 15 s. Cool to 24°C.
Starter	Add 6–10% mesophilic lactic culture. In 3–3.5 h add rennet
Rennet	0.4–2.2 ml rennet per 100 litres to assist coagulation. Temperature 24°C. Curd should set in 3 h.
Cutting	Make wide cuts in the curd with 25-mm blades.
Drainage	When the curds are formed, warm to 32°C. Then pour gently into linen or hessian close-weave cloths or bags, 10–12 kg per bag. Leave for initial drainage. Transfer to cold room at 7°C until required. Note: the drainage of the curds in cloth bags may be replaced by a centrifugal cream cheese separator for large-scale production
Flavouring	Add flavouring to base, plus colour if necessary. Flavours may be fruit (e.g. strawberry, raspberry, banana) or savoury (garlic, onions, chives, ham, etc.) as required.
Packing	Pack in paper foil from mechanical extruder. Use wood-chip box or carton, or plastic film or plastic tubes.
Storage	Temperature 1–4°C.

24. Cheddar

Tall cylindrical cheese, variable size 35.5–38 cm diameter × 33–38 cm height, weight 18–22.5 kg. Truckle Cheddars are 12.5–15 cm diameter, 25–30 cm high and weigh 4.5 kg, Cheddlets weigh 3 kg, plus intermediate sizes. Block shapes are most popular with mechanized systems, typical weight 18 kg. Characteristics:

Colour	Uniform creamy yellow, unless coloured
Body	Firm
Texture	Close texture, uniformly free from slits or round hole. Plug not wet or rubbery (sweet), or chalky (acid), or corky (short of fat).
Flavour	Clean, nutty
Milk	Farmhouse production may use raw milk. Factory: only pasteurized milk, e.g. 72°C for 15 s.
Starter	Mixed culture of lactic organisms, 1.5–3% added at a temperature of at least 21°C. (DVS cultures at 31–32°C.)

Ripening	Leave at rennetting temperature 29–30°C for 1 h or until the acidity is 0.20–0.22%.
Rennetting	Dilute the rennet with five to six times cold clean water. Amount of rennet used, 25–30 ml per 100 litres milk according to season. Temperature 29.5–32°C as above.
Cutting	Allow the curd to firm enough to break cleanly over an inserted rod (6 mm diameter, 250 mm long). Time 30–50 min. Cut both ways with both knives (blades set at 12.5 mm) and then stir uncut curd into the centre and cut with vertical knife to 3–5 mm. (Note: some factory cheesemakers cut to 12-mm cubes to prevent loss of fat in the whey.)
Scalding	Stir gently until curd floats in the whey, then scald at a rate of 0.2°C/min rising to 0.3°C/min until the final temperature is 40°C. (Note that temperatures up to 42°C may be used, but since some starter organisms cannot withstand more than 40°C, high temperatures can only be used with certain starters.)
Pitching	Pitching is a stage which can be used as a control and not a fixed operation depending on acidity. Thus a slow or low acid curd can be pitched early while a high acid curd should be stirred until the whey is removed.
Whey off	Whey is removed when the curd is dry (i.e. shotty to feel) and of the right acidity (i.e. 0.20–0.24% (average 0.22%)). The pH will typically be 6.05–6.10. The curd is drawn or floated to one end of the vat to assist in whey removal. (Note: weights may be required on the curd (via a wooden rack) to assist Cheddaring or matting of the particles in small (farmhouse) vats.) The curd may be dry stirred for more adequate removal of whey especially in factory vats or mechanized systems.
Texturing	The compacted curd is cut into blocks and piled up every 15–20 min. (Weights on the piled curd are needed in small vats.) According to the rate of acid development the curd is turned and kept warm or aerated and cooled until acidity of 0.75–0.85% is reached (i.e. below pH 5.3). At this stage the curd should be dry, firm, mellow and exhibit 'chicken breast' structure. Wet acid curds should be cooled and spread out before milling. Dry sweet curds are kept warm and piled up

	until dry and mellow. Note: cheese from low casein milks may be milled at 0.55–0.65% acidity.
Milling	The curd is milled or chopped in a coarse mill to finger-sized pieces. The milled curd is aerated and cooled to 25–26°C to cool the enclosed fat before salting.
Salting	Approximately 2% salt is sprinkled on the curd (wet curd 2.1%) then well mixed in over a period of 15 min to mellow.
Moulding	Curd is packed into round or block moulds. On a factory scale block formers are now common.
Pressing	For small-scale production, pressure is applied gradually at first then increased, as the curd cools, to 75 kPa for 12–16 h. Turn into fresh cloths and re-press at 200 kPa for a further 24 h. (See notes on finishing cheese and also table of pressures and force per unit area of surface of cheese (Table 12.3).) Most commercial pressing is overnight only.
Storage	Store at ambient temperature to dry off then at temperatures from 7 to 11°C according to marketing requirements for ripe cheese.
Maturation	Depends on storage temperature and rate of acid production. Fully mature at 9–12 months. Block cheese marketed at 4–6 months. (See also Tables 13.3, 13.4 and 13.5 and the following comments.)

Mechanization of Cheddar cheese making

Full automatic mechanization of the Cheddar cheese process involves all the stages from milk to storage shown in the recipe given above. However, mechanization has often replaced manipulation of the curds in the whey off to storage stages of the recipe.

If cheeses are to be of similar quality the biochemistry of cheese made by mechanical or manual means must also be similar. The measures of control normally undertaken by the cheesemaker, through acidity and temperature control and manipulation of the curds, still have to be fed into an automatic controller. Automatic operation follows a similar recipe to the above, except that it must be interpreted according to the knowledge of the mechanical operations to be performed.

A 'New-Way' method for the manufacture of Cheddar cheese has used a thermophilus culture (TS2) along with normal starters in the vat. A time saving of 80 min is claimed by this method. Vacuum pressing is an essential feature of the method.

25. Cheshire

Tall cylindrical cheese, 30.5 cm diameter × 35.5 cm high. Block, film-wrapped cheese and other shapes are made. Weight, 1.3 kg up to 27 kg. Traditionally, three varieties (spring, summer/winter and autumn cheese) have been made according to the chemical quality of the milk produced in these seasons. Recipe 1 covers the main production type, i.e. summer/winter cheese. Recipes 2–4 give the alternatives in respect of acidity, etc. for other cheese. Characteristics:

Coat	No cracks
Texture	Open short silky to 'granular' texture. Medium moist (no free moisture)
Colour	Even, white or coloured
Body	Firm, not hard. Free from gas holes. No internal mould growth.
Flavour	Clean, high acid cheese.

1. Summer/winter Cheshire

Milk	Farmhouse: 182 litres mixed milk (3.5% fat), may be used raw per cheese. Factory: mixed milk, heat treated at 72°C for 15 s; sometimes flash heated to 74°C.
Starter	Mixed culture of lactic organisms. 1–3% starter added at minimum 21°C temperature.
Ripening	The temperature is raised to 30–31°C and held for 40–60 min until the acidity reaches 0.22–0.24%.
Colour	Annatto may be added according to market requirements, 0.016% up to 0.02.%, after the starter has dissolved, on average 18 ml per 100 litres.
Rennet	Quantity, 32 ml diluted with 200 ml of clean water per 100 litres milk. Temperature, 30–31°C.
Cutting	In 40–60 min cut the curd when it breaks cleanly. Cut both ways with the vertical knife (19-mm blades), then cut both ways with the horizontal knife. Whey acidity 0.135–0.155%.
Scalding	Stir until curd floats freely (15 min) then commence scalding, the final temperature depending on the initial whey acidity: Low acid, 0.135%: scald to 32°C in 40 min Medium acid, 0 15%: scald to 33.5°C in 45 min High acid, 0.16%: scald to 34.5°C in 50 min Stir until curd is firm and free from whey pockets inside the curd.

Whey off	When acidity reaches 0.21–0.23% (below pH 6.0) remove whey. Dry stir if the acidity is higher than 0.23%. Then push curd from outlet and channel up the centre for further whey drainage.
Texturing	The curd is not textured but kept free and open as far as practicable. Cut the curd into blocks and turn over. Break the curd blocks by hand every 15–20 min until small in size, then disturb to allow whey drainage until the acidity is 0.65–0.75%.
Milling	Mill the curd through fine mill once. (Sometimes mill after the salting if curd is well matted.)
Salting	Salt the curd either as it passes through the mill or on the milled curd. Then re-mill. Use fine salt at 1.85% (2.1% on wet curd).
Moulding	Fill curds into moulds after mixing in salt. If necessary reduce temperature to 25°C but avoid reducing temperature lower than 25°C. Turn cheese in 2 h. Leave filled moulds unpressed in warm room (Cheshire oven) at 21°C overnight.
Pressing	Press at 35 kPa (about 250 kg per cheese) for 2 h, turn into fine calico cloth and press at 70 kPa (about 500 kg per cheese). Next day turn and increase pressure to 140 kPa (1 T per cheese). Remove on third day, paste bandage and allow bandage to dry off in warm room.
Storage	Store in gradually reducing temperature: 13°C down to 10°C in 1 week. Store at 7°C for 1 month, or until marketed. Fully ripe in 6–9 months.

2. Spring Cheshire (modifications)

Milk	Low quality milk (3.1–3.3% fat).
Starter	1.5–2%.
Ripening	Until acidity is 0.20–0.21%.
Rennet	Quantity, 35 ml per 100 litres milk. Temperature, 31°C.
Scalding	33.5°C.
Salting	At 0.55–0.65% acidity add fine salt at 1.85% (on wet curd use 2.1%). Temperature must be maintained at 21°C in press.
Storage	Should be marketed before the end of July.

3. Autumn Cheshire (modifications)

Milk	High quality milk.
Starter	3.5–4.0%.

Ripening	Until acidity is 0.23–0.24%.
Rennet	30 ml (diluted six to eight times) per 100 litres milk. Temperature 30–31.5°C.
Scalding	33–35°C, 35.5°C if milk is acid.
Salting	1.85% of salt at acidity 0.75–0.85%. Temperature may be reduced to 17–18°C in press room and drying room.
Storage	Can be stored for 9–15 months if well made.

4. Block cheese

Used for cutting consumer portions. Therefore cheese must be drier than traditional cheese.

Scalding	Scald at 33–35.5°C.
Stirring	Dry stir curd after whey is removed.
Milling	Milling acidity 0.5–0.59%.
Pressing	Into mould immediately after milling and press overnight. Next morning film wrap. Curds tend to have closer texture and be less acid than traditional cheese.
Storage	At 7°C, send to market after 2–8 weeks.

26. Colwich

Soft cheese, 7.6 cm diameter × 2.5 cm deep or 15 cm diameter × 3 cm deep, weight 113–227 g. On a small scale, 5 litres milk gives three to four cheeses, depending on size. Often raw milk is used but pasteurized is preferable.

Method 1 (Farmhouse)

Starter	6 drops per 5 litres pasteurized milk; 3 drops per 5 litres raw milk.
Ripening	Ripen at 32°C for 12 h. (For no ripening period add 78 ml starter to 5 litres milk.)
Rennet	3.8 ml per 5 litres at 32°C.
Moulding	When curd is firm, ladle into mould lined with cloth. Fold over cloth. About 2 h later, tie the cloth loosely in a Stilton knot, pulling the curd gently away from mould sides. (Note: moulds have perforated bases in sizes from 7.5 to 15 cm diameter.) On the second day tighten the knot. The cheese ought to be ready to turn out of the mould, but contaminated milk or antibiotics will cause slow draining. Tighten knot if too wet. Turn in mould. Remove from mould when firm. If ripened, turn into dry cloth and keep at 18–21°C for 2–3 days.

Packing	For sale, wrap in greaseproof paper in a card or woodchip box.

Method 2 (Factory)

500 litres milk give 330 cheeses. Milk retained in milk churns of 45–50 litre capacity.

Heat treatment	Pasteurized (often surplus from bottling plant requirements).
Starter	75 ml per 100 litres pasteurized milk.
Rennet	75 ml per 100 litres milk. Dilute rennet with 2 litres of clean water. Leave to set for 4 h and plunge with stirrer enough to pour into moulds lined with cloth. Then treat as for Method 1.
Packing	The cheeses are usually packed for sale in open greaseproof cups, 24 to a wooden tray.
Finished cheese	The cheese should be hollow with raised sides. It is mainly eaten fresh and unsalted. Sometimes with mint or vinegar. A whipped cream whirl of slightly coloured curd may be placed in the cup or hollow.

27. Cotswold

Farmhouse cheese (developed by Gloucester Council as a smallholder cheese). Size variable, weight 0.45–1 kg.

Rind	Thin, yellow to red.
Curd	White to yellowish, soft, but pale red if coloured.
Texture	Close, soft but not too firm, not chalky or crumbly.
Flavour and aroma	Mild, pleasant, not bitter or acid.
Milk	Mixed morning's and evening's milk, clean (3.6% fat).
Heat treatment	None.
Starter	Variable according to acidity of milk. Typically 90–220 ml per 100 litres.
Colour	If used, 11 ml annatto in 100 litres milk.
Rennet	Temperature 32°C, 33 ml rennet extract in 100 litres milk to give firm curd in 60 min.
Cutting	Cut to 9–10-mm cubes. Stir for 10–20 min to float curd freely. Maintain temperature at 32°C.
Scalding	Scald to 40–41°C in 1 h. Stir well.
Whey off	Stir to float curd if settled. Take off whey – dry stir to keep curd free.

Salting	No milling needed – add 2% salt onto the curd and stir to dissolve.
Moulding	Put curd into steel moulds lined with cloth.
Pressing	Press with 13 kg per kg of cheese. Turn into a fine cloth next day and press at 250 kPa (about 250 kg per cheese) for 48 h.
Storage	Store at not more than 15°C. When dry, grease bandage. Turn each day. Ripe in 1–3 months.

28. Cottage cheese

Cottage cheese is made for a number of different purposes. The composition is variable and the manufacturing methods vary for each type (e.g. plain, creamed, flavoured by essences, or by the addition of fruit or vegetable portions).

Method 1: acid-type cheese

Milk	Whole or skim milk; good hygienic quality.
Heat treatment	For small quantities: heat to 63°C for 30 min, cool to 21°C. For continuous process: HTST, 72°C for 15–20 s; cool to 21°C. (Note: a temperature of 75.5°C for 15–20 s has been used.)
Additive	0.02% calcium chloride where the curd does not normally firm up. (Note: calcium chloride may cause a bitter flavour.)
Starter	Temperature: 20°C in summer, 22°C in winter for long set method. 0.5% of active starter for long set method (14–16 h incubation). 3.0% of starter for medium set method (8 h incubation). 5.0% of starter for short set method (5 h incubation). Cultures used include *Lactococcus lactis* sub-spp. *lactis* and *cremoris*, gas producing cultures should be avoided.
Rennet	None needed in long set method.
Cutting	The coagulum should be firm, and the acidity 0.52–0.55% (pH 4.5–4.6) before cutting. Cut to 8-mm cubes with wide blade knife (19 mm) or wire frame. (Note: some markets prefer larger curds of 12 mm.)
Scalding	Stir only sufficiently to float the curd to give even cutting. Leave for 15–30 min then scald. Raise temperature to 49–54°C in 120 min. Stir slowly to keep curd particles free. (Note: excess stirring makes much 'curd dust'.)

Washing	Wash water at 49°C can be used to assist in the final scalding and removal of lactose from the curd. Wash water to be chlorinated (5 ppm), pasteurized or otherwise disinfected to avoid contamination by psychrotrophs. (Note: initial scald can be assisted by layering hot water, 46°C, 50 mm deep on the curd–whey mixture to firm up the top curds.) Stir gently for 15–30 min until curd is firm, or if soft, leave until firm enough to handle.
Whey off	When the curd can be squeezed in the hand, drain off whey and fill with cold water at 7–10°C to the original milk level. Stir to float the curd. Remove wash water and refill with cold water three times. Last wash water preferably should be at 1°C. Leave for 30–60 min for curd to cool below 4°C.
Drainage	Drain off water – trench curds on vat sides for better drainage. Stir gently to assist drainage of the last free water.
Storage	Store curd at 0–1°C until used.

Method 2: rennet-assisted curd from skim milk. Quick set method

Milk	Skim milk, freshly separated
Heat treatment	HTST 72°C for 15–20 s. Cool to 35°C.
Starter	Add up to 10% active starter at 35°C. Mix well for 20–30 min. (Culture as for Method 1.)
Rennet	Add 0.2–0.5 ml rennet (diluted 40 times) per 100 litres milk. Curd should set in about 6 h.
Cutting	Cut to 8-mm cubes. When acidity is 0.53–0.55% (pH 4.6), stir to float the curd. Leave for 30 min for whey to rise. Proceed as for Method 1.

Cream dressing for Cottage cheese

Cream	18% fat cream, salt as necessary (1% normal).
Homogenization	Homogenize the mixture at 14 MPa (first stage) plus 3.4 MPa (second stage).
Heat treatment	Heat to 80°C for 5–20 s. Cool to 4°C for storage or for use immediately.
Dressing	25% addition rate to the blend with Cottage cheese curds, or according to trade requirements.

Marketing

Cottage cheese is marketed in many styles, usually in plastic tubs or waxed cartons, in the following varieties:

(i) Plain, salt free, skim or plain whole milk
(ii) Whole milk or skim milk curd, salted
(iii) Creamed Cottage cheese; 18–40% cream dressing, salt to taste
(iv) Flavoured – savoury with salad, onions, chives, carrots, etc.
(v) Blended with cream to give 8% fat content with a smooth texture
(vi) Various flavours according to market.

Defects

Soft high moisture curd is caused by:

(i) too much rennet,
(ii) too high a heat treatment temperature,
(iii) insufficient scalding.

Too firm or too dry curd is caused by:

(i) lack of acid at cutting,
(ii) prolonged scalding,
(iii) insufficient rennet.

29. Coulommiers

The methods by which Coulommiers cheeses are made are basic for many of the soft cheese varieties. Minimum 40% FDM. Typical size is 12.5–15 cm diameter by 3–4 cm high with a weight of 400–500 g.

Rind	Clean, white, thin, normally with white mould growth.
Curd	Soft, moist, white, close.
Texture	Close, no gas holes, spreadable when fresh.
Flavour and aroma	Clean, slightly acid, not bitter, pleasant.

Small-scale Farmhouse

Milk often raw, 5 litres make four cheeses 11 cm in diameter or three cheeses 15 cm in diameter. All moulds are two piece. The cheesemaking room must be kept at 18–21°C.

Milk	Should be pasteurized at 72°C for 15 s, but if heated and cooled in buckets the milk may be heated to 71°C and cooled to 29.5°C. The fat percentage should

be 3.5%, or adjusted to that value with skim milk or by skimming some cream away. Higher fat percentages are likely to produce wet cheese while a lower fat content produces hard, dry cheese.

Colour	None or slight (0.1 ml annatto per 100 litres).
Starter	None or 0.5 ml per 100 litres at 29.5°C.
Ripening	None.
Rennet	Temperature 29.5°C summer, 32°C winter. 4 ml rennet per 100 litres diluted with 25–30 ml clean water. The rennet should be well stirred in the milk for 1 min and then left for 1–1.5 h to coagulate. (Note: The coagulation has occurred when a drop of clean water on the surface does not spread into the milk.) Traditionally the buckets were covered with a clean cloth to keep the contents warm – note however that a 'clean' cloth is rarely bacteriologically clean and a disinfected stainless lid is more appropriate.
Cutting	When the curd is firm enough to break cleanly over an inserted rod or from the side of the bucket, then four circular 'slices' of curd are cut with a dished skimmer to be retained for topping up the cheese, after ladling, to give a smooth finish.
Moulding	While the curd has been setting, the two-piece moulds are assembled, and along with cane mats or straws and wooden boards or wire mats, are all disinfected with steam or boiling water. Four moulds are then set on cane mats or straws on a board.
Ladling	Note that ladling is the equivalent of cutting the curd with knives. The smaller the pieces of curd ladled, the quicker the drainage (but also the greater the loss of curd and fat). Ladle the coagulum carefully into each mould in turn but do not fill the mould more than one-third full until whey has drained away (20 min). Ladle another third of the curd and leave (20 min) to drain further whey. Finish the ladling and slip one of the undisturbed slices on the top of each cheese.
Drainage	Leave the cheese to drain for 12 h or so until the curd has shrunk into the bottom part of the mould. Turn over by inversion onto a clean mat and board. Leave for 24 h. The temperature should be maintained at 21°C.
Salting	When drained sufficiently to remove the mould, the cheese is rubbed with salt and turned over. The salting

	may be repeated in 12 h. Keep the cheese on clean mats and boards.
Storage	The cheese may be eaten fresh, or stored to ripen for 10 days at 15–16°C and then at 10°C until used. Turn daily. For mould-ripened cheese the period of affinage may be up to 8 weeks.
Packing	Pack in woodchip or card boxes, wrap in greaseproof paper or waxed paper.

Factory method

Since the cheeses are made in small moulds, the curd filled into each mould must be of the same consistency and degree of firmness, otherwise the cheeses will not drain to the same size. The moulds are attached to a base grid (20–100 on a frame) for mechanical handling.

It is usual for manufacturers to arrange that the milk is coagulated at intervals in small batches so that the time lapses between filling each batch of moulds are equal and the coagulum from each batch is filled into the same size moulds. Thus milk coagulated in batches of 15–20 litres can be filled continuously into the moulds.

Method 1

Milk	3.5% fat, adjusted or standardized.
Heat treatment	Pasteurized at 72°C for 15 s. The milk is cooled to 29.5–31°C according to ambient temperatures.
Colour	None.
Starter	Variable – none up to 0.1% (excess starter produces hard shrunken cheese).
Ripening	None.
Rennet	Temperature 29.5–31°C, in winter 32°C. 25–30 ml per 100 litres milk. Diluted one to six times with clean water (150 ml). (Note: $CaCl_2$ solutions are rarely used for soft cheese on account of the bitter flavour and hard dry curds produced).
Cutting	In 60–80 min; when firm, the curds are cut. Vertical knives (12.5-mm blades) are common but wire grids may also be used.
Moulding	The cut curd is run off in channels into the moulds which are fixed to wire grids (20–100 on a frame). Overnight the curd has shrunk. The cheeses are turned onto a plain grid, leaving the grid with attached moulds free for reuse. Perforated plastic sheets/mesh may also be used.

Salting	The cheeses on the wire grid frames are rubbed by hand with salt and turned over. Alternatively, the wire frames in a multi-framed holder are dipped into brine (18–20% strength) for 1–1.5 h before transfer to the store. Brine temperature, 16–20°C, is important. The cheeses are stored to dry off.
Maturation	Cheeses may be packed for local use, fresh, or stored for 3–10 days at 10°C. Alternatively the cheeses are dried off at 15–16°C for 5–10 days and then moved into a humid store (90% RH) at 7°C until sold.

Modification of Method 1

Cutting	The coagulum is cut both ways with both knives (12.5-mm blades) or only one way with the horizontal knife on tender curds. The curds are allowed to settle in the whey until the particles are reduced to half the original size. The whey is removed.
Moulding	The curds are channelled into moulds on wire grids. The same amount of curd is filled into each mould. (The moulds are not so tall as those used in Method 1.) The final cheese made by this method tends to be more grainy than that from Method 1.

Coulommiers from UF milk

Coulommiers cheese can be made from UF concentrated milk by same method as for Camembert cheese, omitting the inoculation with *Penicillium camemberti* spores if an unripened cheese is desired. Ripen at 10°C, as required for sale.

30. Cream cheese

The method given is for small-scale production. For larger quantities of cheese a special form of separator can be used to produce the cream cheese from ripened cream (pH 4.4–4.8) containing between 9 and 11% fat.

Cream	15–20% fat separated from clean milk or from diluted 42 or 40% cream.
Heat treatment	Variable: (a) Heat in open vat to 88°C and cool immediately. (b) Heat to 82°C, cool, stir slowly. (c) Heat to 63°C, hold for 30 min then cool.

Starter	Add 0.13% mixed culture starter to 100 kg cream at 16°C. Use 0.5% starter for more rapid coagulation.
Ripening	Usually 18–24 h. (Note: rapid coagulation may cause bitter acid flavour.) Remove to cold room at 10°C. (Note: for low fat cream (15%), 25–40 ml of rennet per 100 kg assists more rapid coagulation.)
Drainage	Pour coagulated curd into linen or hessian cloth laid over hollow former on the drainer, 4.5–7 kg per cloth. Leave 2–3 h, then scrape the curd off the cloth. Leave for another 2–3 h, scrape the curd from the cloth surface then tie the four corners of the cloth together.
Pressing	Press weight the cloths, 1–2 kg to assist drainage.
Salting	When the curd is firm enough add 2% salt. Mix well and leave to drain.
Packing	Use extruder or fill steel hoops with curd. Pack cheese in greaseproof paper, foil or plastic film. Enclose in carton or woodchip box.
Storage	Store at 1–4°C. Shelf-life according to hygiene standard 4–14 days.

31. Danablu

Blue-veined cheese. More moist than some blue-veined cheeses and so needs to be kept in cold storage before sale. Packed in metal foil to reduce evaporation of moisture. Flat round shape. Size 18 cm × 10 cm, weight 2.6–3.2 kg. Normally 1 kg of cheese needs 7.5 litres of milk but only 5.3 litres of milk are needed when the UF method is used.

Rind	Slightly rough, packed in metal foil.
Curd	White with blue veins well distributed, moist, soft.
Texture	Open with mould growth in open spaces.
Flavour and aroma	Piquant to strong flavour. Aromatic ketonic aroma, aroma tends to have a salty, sharp flavour.

Methods of manufacture vary but typical method is below:

Milk	Good quality, 3.9–4.0% fat, good hygienic quality. Cream is separated at 15–20%, homogenized at 15 MPa single stage (or with a second stage at 3.4 MPa) then remixed with warm milk.
Heat treatment	The blended milk should be immediately pasteurized at 72°C for 15 s or 74°C for 10 s. Cool to 31–32°C.
Additives	10 g sodium nitrate (0.01%) per 100 litres milk, 2–5 g calcium chloride per 100 litres milk, 8 g decolourizer (chlorophyll extract) per 100 litres milk.

Starter	100–200 ml of starter culture per 100 litres milk (i.e. 0.1 to 0.2% active culture). Also add 3 g of *Penicillium roqueforti* powder and two plates of grown *Penicillium camemberti* culture. Squeeze spores from mat on the plates into the milk.
Rennet	Add 30 ml of rennet extract per 100 litres milk at 29–31°C to give firm curds in 75–100 min.
Cutting	Cut the curd to 12.5-mm cubes. Stir slowly.
Scalding	Remove 25–30% of whey and add warm water or use the vat jacket to raise the temperature by 2°C. Stir intermittently for 1 h.
Whey off	(1) Some cheesemakers take off 30% of the whey and mix into the curd–whey 2 g of salt per 100 litres at this stage. Stir then run off whey. Fill curd into plain hoops on drainer. (2) Others take off 30% of whey and then run off the remaining curd–whey onto a perforated tray, covered with open weave cloths. Make cloths into bundles and periodically open bundles and disturb for whey drainage.
Moulding	Metal hoops 20 cm diameter × 20 cm high are placed on the drainer and filled with the drained curds. (Note: during filling of the hoops a powder culture of *Penicillium roqueforti* may be blown onto the curd, if not already inoculated into the milk.)
Pressing	No applied pressure, gravity only. The cheeses are turned five to six times in the first 24 h.
Salting	The cheeses are stored at 23–24°C and 95–100% RH for 24 h. The temperature is then reduced to 18–20°C and the cheeses are dry salted top and bottom. The hoop is removed and dry salting continues for 3 days. If the cheeses are very dry, brine is poured over them.
Maturation	The cheeses are washed and stored in a warm room to dry for 4–8 h. They are transferred to the ripening room at 14–18°C and are pierced top and bottom with wire skewers; the cheeses are returned to the shelves and put on edge in cradles cut into the shelves. The temperature is reduced to 5–10°C for 3–5 weeks with the humidity at 90–95% RH and the cheeses are turned regularly. The mould growth on the cheese turns slimy as the cheese ripens in 6–8 weeks.
Packing	The cheeses are scrubbed clean and then wrapped in metal foil. Storage The cheeses need to be stored at

2°C until sold and also while in transport. The quick ripening cheese will soon become wet if held in warmer temperatures.

32. Derbyshire

Flat, round or block shape cheese 41 cm diameter × 11.5 cm high. Almost a quick-ripening Cheddar. Characteristics:

Rind	Mature cheese becomes covered with grey green mould.
Texture	Close, not pasty or chalky.
Flavour and aroma	Medium acid, full milk flavour at 4 months.
Milk	Farmhouse cheese from good quality mixed raw milk, 136 litres per cheese.
Heat treatment	Factory: mixed milk, pasteurized at 72°C for 15 s, cooled to 29°C.
Starter	Mixed culture of lactic organisms, 1.0–1.75% added at 21°C or more.
Ripening	Temperature 29°C, for 45–60 min according to amount of starter added. Acidity 0.165–0.18%, or a rise of 0.01% (a slight drop in pH).
Rennet	Dilute rennet with five to six times amount of water. 25–30 ml per 100 litres at 29°C.
Cutting	After 40–50 min when curd breaks cleanly, cut both lengthwise and crosswise with 12.5 mm blade. Stir and then cut crosswise. Curd larger and firmer than Cheddar but more fragile. Acidity 0.12–0.13%.
Scalding	Stir for 15 min then scald to 35–36°C, taking 1 h. Keep the curd particles free until broken curd shows no free whey pockets inside the cubes.
Running whey	If the curd is firm, it may be pitched until the acidity is 0.15–0.18% at which point the whey is removed.
Dry stirring	The curd is still fragile hence stir carefully to allow all free whey to drain off. This is important if block cheeses are being made.
Texturing	Pile curd into half the vat. Cut into four blocks and turn each over. At 10–20-min intervals turn over and pile the blocks double until the curd shows leafy texture, moist but with little free whey.
Milling	Mill once through coarse mill.
Salting	At acidity of 0.45–0.55% salt the curd using 1.8% on dry curd, 2.0% on wet curd. Temperature 25–26°C. Wet

	curds lose salt in pressing hence apply more. Allow at least 15 min for salt to dissolve before moulding.
Pressing	Moulds 41 cm diameter × 15 cm deep with split collar. Press in coarse hessian cloth at 20 kPa (about 250 kg per cheese). Turn in 2 h to calico cloth and press at 40 kPa (500 kg per cheese) increasing to 60 kPa (750 kg per cheese) overnight. Next day turn into tubular bandage and press at 95–110 kPa (1.25–1.5 T per cheese) for 24 h. Iron on a paper cap and convey to store. Alternative is to paste on a calico bandage. Storage temperature 10–13°C. Some cheese may be in block form and a vacuum pack is used inside a cardboard container.
Faults	High acid milk or curds produce dry acid crumbly cheese. Wet curd at moulding provides soft cheese which dries out to give cracked coats growing mould. Poor quality milk (3.1% fat) gives chalky cheese. Sweet (low acid) curds favour taint-producing bacteria. Some produce discoloured curds. Weed and feed taints are transmitted to the curd.

33. Domiati

White brine pickled cheese. The cheese curd is pickled in layers in brine. Square cheeses are typically 8 cm square × 3.5 cm in size, and round cheeses are 12 cm diameter × 3.5 cm deep.

Rind	None, white coat.
Curd	Softer than Coulommiers; white turning brownish.
Texture	Close, no gas holes, flaky and brittle when old.
Flavour and aroma	Distinct flavour, slightly salty, clean, acid; when old, strong flavour.
Milk	Whole buffalo and cow's milk or mixed in proportion. Buffalo milk (7% fat) preferred.
Heat treatment	65.5°C for 15 min. Cool to 35–40°C.
Salting	5–15% salt added to milk to counteract the microbial population (average salt 10%). Alternatively add the salt to two-thirds of the raw milk and heat the remaining third to 77°C and add to the remainder to give a temperature of 35–46°C ready for renneting.
Rennet	Temperature variable, 35–46°C. Rennet extract 22–30 ml per 100 litre milk. Set for 2–3 h.

Moulding	Ladle the curd into hoops of suitable size, e.g. 12 cm diameter × 11 cm high. The hoops are placed on a porous mat of straws or cloth. Alternatively the hoop is made long (50 cm) and wide (50 cm) with a capacity of 50 kg of curd and factory systems use tables holding 500 kg of curd. Drainage takes 12–40 h.
Whey drainage	The curd is enclosed in cloth and is pressed by weights to expel more whey. When the curd is firm it is cut into pieces 8 cm square.
Pickling	The pieces of curd are put in layers into large tin cans or earthenware vessels. The spaces between the curds are filled with salted whey (10–15% salt) and the vessels made airtight.
Storage	Normally the containers are cold stored until required.

34. Double cream cheese

Cream cheese, 57–113 g, packed in paper or muslin in cartons or woodchip boxes.

Cream	50–60% fat, separated from clean milk.
Heat treatment	Heat to 71°C and cool immediately to 10°C. Stir gently to prevent fat loss.
Starter	For ripened cheese: Not more than 50 ml of mixed culture starter per 100 kg of cream at 10°C. Mainly fresh cream without starter is used.
Rennet	None needed.
Drainage	Pour into linen close weave cloths (Huckaback) after 2 h at low temperature, 5°C. Drain for 4–6 h. Scrape the curd from the cloth and mix into the centre. Press cloth with 1–2 kg weights to assist drainage.
Salting	When the curd has formed, mix in 0.5% salt. Leave to dissolve. (Note: some cheesemakers add 0.25% salt to the cream before drainage to keep acidity low.)
Packing	Pack in paper or muslin – using steel forms for measuring weight of curd. Pack in cartons or woodchip boxes.
Storage	Store below 4°C (preferably at 1°C) until sold. Shelf-life variable, 2–7 days or longer where good hygienic standards prevail. Drying out of curd or melting of fat occurs in unsuitable store temperatures and humidities.

35. Double Gloucester

Flat round cheese 46 cm diameter × 12.6 cm high; weight 18–20 kg. Approximately 180 litres milk per cheese. Some farmhouse cheeses have extra added cream, 4.5–9 litres per batch.

Milk	Typically 3.75–3.9% fat.
Heat treatment	Farmhouse: mainly raw milk is used. Factory: heat treated at 72°C for 15 s.
Starter	Mixed culture of lactic organisms, 1–3% according to type of cheese and chemical quality of the milk. Temperature not less than 21°C.
Colour	Annatto, 15–30 ml per 100 litres according to market requirements.
Ripening	Temperature 30°C for 30–45 min to an acidity of 0.21–0.22%.
Rennet	Temperature, 30°C. Amount, 25–30 ml per 100 litres, dilute six to eight times with cold water. Setting time 45–60 min.
Cutting	Allow curd to firm for 40–60 min to give clean break over inserted rod. Knife with 12.5-mm blades. Cut lengthwise and crosswise. Stir curd from bottom and sides of vat. Cut crosswise until the curd is as fine as small peas. Acidity 0.14–0.16%.
Scalding	Stir until curd is floating free in whey. Scald to 37.5°C in 40–45 min. Continue to stir until curd is firm. Pitch until acidity is 0.185–0.23% according to season.
Drawing whey	The whey is removed at acidity 0.185–0.23%. The curd (especially acid curd) is dry stirred to remove free whey.
Texturing	The freely drained curd is piled into two-thirds of the vat. Cut into blocks 20 × 30 cm. Every 10–20 min the curd is cut and piled double. Acidity approximately 0.55–0.65% (using the hot iron test), the curd blocks are sliced to 2.5–4 cm ready to mill.
Milling	Mill through coarse mill once only.
Salting	1.9% mixed well until dissolved in curd. Acidity 0.60–0.75%.
Moulding	Fill curds into moulds in coarse cloths.
Pressing	Press, gradually increasing pressure to 15 kPa (about 250 kg per cheese) for 30 min. Turn and press for 16 h at 30 kPa (about 500 kg per cheese) for 24 h. Acidity of whey 0.8% (0.75–0.95% according to season). Next day turn cheese into fine calico cloth and press at

	45 kPa (about 750 kg per cheese) for 24 h. Then cap each end with calico cap and press with 60–75 kPa (1–1.25 T per cheese). Then paste bandage.
Farmhouse cheese	Grease bandage is often used for the inner bandage. Paste bandage or sewn-on bandage for outside.
Storage	Temperature 10–13°C.
Characteristics	Acid curd, close texture, moist flavour, mild body, silky, mellow. 4–5 months storage. Difficult cheese to make – acid curds discolour rapidly, sweet curds develop off-flavour and discolour.
Factory cheese	Factory methods have modified recipe to give a higher rate of acid development. Starter 3%. Scald 39°C. Drawing whey 0.23–0.25%. Milling through fine mill, salting at 0.60–0.70% acid. Note: film-wrapped, block-shaped cheeses are made in place of the flat, round, bandaged farmhouse shape. In factory practice the cheeses tend to be firmer in body than traditional cheeses, and less moist.

36. Dunlop

Flat, round cheese, 35–40 cm diameter × 20 cm, weight 13.6–15.8 kg. (Almost a quick ripening Cheddar cheese.) Characteristics:

Texture	Smooth close
Flavour	Mild, free from acidity or bitterness.
Milk	Fresh whole milk (3.75% fat), acidity 0.15–0.19% (not over 0.2%).
Heat treatment	Pasteurized at 72°C for 15 s, or at 70°C for 15 s. Cool to 30°C.
Starter	1–1.5% starter ripened for 1–1.5 h. (Rise in acidity 0.02%.)
Colour	Sometimes none, more often 12–32 ml per 100 litres according to trade requirements.
Rennet	Temperature, 30°C. Quantity, 22–28 ml per 100 litres according to season. (Curd should set in 40–50 min.)
Cutting	Cut with 6-mm blades to 6-mm cubes (i.e. larger than for Cheddar cheese). Acidity 0.13–0.14%.
Scalding	Stir to float curd freely. Raise temperature to 35.5–37°C at a maximum rate of 0.2°C per min. Stir for 20 min after temperature is uniform. Leave to settle for 20–30 min.
Whey off	When acidity of whey is 0.22%, run off whey. (Note: if acid development is slow, run whey at 0.25%.) (Hot iron test: 3–13 mm threads.)

Texturing	Open a channel in centre of the vat for whey drainage. Cut curd into 20-cm slabs. Turn over every 20–30 min, keeping slabs separate at turning but piled two to four deep on sides of vat.
Milling	Acidity 0.55–0.65%. Curds should have developed a 'chicken breast' texture. Use coarse mill. (There is a preference for peg mills but chip mills tend to give a closer textured curd.) Temperature 25.5–26.5°C. Acidity 0.75%.
Salting	3–3.5% salt applied to curds (1.9–2.0% in pressed curd). Well mixed and allowed to dissolve (5–15 min.)
Moulding	Curds will cool to 23–25.5°C. The hoop is lined with hessian cloth, filled with curd, and the cloth folded over the top and the follower applied.
Pressing	Press lightly for 15 min, then more pressure 50 kPa (about 500 kg per cheese) for 3 h. Turn over the curd and apply 150 kPa (about 1.5 T per cheese) overnight. Acidity of whey from press 0.8%.
Dressing	Take cheese from hoop, immerse in water at 60°C for 1–2 min. Re-press overnight. The next day remove cheese and grease with unsalted lard, bandage and cap. Then return to press at 100 kPa (about 1 T per cheese) overnight. (Note: a stocking type bandage is often used.) Paste on a bandage before removing to store.
Storage	Temperature 10–15°C. Humidity 85% RH. Store for 4–12 months according to trade.
Farmhouse cheese	At the texturing stage the curd is cut into 76-mm blocks which are gathered into cloths. The cloths are folded over the cheese and a 25 kg weight applied. The parcels of cheese are turned over at 25–30-min intervals and the individual curd blocks torn apart and re-piled in the cloth parcels until ready for milling. (Factory methods dispense with the cloth.)

37. Edam

Though originating in The Netherlands it is now made in many countries. There are many variations of this cheese produced, especially in respect of the fat content, i.e. 30, 40, 45 and 50% FDM. The milk used must be standardized for fat content. Typical examples are:

30% FDM cheese, milk should have 1.6% fat,
40% FDM cheese, milk should have 2.6% fat
45% FDM cheese, milk should have 3.0% fat.

Flattened spherical shape and square block forms are made. Typical sizes 30 cm diameter × 13.5 cm, weight 4–4.5 kg and 25 cm diameter × 11 cm, weight 2–2.5 kg. Other sizes and weights are also made from 0.6 to 5 kg.

Rind	Smooth, even, thin. During ripening may become grey–green with mould. For sale the cheese is coated with polyvinyl acetate or, for export, with red wax coat.
Curd	Ivory white to yellowish, dull cut surface.
Texture	Elastic, pliant, softer than Gouda. Gas holes few, round to oval, size of peas.
Flavour and aroma	Very mild, not acid, clean flavour.
Milk	Due to high pH of cheese, high hygienic quality milk is required.
Heat treatment	Milk is heat treated, HTST 72°C for 15–30 s. Cool to 30°C. Lower temperature in summer.
Additives	0.015–0.02% calcium chloride (added in solution) and 0.005–0.02% sodium nitrate.
Colour	None or 0.001% if needed
Starter	0.3–1% starter added at 30°C and ripened for 20–30 min. Acidity 0.155–0.165%.
Rennet	Powder rennet, 2.7–3.0 g, or rennet extract, 18–22 ml per 100 litres milk. Temperature 30–31°C. Curd is firm to cut in 30 min.
Cutting	Cut into 1-cm cubes. Stir to float curd in whey.
Stirring	Stir for 15–30 min. Settle curd and drain off third to half of whey.
Scalding	Stir while adding water at 50–60°C to approximately original liquid level. Alternatively spray in hot water to give a final temperature of 36–37°C. (Note: the hot water must not be allowed to plasticize the curd.) The acidity of the whey is reduced to 0.07–0.09%. Stir until curd is firm and pH is 6.0–6.1 (approximately 40 min stirring). Allow to pitch.
Whey off	Place flat plates to consolidate curd. Take off whey.
Moulding (hooping)	Line special shaped moulds with cloth. Cut curd to fit size of moulds. Fill mould and put top piece on curd ready for press. (Note: traditional moulds are two piece with spherical curd space (15 cm diameter) The bottom mould is perforated. Plastic moulds have replaced the wooden moulds.)
Pressing	Press cheese (usually in groups) at 10–15 kPa for 3 h. During pressing the cheese is turned in the mould (also sometimes dipped in hot whey at 48°C). The cheese is

	turned and re-pressed overnight (15–25 kPa). Temperature maintained at 15–20°C.
Salting	The cheeses are immersed in brine (22–25% strength) at 16°C for 2–3 days or at 12–14°C for 3–4 days. The coat is allowed to dry off in a warm room.
Maturation	The cheeses are matured for 3–4 weeks at 12–14°C. The cheese rinds are washed ready for coating.
Coating	For export coat in red wax at 120–140°C. Dry off in the store at 8–10°C. For sale the cheese may be wrapped in film or foil. Mature in 3–8 weeks according to size and temperature.

38. Edelpilzkäse

Mould ripened cheese. Cylindrical cheese, various sizes 11–17 cm diameter × 4 cm deep. Weight from 0.85 to 5 kg. Characteristics:

Rind	Yellowish with darker green mould, pierced holes on the top surface.
Curd	Soft yellowish with dark green or blue veining due to mould growth.
Texture	Open curd spaces with mould growth. Slightly crumbly but with adequate fat.
Favour and aroma	Piquant to strongly piquant, aromatic.
Milk	Cow's milk or sheep's milk or both.
Heat treatment	Normally heat treated. Flash 75°C or HTST 71.6–72°C for 16–40 s.
Starter	1–2% of active lactic culture. Also add 5 ml of liquid *Penicillium roqueforti* culture per 100 litres milk.
Rennet	Temperature 30–35°C. Acidity 0.17–0.185%. Rennet extract 20–25 ml per 100 litres milk. Curd should be firm in 50–60 min.
Cutting	Cut to 10- to 15-mm cubes (hazelnut size). Float curds and leave until whey is 0.21–0.22% acid. Time from adding rennet to hooping, approximately 90–100 min.
Moulding	Ladle curd into metal hoops. Turn each hour for first day.
Salting	Immerse in salt bath (23% salt) at 14–15°C for 48 h. Then dry the cheese.
Maturation	Storeroom 12°C. Pierce the cheese many times, top to bottom, after 5 days to encourage mould. Mature in 6–8 weeks. Reduce temperature to 6–8°C.

Note: A similar cheese curd but using the white mould *Penicillium candidum* produces Edelschimmelkäse with a white surface-mould growth.

39. Emmental

Since the approximate weight of the Emmental cheese is 100–110 kg, the cheese vat or kettle is required to be at least approximately 1400 litres. Traditionally the Emmental cheese was made only in copper-lined vats. Stainless steel is now used in the newer installations but may still be copper lined. For factory production the one cheese per copper kettle manufacture has given way to the large capacity stainless steel cylindrical vats producing curd for 10–12 cheeses (approximately 12 000 litres). Cheese size 70–100 cm diameter × 13–25 cm thick. Emmental is also produced in a block form, size 66 × 40 × 20 cm and weight 38–45 kg, as well as the standard 18 kg block. Characteristics:

Rind	Blue–grey at first, turning to golden brown.
Body	Firm.
Texture	Numerous eyeholes 1–2 cm diameter.
Flavour and aroma	Mild, sweet, slight aroma.
Milk	Quantity from 1000 to 1136 litres, according to quality of the milk, for a single cheese. Evening milk skimmed, and morning and evening milk at 3.0–3.1% fat. Factory cheese: Standardize the milk at 3.0% fat through a standardizing separator (which also acts as clarifier) at a temperature of 35°C. Then heat treat at 70°C for 15 s. (Milk heated up to 72°C may not always be used for Emmental in view of a desire to preserve some milk enzymes while destroying coliform and other harmful micro-organisms.) Convey to the cheese vat or kettle at 35°C.
Colour	None.
Additive	45–120 g of calcium chloride (dissolved in water) added per 1000 litres milk if the curds are normally weak.
Starter	Three cultures are used as follows. To every 1000 litres of milk add: 330 ml *Streptococcus thermophilus*, *Lactobacillus lactis* culture. 660 ml *Lactobacillus helveticus*, *Lactobacillus lactis* culture. 1–5 ml *Propionibacterium freundreichii* sub-sp. *shermanii* culture. Stir the cultures into the milk for 5–10 min to give even inoculation. (Note: some cheesemakers also add a culture of *Lactobacillus delbrueckii* sub-sp. *bulgaricus*.)

Rennet	To every 1000 litres of milk add 155 ml (1 in 10 000 strength) rennet diluted 1:40 with clean water. Mix in well for 3 min (with acid milk use more water). Temperature 31–32°C.
Cutting	After 25–35 min the curd should be firm enough to cut. Wires spaced at 6 mm on a frame are used for cutting. Since the vat is often circular and hemispherical a 'figure of eight' motion is used. Cut fine to about the size of rice grains (cf. Cheddar curds). Allow the curd to settle for 5 min.
Scalding	Stir for 30–40 min at medium speed (to avoid loss of fat). Cook slowly from the jacket of the kettle to 50–53°C (max 57°C) allowing at least 30 min. Stir without further heat for 30–60 min until the curd is firm and the pH is approximately 6.3–6.4. The curds should separate freely at this stage.
Pitching	If the pH is still high (6.4) leave for acid to develop but not long enough for curd to mat.
Whey off	Stir the curd vigorously with circular motion to bring the curd into the centre of the kettle. Then with a coarsely woven cloth (linen or hessian) overlapped on a light steel strip gather the curds by scooping the curd into the cloth and, by taking the four corners of the cloth, lift the curd from the vat (by block and tackle) into a round wooden (or steel) hoop on a square board under the press. Run off the whey and collect remaining curds.
Moulding	Knead the curd in the cloth to fill the circular hoop and to incorporate the curds recovered from the whey. On a large scale the curd–whey mixture is run directly into perforated steel moulds.
Pressing	Place board on cheese. Press lightly with the vertical screw, 3 kPa (100–250 kg per cheese depending on size). After 5–15 min remove board and circular hoop. Open up the cloth. Place new finer woven cloth on cheese then invert the cheese onto a clean board. Remove coarse cloth and fold over the finer cloth, then replace circular hoop and tighten to contain the cheese. Re-press for 2 h. Repeat three times at 2-h intervals. Finally remove coarse cloth and press overnight at ≈20 kPa (about 0.75–1 T per cheese). (Note: If there is a tendency for the press cloth to stick to the curd, a square of thicker material is put under the press boards.)

Salting	Take the cheese out of press (still on a board) and slide the cheese into a brine bath. 23% salt at 10°C for 2–3 days. Rub salt onto the surface of the floating cheese each day.
Drying off	Using a circular board take the cheese from the brine bath. Wipe the cheese dry with cloth dipped in salt. Turn the cheese onto a clean board and store at 10–16°C, 90% RH for 10–14 days. Rub salt over the cheese surface when turning each day.
Warm maturation	Remove cheese (still on its board) to warm room at 20–24°C, 80–85% RH, for 3–6 weeks until secondary maturation takes place (eyehole formation). Turn two to three times weekly. Wipe cheese surfaces with salt cloth when turning.
Cold maturation	After 4–8 weeks remove to room at 7°C, or lower if trade or other reasons demand longer maturation. Normally mature in 6–12 months (pH 5.4 at 3 months, 5.75 at 12 months, 85–90% RH).
Packing	The cheese (wheels) are cleaned and stamped for identification and then packed in wooden barrels or tubs for shipment.

Block (rindless) Emmental

Milk	2.5–2.6% fat. Fat/protein ratio 1:0.93. pH 6.6–6.7, TA 0.16%
Heat treatment	72°C for 16–20 s and cool to 31°C.
Starter	1.25% of starter culture of lactococci and lactobacilli including *L. helveticus* and *Propionibacterium freundreichii* sub-sp. *shermanii*. pH 6.5–6.6, TA 0.175%.
Rennet	20 ml of rennet extract per 100 litres milk. Temperature 31°C. pH 6.4–6.5, TA 0.18%.
Cutting	Curd cut in 25–35 min to 5 mm (whey pH 6.3–6.4, TA 0.11%). Stir to float curd.
Scalding	Take off one-quarter to one-third of whey as scalding proceeds. Scald to 39–40°C with hot water. At pH 6.3–6.4, TA 0.8% leave to pitch on the vat bottom.
Whey off	Take off whey at correct acidity (pH 5.9–6.1, TA 0.14%).
Moulding	Cut the curd to fit the hoop size.
Pressing	Press overnight at 29 kPa increasing to 44 kPa. Whey pH 5.8–5.9, TA 0.25%.
1st maturation	Mature for 2 days at 7°C.
Salting	Immerse in salt bath until salt content of cheese is 1.25%.

2nd maturation	Dry off cheese and mature for 4 days at 7°C. Then when dry, pack in film (vacuum pack), Saran, Cryovac or similar. Cheese pH 5.1–5.2.
Storage	Store in room at 20–22°C for 10–15 days until eyeholes have formed (1–12 mm diameter). Store at 8–12°C until ripe. (Note: humidity control is not necessary.)

40. Estepe

Cheese similar to Prato cheese, a Gouda-like cheese. Rectangular shape, weight approximately 5.5–6.5 kg.

Rind	Smooth, tough.
Texure	Close with oval eyeholes (3–6 mm).
Flavour and aroma	Mild, more pronounced when ripe.
Milk	Good quality, 3.6% fat. Filtered before heat treatment.
Heat treatment	HTST or similar treatment.
Starter	1.5–2.0% pure culture starter (mesophilic) plus 0.5% *Lactobacillus delbrueckii* sub-sp. *bulgaricus* culture.
Colour	Sufficient to give a pale yellow colour.
Additive	25 g calcium chloride per 100 litres milk.
Rennet	Temperature 30°C. 45 ml rennet extract per 100 litres milk. Curd should be firm in 45–50 min.
Cutting	Cut to 5- to 6-mm cubes. Stir to float curds.
Scalding	Scald to 34–40°C by adding water at 80°C.
Whey off	Take off whey when curd is firm.
Moulding	Press curd in the vat, then cut into blocks to fit the hoops.
Pressing	Press curd in the hoops using 60–70 kg per cheese for 30 min. Turn and wash in brine, then press for 3 h at 90–100 kg per cheese.
Brining	Immerse in brine (18–20%) for 2 days.
Maturation	Mature at 11–13°C and 80–82% RH to dry off. Transfer to 13–16°C and 85–90% RH for 2–3 months. Mature in a cold store until mature.

41. Feta

Soft, brine pickled cheese made in many countries from local milk (i.e. cow's, sheep's or goat's milk). Methods of manufacture vary slightly according to type of milk used. The cheeses are cubical or rectangular in shape and made in various sizes and weights. Sheep's milk is preferred since the cheese is white. Cow's milk imparts a yellow colour to the cheese.

Rind	None.
Curd	Soft and white.
Texture	Soft and not firm enough for slicing. No gas holes but with mechanical holes between curds.
Flavour and aroma	Piquant, slightly acid and salty but not bitter.
Milk	Sheep's or sheep's and goat's milk mixed. Cow's milk occasionally. Fat 7%.
Heat treatment	None in farmsteads but collected milks are heat treated. Low temperature, flash or HTST treatments are used, typically 65°C for 30 min. Cool to rennet temperature 32–33°C but also temperatures between 27 and 40°C have been used.
Starter	None to 1–3% starter in heat treated milks. *Lactococcus lactis* sub-spp. *lactis* and *cremoris* based starters have been used. Milk may be ripened before rennet is added.
Additive	0.02% of calcium chloride may be used in heat-treated milks.
Rennet	20–30 ml of rennet extract or 50–60 ml of locally made lamb or kid vell extract per 100 litres of milk. Powder rennet (2 g) may also be used. Temperature variable but normally 33°C. Curds should be firm to cut in 50–150 min according to temperature. Longer setting time in winter.
Cutting	Cut into 2–3-cm cubes. Leave for cubes to become firmer (10–15 min).
Drainage	Method 1: Ladle curds into large, cloth-lined metal hoops 20 cm deep. Next morning turn the cheese and leave overnight (if necessary cut the cheese into 20-cm blocks). Pack the cheese in a drainer to retain shape. Method 2: Ladle curds (uncut) in slices into cloths. Hang the cloths by the corners to drain for 2 h. Press the curd (in the cloth) into balls (cubes) on the drainer.
Salting	The cheeses are dry salted with coarse (sea) salt for 2 days. 6% of salt is incorporated in the curd. Alternatively, immerse in salt brine (16% salt) for 24 h.
Maturation	(1) Mature the cheese on a drainer for 8–15 days, or (2) Mature for 1–2 days then pack in barrels, canisters or drums (fill spaces between curds with broken curds),or (3) Fill the cheese into animal skins (sheep). Brine collects between the curds, and also gas. Therefore the barrels, canisters or skins are moved or rolled to distribute liquid and allow gas to rise. Mature the cheese for about 30 days at 9–10°C.

Storage	The containers are usually collected into a central store at 2–5°C.

Note: Sheep's milk is the preferred milk for Feta cheese. Goat's milk makes a harder curd with stronger flavour. Cow's milk gives a yellow tint to the curd which may be treated with chlorophyll to mask the colour.

42. Feta type cheese from UF concentrated milk

Milk standardized and concentrated by ultrafiltration.

Milk	Standardized to 3–5% fat, 3.4% protein and 12.5% total solids. Temperature adjustment to 50°C as part of the standardizing process.
Ultrafiltration	Temperature 50°C. Product ratio 1:5.1. Retentate to be 39.5% total solids (17.8% fat and 16.4% protein).
Heat treatment	Temperature 72°C or 77°C for 1 min.
Homogenization	Integral with the heat treatment. Pressure of 5.2–7 MPa. Cool to 27°C.
Starter	Add 2–3% of active starter. Leave to ripen for 16–24 h.
Rennet	Add 0.005% of rennet (i.e. 20% of normal) and allow to set for approximately 20 min in shallow tanks.
Salting	Cut the curd into blocks and add salt (3%). Pack in containers and leave for 24 h. Turn once during this time. Drainage, if any, will occur at this stage. Acidity develops to pH 4.8.
Maturation	Final maturation at 8°C for 9–12 days, the blocks of cheese curd are placed in plastic drums or tinned steel boxes and the containers filled with 18% brine. During this maturation, the salt content is adjusted (i.e. cold temperatures allow the cheese to take in water, warm temperatures cause the cheese to lose water).
Storage	Store at 2–5°C. Mature in 8–10 weeks.

43. Fromage Frais

The term Fromage Frais describes a family of cheeses, mainly unripened, which includes Fromage à la Pie (skimmed milk cheese) and cheese with FDM from 20 to 75%, including Boudon, Demi-Sel, Double and Triple Cream Suisse, Spoisi, Gerome, Gornay, Neufchâtel, etc. These cheeses vary in size, shape, rind, fat and salt content and method of packing.

1. *Triple Cream Suisse*

The simplest manufacturing procedure is for the group of Suisse cheeses. Farmhouse methods have used raw milk but most Suisse cheese is now made with heat treated milk.

Milk	Cheese milk for triple cream cheese is prepared from 100 litres whole milk (4.0% fat) and 35–38 litres of 30% cream or equivalent.
Rennet	3.7 ml of rennet extract are added at 16–18°C (for double cream cheese, 2–3 ml of rennet at 18–21°C). Leave to set for 36–48 h.
Drainage	Cut the coagulum sufficiently to pour into bags of hessian or linen cloth. Disturb curd periodically to present a new surface to the cloth.
Homogenization	When curd is well drained pass it through granite rollers to produce a smooth paste.
Packing	Produce short cylinder shaped curds and pack in waxed paper or aluminium foil.

2. *Demi-Sel cheese*

Similar method to Suisse but addition of 1–2% salt.

	Demi-Sel	Double Cream Suisse	Triple Cream Suisse
FDM (%)	40	60	75
DM (%)	28	30	30
Salt (%)	1–2	0	0

Note: Starter may be used to assist the coagulation process (0.5–5 ml per 100 litres) especially in colder weather. However, the Fromage Frais group of cheeses are traditionally slow to coagulate and drain.

3. *Neufchâtel*

Soft cheese from Normandy in France in varying forms, e.g. cylindrical form: Boudon; square form: Carré; rectangular form: Briquette. Composition typically 45–60% FDM: 25% when unripe but when ripe about 50%. Salt content from 2% to 3%. Characteristics are:

Rind	Slight mould growth.
Curd	Creamy white to light yellow, close textured without gas holes.
Flavour	Sour, aromatic and slightly salty.
Milk	Full cream milk adjusted, if necessary, to produce 50% FDM cheese.

Heat treatment	Farmhouse – none; creamery – pasteurized.
Starter	Normally none but sometimes a few drops per 100 litres.
Rennet	Temperature 15–20°C, 1–1.5 ml per 100 litres. Setting time 20 h.
Drainage	Curds ladled into linen cloth in drainer. Parcels of curd rearranged to release whey. Sometimes weighted down with boards. Temperature 16°C. Drainage time 15–20 h.
Curd treatment	Curd passed through granite rollers to produce smooth paste.
Packing	Curd moulded to shape and packed in cellophane parchment or waxed paper. Weight variable according to type (approximately 120g).

44. Gamelost

Skimmed milk cheese with surface mould growth. Cylindrical cheese 10–20 cm diameter × 12–20 cm high. Weight 1–3 kg. Old Western Norway cheese (1774) with typical composition: DM 51%, FDM 2%, salt 0.2%, protein 46%. Two types: Hardanger and Sogn.

Rind	Firm, rough, brown colour, with mould growth of *Mucor racemosus*.
Curd	Firm, brownish colour.
Texture	Granular, short, no gas holes.
Flavour and aroma	Very sharp and aromatic.
Milk	Skimmed milk is used.
Heat treatment	Skimmed milk, HTST 70–72°C for 15 s.
Starter	0.5% starter culture added at 19°C. Ripened for 18–24 h until acid is 44°SH (0.9% TA).
Breaking coagulum	Break up coagulum slowly.
Scalding	Drain off part of the whey, heat and return it to the curd. Warm slowly to 60–63°C for 2 h to collect and firm the curd.
Pressing	The curd is filled into cloth press bags in a perforated stainless steel mould. The curd is pressed hydraulically to remove surplus whey.
Milling	The compacted curd is milled into a special mould lined with cloth.
Cooking	The cheese and mould are heated in whey to 92–93°C for 2–3 h. The curd is then left to cool and drain on open strip shelves.

Inoculation	When drained, the cheese, now firm, is removed from the mould and may be sprayed with a suspension of *M. racemosus* spores, although there are usually enough spores in the air for inoculation purposes.
Maturation	The cheese is transferred to store, temperature 12–13°C for 5 days for the mould to develop. The temperature is reduced to 8–10°C for 4 weeks.
Storage	Store at from 3 to 10°C.
Packaging	Pack in aluminium foil in cartons for sale.

45. Gorgonzola

Originally made from sheep's milk but now goat's and cow's milk are used. Traditionally made from mixing two curds made from separate milkings to produce open curds in which moulds grow. Flat round shape. Size of cheese variable: 25–30 cm diameter × 16–20 cm high; weight 4.3–12 kg. Note that several variants exist. Typical characteristics are

Rind	Firm close uniform reddish colour with small holes.
Curd	Whitish to straw colour with mould growth evenly spaced.
Texture	Elastic, soft, nearly spreadable, irregular mechanical holes and mould growth.
Flavour and aroma	More or less piquant, spicy.
Milk	Sheep's, goat's or cow's milk individually or mixed. Fresh milk used but mainly heat-treated milk.
Heat treatment	HTST, 72–75°C for 15–20 s. Cool to 30°C, in winter 34°C.
Starter	1–2% starter culture (*Streptococcus thermophilus*, *Lactobacillus delbrueckii* sub-sp. *bulgaricus*) at acidity 1.23%. Also culture of *Penicillium roqueforti* added separately (2 ml per 100 litres). Ripen for 30–60 min for acid to increase by 0.02%.
Rennet	25 ml per 100 litres milk. Temperature 30°C, or 32°C in winter. Rennet volume variable, 18–27 ml but sufficient to give firm curd in 20–25 min.
Cutting	Curds cut into 2.5- to 4-cm cubes using wire knife or special curd harp ('panarola').
Stirring	Curds stirred and rested at intervals until acid increases 0.1–0.12%.
Drainage	Whey is removed to curd level. Curd is scooped into coarse cloth (hessian) on a draining tray. Each cloth holds equivalent of 33 litres milk. The curd is cut crosswise and the cloth tightened in 15–20 min.

Moulding	The drained curd is broken and filled into the hoops in layers. The prepared mould powder is sprinkled on each layer or over all the broken curd pieces. The hoops are on straw or cane or mesh mats 'persianina' for whey drainage. The cheese is turned five to six times in the first day to close up the holes in the rind. The slab 'spersole' on which the cheeses are drained is sloped to allow whey drainage. The cheeses are not pressed but maintained at 22–24°C and 90% RH.
Salting	Salt is rubbed into the rind of the cheese when the hoop is removed. Salt is rubbed into the rind for 2–3 days as the whey is released from the curds.
Maturation	After 3–4 days in the salting room (22–24°C) the cheeses are turned regularly and are then transferred to a store room at 4–5°C, 90% RH for 15–20 days.
Piercing	The cheeses are then pierced from one end and replaced on the shelves. After 10 days the other end is pierced. The blue mould takes 20–30 days to begin to grow. Ripening takes 50–60 days for the blue mould to grow well.
Final maturation	The cheese is transferred to a temperature of 4°C for final maturation of up to 5 months.
Packaging	Formerly the cheese was coated with barium sulphate or calcium sulphate to give a fine coat and then put into a wicker basket. Today the cheeses are packed in aluminium foil and then in tissue paper.
Faults	Curd too hard and dry. Rind cracked and open. Bitter flavours. Uneven mould growth.

The cheeses today are made more quickly and the curd is softer. The flavour is less piquant and the maturation time is reduced to 2 months. Formerly a two-curd system of manufacture was used. The warmer curds from immediate production were placed on the outer part of the hoops and the colder previous day's curd on the inner part of the hoops. The two curds of different acidities provided the places for moulds to grow. Temperatures for maturation were lower than today, 13–15°C as against 22–24°C, and the humidity was 75–80% RH as against today's 90% RH. The total maturation time of up to 8 months has been reduced to 2–3 months.

46. Gouda

Originally from The Netherlands, but now made in many countries. Gouda cheese resembles Edam but has a firmer curd and larger size. The cheeses are made in various shapes and qualities in respect of fat (30, 40, 45 and 50%

FDM). In some areas full cream cheese is mainly made. Flat, round shapes and block forms are made; weights vary from 'baby' size, 600 g to full size 20 kg. Sizes vary, 24–50 cm diameter × 6.5–12 cm high. Variants of Gouda are made spiced, e.g. Kanter cheese with clove or Leyden cheese with cumin.

Rind	Thin, dry smooth rind, light grey to grey–green colour. Coated with linseed or other oil for sale (or with yellowish wax).
Curd	Ivory white to yellow colour, not waxy but dull.
Texture	Firm and flexible. Round or oval gas holes scattered throughout the curd.
Flavour and aroma	Mild becoming piquant but not sour.
Milk	Whole milk (3.75–3.9% fat for 50% fat cheese). Good hygienic quality.
Heat treatment	HTST 72°C for 15–20 s. Flash heating to 75°C is used in some areas. Cool to 31°C.
Additives	0.02% calcium chloride (added as 35% solution), also, for control of gas producing bacteria, 0.005% sodium nitrate added.
Colour	Mainly in winter, early spring 1–2 ml of annatto or carotene per 100 litres. For more highly coloured cheese add 20–25 ml of annatto per 100 litres.
Starter	0.5–1% starter (average 0.8%) culture. Cultures of *Lactococcus lactis* sub-sp. *lactis, L. lactis* sub-sp. *lactis* biovar. *diacetylactis* and *Leuconostoc mesenteroides* sub-sp. *cremoris* are used in some creameries. Temperature 31°C.
Rennet	Rennet sufficient to give a curd to cut in 25–30 min, typically 22–28 ml of rennet per 100 litres milk. Temperature 28–30°C. Acidity 0.19–0.20%.
Cutting	Steel bladed knives or wires in frames to cut curd into 0.5–1.5-cm cubes, taking 10–15 min. Stir curd to float in whey.
Stirring	Stir for 20–30 min then allow whey to rise and run off 30% of whey.
Washing	Add water at not more than 80°C to give final temperature of 36–38°C. (Variable 32–40°C.) (Note: the hot water must not plasticize the curd. Therefore, the water is often sprayed into the vats; some modern vats can remove the whey slowly and replace with hot water at 36°C.) Stir continually while hot water is added (i.e. for 15–20 min) and then for 20–30 min thereafter. Allow to pitch and press curd with steel plates.
Whey off	When curd is consolidated, take off whey.

Moulding	Wooden or plastic moulds (two-piece) are filled with curd at pH 5.85–6.05. Cut to size from vat. Cheeses are turned into cloth and replaced in the mould and top of mould placed on the cheese ready for pressing.
Pressing	The cheeses are pressed for 5–8 h. During this time the cheese is turned and dressed to produce required shape (rounded corners). Increase the press weight when the cheeses are returned from 95 to 195 kPa. Pressed curd pH 5.1–5.2 and acidity of whey from press 0.35–0.40%.
Salting	Immerse cheese in brine at 15°C for 3–5 days. Brine strength 20% and pH 4.8. To give 1.5–1.8% salt in cheese at pH 5.15–5.25. (Note: the salt takes more than 1 month to penetrate to the centre of the curd.)
Coating	The cheeses are dried off at 15°C and 80% RH.
Maturation	Temperature 15°C for 4–6 weeks, pH at 2 weeks 5.2–5.3.
Storage	Temperature 10°C for 6–12 months for mature cheese (pH 5.3–5.5 at 8 weeks).
Final coating	For sale the cheese may be oiled (linseed oil) on the rind and packed in film. Some cheeses are waxed (black).
Faults	Excessive mould growth on rind (especially farmhouse cheese). Gas holes in curd too numerous or uneven and too large. Bitter flavours. Soft or too dry rind. Cracked rinds.

47. Gruyère

The Gruyère cheese, often confused with Emmental, is a smaller mature flavoured cheese with fewer and smaller holes. Diameter 40–64 cm × 8–13 cm, weight 35–40 kg. Cheese characteristics are a dry body, waxy with hazelnut-sized holes, flavour mature, aromatic and sharp.

Milk	Cow's milk (3.2% fat) not standardized. Sometimes evening milk is skimmed before mixing with morning milk. Maximum acidity 0.18%.
Heat treatment	Farmhouse none – flash treatment rarely.
Colour	None.
Additive	Calcium chloride rarely needed.
Starter	0.5–0.7% starter culture (*Streptococcus thermophilus* and *Lactobacillus helveticus* mixture, 1:2) with propionic bacteria in addition. Temperature 31–32°C.
Rennet	Temperature 31–33°C. Quantity 16–20 ml per 100 litres, sufficient to give curd in 25–30 min.

Cutting	Curd cut to 0.5–1-cm cubes with harp or steel or wire knives (i.e. larger than for Emmental).
Scalding	Stir gently at first then scald to 52–54°C. Stir for 40–50 min.
Whey off	Curds are lifted from the whey by a steel strip covered with cloth (as in Emmental). In larger vats the whey curd mixture is run through perforated sieves, the curd being collected in the hoops directly.
Pressing	Cheeses covered with cloth are pressed and turned with increasing pressure to 60–70 kPa for 2–3 days. The final whey from press should have 0.9% acidity. The cheese edges should be sharp and the coat smooth. The cheese is allowed to dry partially before salting.
Salting	Salt is rubbed into the cheese surface. The cheese may be immersed in brine (9°SH). Temperature 10–13°C for 2–6 days. Excess salt is undesirable.
Primary maturation	The cheeses are matured at 10°C for 3 weeks and then at 15–20°C for 2–3 months to form the eyeholes. Humidity 90–95% RH.
Secondary maturation	While maturing, the cheeses are turned regularly and rubbed with cloth moistened with brine. This rubbing also aids in establishing a smear coat of the red–brown *Brevibacterium linens* which aids ripening.
Storage	Final store temperature 12–15°C, RH 85%. Ripe in 8–12 months.

48. Halloumi

Firm pickled cheese made in Cyprus only.

Milk	Sheep's or goat's milk or a mixture of both. High fat content, 6% fat.
Heat treatment	None but sometimes low temperature, long hold where equipment is available.
Starter	None.
Rennet	Rennet extract 27–30 ml per 100 litres milk; or from lamb or kid extracted vells. Sufficient to give firm curd in 40–45 min. Temperature 30–34°C.
Cutting	Curd cut to 2–4-cm cubes with knives or wires on a frame.
Scalding	Stir to float curds gently at first. Raise temperature to 38–42°C by hot water in vat jacket. Stir for 20 min after temperature is reached. Allow curd to settle, then press curd by hand to bottom of vat (cylindrical).

Whey off	Ladle whey off the curd and collect whey for use later.
Drainage	Cut the curds into 10-kg blocks and put onto large cloths which are folded over the curd.
Pressing	Press the curd in the cloths – pile the curds to assist drainage. Re-press with 24 kPa (about 500 kg per cheese) pressure. (Whey treatment: heat the collected whey to 80–90°C (according to acidity); skim off whey proteins which are pressed dry and sold as Anari.)
Scalding	Cut the curd from the press into 10 × 15 × 5 cm curd blocks and put the blocks into boiling whey. Leave for 40–80 min until the curd assumes a chicken breast texture and is plastic throughout. Blocks rise to surface at this stage.
Salting	Cool the blocks on a drainer for 15–20 min. Sprinkle 5% salt onto the curd and leave for 2–4 h. During this period the curd cools.
Packing	Put the curd blocks into containers, i.e. cans, drums, etc. and fill the spaces with 30% brine.
Ripening	The cheese is eaten fresh or after storage in cool store.

49. Herregaardsost

Hard cheese made in Sweden. Flat, round in shape. Variable sizes, typical 35 cm diameter × 14 cm high, weight 14 kg. Three types of cheese, 30, 40, and 45% FDM.

Rind	Hard rind, tough and dry.
Curd	Colour yellow, soft and greasy.
Texture	Fairly firm but pliable with round gas holes, 1–2 cm size, distributed in the centre, not at the rind.
Flavour and aroma	Mild, full flavour, aromatic, nutty.
Milk	Cow's milk, fat content adjusted to type of cheese (e.g. 1.5, 2.2, 2.8% fat).
Heat treatment	HTST 72°C for 15–20 s.
Starter	0.7–1 litre starter culture per 100 litres milk. Temperature 30°C.
Additive	20 g of sodium nitrate per 100 litres milk.
Rennet	Temperature 30°C, 20–30 ml rennet extract (according to milk) per 100 litres milk. Curd should set firm in 30–35 min.
Cutting	Cut curd to 8-mm cubes. Stir to float.

Scalding	Scald the 45% FDM cheese curd to 42–45°C taking 60 min, the 40% and 30% FDM cheese curds to 38–41°C taking 45 min. Stir until curd is dry (i.e. no internal whey).
Whey off	Press curd under the whey with light pressure plates, then take off the whey.
Moulding	Cut curd to fit the moulds and fill into the moulds.
Pressing	Press lightly and allow curds to cool.
Salting	Immerse in salt bath for 2–3 days at 10°C. Drain and dry for 14 days at 10–12°C.
Ripening	Store for 3 weeks at 16–18°C to allow gas holes to form. Then store for 3 months at 12°C or until ripe.

50. Hervé

Soft cheese with smear coat. Mainly farmhouse production. Typically a 7.5 cm cubical cheese, weight 400 g. Similar to Remoudouis, a rectangular cheese, size 10 × 4 × 3 cm, weight 220 g.

Rind	Thin, smooth, moist, slightly greasy. Dark yellow, brownish colour.
Curd	Soft to firm yellow colour
Texture	Soft becoming firmer; either no gas holes or very few small ones.
Flavour and aroma	Mild but tending to be sharp in young cheese. Older cheese flavour strong, slightly salty, penetrating odour.
Milk	Cow's milk.
Heat treatment	Farmstead, none. Creamery, HTST 71–72°C then cooled to 30–34°C.
Starter	Farmstead, none. Creamery, 5–10% of a mixed culture.
Rennet	10–15 ml rennet extract. Sufficient to give firm curd in 1–2 h according to temperature of setting, 30–34°C.
Cutting	Cut using stirrer (harp) or six-bladed knife to 0.5–1 cm size. Stir to float the curds.
Drainage	Ladle curds into long wooden troughs on draining table. (Troughs 12 cm wide × 12 cm high). After 1 h cut the curds into 7.5-cm pieces. Turn the cheese between the wooden divisions for 6–8 h. When the curds are firm enough to keep shape, 1–3 days, the cheeses are removed to shelves for salting.
Salting	The cheeses are dry salted by hand. Note: in creamery practice using HTST heat treatment and 0.1–1.0% starter to give quicker maturing cheese, only 20 h are required before salting the cheese.

Smear coating	Temperature 14–15°C and 85–95% RH. The smear coat is inoculated from cheese to cheese during the handling and turning process.
Packing	The cheese is ripe in 4 weeks and is packed in film, parchment paper or aluminium foil.

51. Italico

Italico is a family of cheeses, of which Bel Paese (made from 1929) is the best known. Flat, round form 15–20 cm diameter × 5–6 cm, weight 1.8–2.2 kg.

Rind	Straw coloured, uniformly smooth without cracks.
Curd	Straw to yellow colour; soft, pliable, compact. No eyes or gas holes.
Flavour and aroma	Mild, delicate, slightly acid. Butter-like and melts easily on the tongue.
Milk	Cow's milk 3.5% fat. Clean and free from taints.
Heat treatment	None or low temperature pasteurization (61°C for 20–30 min) in small-scale manufacture. HTST (75°C for 15–20 s) in larger quantities.
Starter	0.25–1% lactic culture. Includes *Streptococcus thermophilus* so incubation at 45°C.
Rennet	30–50 ml per 100 litres milk. Temperature according to season 40–43°C. Curd sets in 15–25 min.
Cutting	Cut to 1-cm cubes. Stir between cuts. Stir to firm curd then take off whey and press lightly in vat.
Moulding	Curd cut and placed in hoops on straw or wire mesh mats or drainers. Turn repeatedly for 6 h. at 28–30°C, RH 9%.
Salting	Cheese immersed in brine (18% salt) at 13°C overnight.
Maturation	Mature at 5°C (85% RH). Turn frequently and wash with salt water for 3 weeks. Dry salt occasionally.
Packing	Foil wrap.

52. Kashkaval (Kaschkawal)

Typical of hard kneaded plastic cheese family. Sizes variable, typically 31 cm diameter × 12 cm, weight 6–9 kg. Flat, round form but also other shapes, e.g. cylindrical shape.

Rind	Yellowish, amber colour, smooth, shiny.
Curd	Yellowish, uniform consistency.
Texture	Semi-hard, pliable, some small gas holes.

Flavour and aroma	Pleasant, mild, aromatic but slightly sweet or salty.
Milk	Sheep's milk, 6–10% fat according to time of year.
Heat treatment	Normally raw milk but some cheesemakers heat treat the milk.
Additive	Calcium chloride 0.01–0.015% according to milk quality.
Starter	Culture containing *Lactococcus lactis* spp., *Lactobacillus casei.* Add 0.1–0.5% culture. Temperature 31°C (some cheesemakers prefer 28°C for ripening).
Rennet	30–40 ml of rennet extract per 100 litres to give a firm curd in 45 min at 31–32°C.
Cutting	Cut to 2 cm size. Stir to float the curd then for 5–10 min and reduce curd size to 6–8 mm (maize) with stirrer (harp or lyre).
Scalding	Raise the temperature to 36–38°C, occasionally 40–42°C. Stir until curds firm then allow to settle.
Drainage	Take the whey off and ladle the curds into cloths.
Pressing	Press the curds in the cloths using weights, 1 kg rising to 15 kg per kg of curd over a 1–2 h period. Temperature 15–20°C.
Milling 1	Slice the curd into pieces (25 × 20 cm) and repress for 6–8 h to give a texture as for Cheddar cheese.
Texturing	Soak the curds in hot water at 72°C until the curd is plasticized and elastic. Acidity 1.25–1.4%, pH 5.2.
Milling 2	Slice the curd into 0.5-cm strips in a rotary knife mill and put the strips into wooden baskets or tubs. Turn the curd over once or twice.
Salting	Immerse the curd in brine (5–6% salt) at 72–75°C for 5–8 min. Then kneed the curds on a table and press into cloth lined metal hoops. Turn many times in the next 24 h. Dry salt the cheese for 12–18 days at 90% RH, 16–18°C. Turn regularly.
Maturation	After salting, wash the cheese with a cloth wrung out in salt water. Reduce the temperature to 12–16°C and the humidity to 85% RH. The mature cheeses are washed, dried and waxed.
Storage	The cheeses, packed in clear plastic film, are stacked in groups of five at 2–4°C. Ripe in 10 months.
Faults	Slits and holes in curd due to *Clostridium* spp. Salt-resistant yeasts, moulds and bacteria produce off-flavours, e.g. strong, bitter or sharp. Red or yellow mould growth on rind due to *Oidium crustacea* or *O. sulfurea.*

	Mite attack on rind, *Tyroglyphus siro* or *T. longior*, due to low humidity. Blown curds due to coliform bacteria.
	Acid curds hard and dry when the cheddaring process is too long. Oxidized fat and ketonic flavours due to mould growth on the rind.

53. Kefalotyri

Strong flavoured, salty, hard cheese (3–7% salt). Flat cylindrical shape but various forms. Sizes variable but typically 30 cm diameter × 12 cm, weight 5–10 kg.

Rind	Smooth, polished.
Curd	Whitish
Texture	Very firm, small gas holes with some large slit holes.
Flavour and aroma	Strong, piquant flavour, salty, typical of the animal milk (sheep or goat).
Milk	Whole sheep's milk but may be mixed sheep's and goat's milk (sheep's milk may be mixed with 25% of skim milk). Note: butter is produced from the cream skimmed off the milk.
Heat treatment	None. Warmed to rennet temperature 32°C.
Starter	Usually none.
Rennet	25–30 ml per 100 litres at 32°C. Sufficient to give firm curd in 30–40 min.
Cutting	Curd is stirred rather than cut with wooden knife/rake (this causes some fat loss).
Scalding	Scald to 42–45°C in 30–40 min. Stir and then pitch curd for 10–15 min.
Whey off	Push curds to end of vat and remove whey.
Moulding	Press sufficient curd into each hoop to fill it, and to press out some loose whey.
Pressing	Press cheese with low pressure at 15 kPa (100 kg per cheese). Ensure even pressure if more than one cheese is in the press. In 5–6 h the cheeses are turned into a cloth and re-pressed to produce a hard smooth coat.
Salting	Rub dry salt into cheese coat for 6–8 weeks. If the cheese is dry use wet salt.
Maturation	The cheeses are piled four to six high as they harden.
Storage	Ambient temperatures may be used but a cool store is used in summertime.

54. Kesong Puti

An acid coagulated cheese. Rectangular in shape 7 × 9 × 2 cm, weight 150 g.

Rind	Whitish, smooth.
Curd	Rubber-like, firm, rough, crumbly when salted.
Flavour	Mild to strong, salty in young cheese. Sour and bitter in older cheese.
Milk	Water buffalo (fat 9.7%, protein 5.8%).
Heat treatment	Heat the milk to 85–90°C then coagulate with acid (vinegar).
Additives	Vinegar is produced by fermenting the juice of the water palm, *Nypa fruticans*.
Coagulation	Mix in enough vinegar to give coagulum.
Moulding	Fill curd into cloth in a wooden hoop.
Salting	When cheese is solid immerse in (15% salt) brine for 2–3 h. Temperature 25–30°C.
Packing	The cheese is packed by hand into banana leaves and sold fresh.

Note: other methods, e.g. for Cebu and St. Cruz cheese, use calf or buffalo vell extract for coagulation.

55. Kingston

Farmhouse cheese, 0.45–2.2 kg weight. Highly coloured cheese, mild flavour, firm body; texture slightly open, no gas holes.

Milk	Mixed milk (3.5% fat).
Heat treatment	None, occasionally 63°C for 30 min. Cool to 29°C.
Starter	Mixed culture starter 75–100 ml starter per 100 litres. No ripening time. Temperature 29°C.
Colour	23 ml annatto per 100 litres.
Rennet	27–33 ml rennet per 100 litres. Temperature 29°C, acidity 0.19%.
Cutting	Cut with 6 mm blade knife to about 6 mm (size of maize grain).
Scalding	Stir for 10 min to float curd. Raise temperature slowly to 33–33.5°C.
Whey off	Drain whey off at acidity 0.16%. Dry stir, then allow curd to mat.
Texture	Repeatedly cut matted curd to smaller cubes for 2 h until ready for milling (see Leicester cheese).
Milling	Mill through fine mill at acidity 0.35–0.45%. (Small quantities may be cut fine with knife.)

Salting	Salt at rate of 2% (salt through mill to mix well).
Moulding	Fill into steel hoops but no cloth in hoop.
Pressing	Press for at least 4 h at 10–15 kPa (about 250 kg per cheese).
Packing	Scrape cheese surface with knife to fill holes in the curd, then paste paper on the cheese surface.
Storage	Store at 12°C for 1–4 weeks. Avoid mould growth on paper coating, wipe with salted cloth. Dip in wax if necessary for longer storage. (Sometimes liquid wax is brushed on the coat.)

56. Lactic curd cheese

Small dairy or farmhouse cheese which is made by the acid coagulation of whole or skim milk, without the addition of rennet. (Note: surplus starter may also be drained for lactic cheese curd.) 5 litres whole milk will make approximately 1 kg cheese while 5 litres skim milk will make about 0.8 kg low fat lactic curd cheese.

Heat treatment	Heat the milk to 82–90°C. Cool quickly to 24°C.
Starter	Add 3–4% of active starter.
Coagulation	Leave the milk to coagulate to a solid curd at 21–24°C.
Drainage	When firm put the curd into a thick linen cloth and hang it up to drain. Each day until the required consistency is obtained, scrape the harder curd from the sides of the cloth and stir. This gives a more even and creamy texture to the finished cheese. To hasten drainage, wrap the curd in the cloth and cover with a board and weight.
Salting	When ready, mix in salt to taste.
Packing	Pack in plastic cups or waxed paper, tin foil or greaseproof paper in cardboard boxes. Farmhouse cheesemakers often use butter muslin to pack a 113 g (4 oz) cheese using a woodchip box or basket for handling.

57. Lancashire

Cylindrical cheese 35 cm diameter, 18–22.5 kg weight. Rarely block shaped. Toasting cheese (sometimes with sage in the curd). Characteristics:

Body	Soft silky body. Cheese should spread like butter but it must retain its shape.
Texture	Loose open texture.

Flavour	Clean, mild, mellow but slightly acid.
Milk	Good quality raw milk, 3.4–3.6% fat.
Heat treatment	Rarely heat treated for farmhouse production. HTST used for creamery production.
Starter	Farmhouse: 0.12% (no starter in warm weather). Factory cheese should have 0.12–0.15% starter. Temperature 21°C.
Ripening	Ripen for 45–50 min at 30°C until acidity has risen 0.005%. (Average acidity 0.16–0.185%.)
Rennet	Temperature 30°C. Quantity 30 ml per 100 litres. Dilute six to eight times with water. Setting time 50–60 min.
Cutting	Cut both ways with both knives (12.5-mm blades) till particles of curd are the size of small beans.
Scalding	None but float curd before allowing to settle so that uncut curd can be recut.
Pitching	Pitch after short stir to float curd.
Drawing whey	Draw off whey at acidity 0.13–0.14%. Drain thoroughly.
Texturing	Cut drained curd into blocks and scoop blocks into a coarse hessian/linen cloth in a drainer. Cut and break curd at 15–20 min and repeat until dry and fairly firm (not hard), acidity 0.2%.
Storage of curd	Keep at 20°C in large rectangular vats until required. Acidity should rise in 24 h to 1.2–1.4%.
Mixing of curds	Two-curd system: curd from drainer, acidity 0.2%, plus curd from 24-h vats at 1.3% mixed in proportion to give 0.9% acidity. Three-curd system: to curd from drainer, first day acidity 0.2%, add curd from vats, second day acidity 1.2%, and add curd from vat, third day acidity 1.6%, in proportion, to give 0.9–0.99% acidity.

Formula for 2-curd system:

weight of old curd × (old curd acidity − average acidity (0.9%))/average acidity (0.9%) − acidity of (new) drainer curd = weight of (new) drainer curd required

Milling	Mill the two curds together through peg mill.
Salting	Salt at acidity 0.9% summer and 0.99% winter with 2% salt (2.12% salt for slower ripening or wet curd cheese).
Moulding	Fill moulds lined with fine calico. Invert once and leave to drain overnight.
Pressing	Second day, press at 25 kPa (about 250 kg per cheese) increasing to 100 kPa (about 1 T per cheese). Third day,

	immerse cheese in water at 60°C for 30 s and press at 125 kPa (125 kg per cheese). Fourth day bandage after grease capping with calico caps. Tubular bandage stitched on and press for 12 h at 50 kPa (about 500 kg per cheese).
Storage	Avoid draughts – grease with lard or whey butter as necessary. Temperature 16–18°C for quick ripening, 13°C for slow ripening Marketed in 3 weeks to 3–4 months.
Faults	Too little acidity before rennetting may give acid, sour cheese which leaks whey. 'Heaving' cheeses are due to low acidity curd and/or too low salt, or high temperatures in storage. Acid curds produce hard dry shrunken cheeses which exude sticky whey.

58. Leicester(shire)

Traditionally a flat round cheese, 45 cm diameter × 11 cm high; weight 18–22.5 kg but now cheese also in film packs 4.5 kg to 9 kg. Characteristics:

Curd	High, even colour.
Texture	Close, dry and buttery but with some granular openings.
Flavour and aroma	Medium acid cheese but free from taint.
Milk	Farmhouse: about 180 litres per cheese, rarely pasteurized therefore high quality milk is needed.
Heat treatment	Factory: mixed milk, heat treated at 72°C for 15 s, cool to 29.5°C.
Starter	Normally a low acid cheese. 0.5–1.25% starter (normally 0.75%). Temperature of milk should be at least 21°C.
Ripening	Ripen the milk for 15–45 min to reach acidity 0.19–0.195% at 29.5°C.
Colour	Annatto at 0.03% to give highly coloured curd. Add when starter is well distributed. Do not add rennet for 10 min (annatto is alkaline).
Rennet	Add 30 ml of rennet diluted with 200 ml water per 100 litres at 29.5°C (29°C in hot weather).
Cutting	After 45–60 min, cut when curd breaks cleanly with green whey showing. Whey acidity 0.12–0.125%. Cut with both knives both ways. Blade width 6 mm. Stir to free the curd and recut with vertical knife until fine (slightly larger than rice).
Scalding	Stir until curd floats freely. Scald slowly to 35–37°C in 60–70 min. Stir until the curd is very firm, free but 'shotty' to feel.

Pitching	Pitch low acid soft curds early but stir high acid curds until whey off. Acidity 0.145–0.16%. Press curd in bottom of vat with wooden rack and weight (small vats only). (Note: factory vats, due to curd mass, rarely need the rack and weight.)
Drawing whey	Acidity 0.155–0.17%.
Texturing	Cut pressed curd into even 15-cm blocks and turn over. In 15 min turn and pile blocks. Cut blocks in half and turn every 10–15 min. When blocks are small, turn every 10 min until acidity is 0.4–0.5%..
Milling	Mill once through fine mill. Apply salt onto milled curd, then re-mill.
Salting	Quantity 1.8–2.0% according to moisture in curd. Acidity 0.45–0.55%. Salt must have dissolved in curd before second milling.
Moulding	Fill moulds lined with coarse hessian cloth with curd at 25–26°C.
Pressing	Press at 16 kPa (about 250 kg per cheese) at first, increase gradually to 32 kPa (500 kg per cheese) in 2 h. Then turn into fine calico cloth and press overnight at 50 kPa (750 kg per cheese). Next day cap both ends with calico caps and re-press at 63–95 kPa (1–1.5 T per cheese).
Storage	Store at 11°C reducing to 9°C. Ripe in 4–6 months. Avoid draughts in store, which cause cracked coats.
Faults	This cheese is prone to discoloration of the curd, taints, and cracks with mould growth. The cheese has traditionally been grease bandaged to prevent cracking. Low acid at salting tends to produce unpalatable 'blown' cheese.

59. Limburger

Soft cheese with smear coat. Family of cheeses of various shapes and sizes ripened with smear coat (*Brevibacterium linens*) including Romadur, Weinkase, Herve, etc. Cheese fat qualities, from 20% to 50% FDM. Shapes variable from square to rectangular; rarely round. Sizes and weights variable.

Limburger, Belgium – 15-cm square × 8 cm high, weight 1.25 kg.
Limburger, Germany – 10–18 × 4–6 × 4.5–5.5 cm, weight 0.5 kg.
Limburger, Germany – 6-cm square × 5 cm high, weight 185 g.

Semi-hard to soft cheese according to manufacture.

Rind	Yellow–brown to reddish coat; no rind, greasy coat.
Curd	Weak, shiny curd; whitish to yellow, not dry or crumbly.
Texture	Pliable, soft, fatty. Some curd spaces in unripe cheese, no gas holes.
Flavour and aroma	Spicy, piquant, fragrant, aromatic becoming buttery on ripening.
Milk	Cow's milk (3.5–3.7% fat). Adjusted for the various FDM qualities.
Heat treatment	Formerly mainly flash heating to 75°C but more HTST treatments at 72°C for 16 s are being used. Cool to 31–34°C.
Starter	Low acid development required; 0.5–1% starter.
Colour	None.
Rennet	25–30 ml rennet extract per 100 litres. Temperatures 31–33.5°C, but 30°C and 37°C have also been used. Curd should be firm to cut in 30 min.
Cutting	Cut curd to 2-cm cubes and later to 1-cm cubes. Stir to float the larger curds.
Scalding	None, but some cheesemakers prefer to warm up to 35.5–37°C, especially when curds are acid.
Whey off	Draw off the whey to curd level.
Moulding	Fill the curd into rectangular wooden or metal multiple moulds or hoops on a draining table. Press the curd (by hand) to give even-size cheese.
Ripening	Turn the cheese on the drainer after 1 h and then frequently in the next 6 h. After 24 h at acidity 0.40% (pH 5.1), take out of mould and if necessary cut to size.
Salting	Immerse cheese in 16–18% salt solution at 16–18°C for 24 h for the larger Limburger, 8–12 h for the smaller Limburger, 3–4 h for Romadur cheese. Alternatively dry salt for 2–3 days.
Storage	Mature in store 15°C, humidity 90–95% RH. Turn daily. Wipe the cheese with cloth soaked in brine to coat the cheese (inoculation from cheese to cheese). First coat is yellowish turning to red–brown and moist due to the growth of *B. linens*.
Packaging	Cheese mature in 3–4 weeks and is packed in film for sale.

60. Liplovska Bryndza

A 'Carpathian' cheese, typical of many other members of the group (two-stage manufacture), Liptauer, Brindsa, Brinse. Typical characteristics are:

Rind	None.
Curd	White colour typical of sheep's milk, turning to ivory colour later.
Texture	Tender curd, not spreadable, but with small grain, not rubbery or streaky.
Flavour and aroma	Mild acceptable flavour, slightly acid, typical of sheep's milk cheese. Some cheese of this group (i.e. Zimna bryndza) may be salty and stronger flavoured.

Stage 1 – manufacture of Hrudka cheese (Hrudkovy Syr)

Milk	Sheep's milk (6–7% fat).
Heat treatment	None, but cool to 30°C.
Starter	None, but ripen with natural flora.
Rennet	Extract of lamb vells with salt. Sufficient to produce a firm curd in 30–40 min. Temperature 28–32°C.
Cutting	Cut the firm curd to bean sized pieces (0.5–1.5 cm) and then leave in the whey to mature.
Whey off	Take off part of the whey.
Curd forming	Form the curd by hand under the whey into clumps (spherical or cubical) of about 3–5 kg weight; pH 5.2 in 24 h (1.55% TA).
Maturation	Hang the clumps of curd in cloth to drain and mature. Optimum temperature 18–20°C. Maximum pH 4.5, reached in 7–10 days (acidity 2.2–2.4%) while the lumps of cheese ferment on shelves. (Note: modern production utilizes pure culture starters, including *Lactococcus lactis* sub-spp. *lactis* and *cremoris* plus *Leuconostoc dextranicum*, *Lactobacillus casei* and *L. plantarum*. The surface growth of *Torula* spp. and *Oospora* spp. is obtained by inoculating the rind with such cultures.

Stage 2 – manufacture of Bryndza cheese

Hrudka or Krudka cheese from the dairy is graded and if necessary re-ripened for the manufacture of Bryndza cheese. The rind of the Hrudka is stripped off and is used to make a special type of cheese when mixed with Quark and 4% of salt.

The cleaned up and matured Hrudka curd is ground up, or milled through granite rollers, to make a firm homogeneous paste incorporating a further 1–2% of salt. The milled curds are packed in softwood (spruce or pine) moulds in quantities from 1 kg to 20 kg.

For sale, the wholesale packs are film-lined wooden (beech) tubs or boxes, but modern methods use machine packing of the curds.

Faults	Too acid. Too chalky through lack of fat. Too sweet or too soapy due to lack of acid. Mouldy, putrid, yeasty, and the flavours sharp, bitter or burnt.

61. Livarot

Soft cheese made mainly by small producers. Strong flavour. Cylindrical shape 13 cm diameter × 5 cm, weight 450–500 g.

Rind	Washed free of smear coat then coloured with annatto.
Curd	Yellow to golden yellow, soft.
Texture	Close, soft, pliable with some open curd spaces.
Flavour and aroma	Strong aromatic, often rather sharp (similar to Limburger).
Milk	Good quality 3.5–3.8% fat.
Heat treatment	None in farmhouse. Flash or HTST in creamery.
Starter	None. Some traditional cheesemakers keep milk for 16–24 h at room temperature. Acidity 0.2–0.22%.
Rennet	Rennet extract 20–25 ml per 100 litres milk. Temperature 28–32°C, average 30°C. Curd should be firm in 75–100 min.
Cutting	The curd is sliced with a long knife (sabre) and then poured onto a fine mesh cloth.
Moulding	The curd is drained for 15–30 min and then filled into metal hoops for 24 h. The cheeses are kept at 18–19°C to develop acidity, 1.10–1.20%.
Salting	When the hoop is removed the cheeses are washed with weak brine. In a temperature of 16–18°C the cheeses are dry salted for 3–4 days. Humidity 90% RH. Smear coat develops (*Brevibacterium linens*) and the cheeses are rubbed to smooth the coat for 3–4 weeks.
Maturation	The cheeses are matured at 12°C, 95% RH and are ripe in 2–3 months.
Packing	The rind is washed and coloured orange–brown. The side of the cylinder of cheese is bound with reed or sedge grass to keep the shape. No other packing is used.

62. Manchego

Important Spanish sheep's milk cheese. Roncal is a similar cheese. Cylindrical shape, size approximately 21 cm diameter × 8–10.5 cm, weight 3.3 kg.

Rind	Dry straw colour with mould growth (dark green). When ripe may be oiled or waxed.
Curd	Ivory white colour.
Texture	Slightly pliable, firm to hard. No gas holes usually, but may have many small ones.
Flavour and aroma	Typical of sheep's milk cheese, strong, distinct when ripe.
Milk	Sheep's milk, well filtered. 0.03–0.06% hydrogen peroxide added to preserve milk. Half an hour later, catalase added.
Heat treatment	HTST, 70–75°C for 16–20 s. Cooled to 30–32°C.
Starter	2% of lactic starter culture (*Lactococcus lactis* sub-spp. *lactis* and *cremoris*) plus 0.5% thermophilic culture (*S. thermophilus*).
Additives	Add 25 g calcium chloride (or calcium hydrogen phosphate) in solution per 100 litres, and before rennetting add 0.5 g lipase preparation per 100 litres milk.
Rennet	20–25 ml of rennet extract or equivalent per 100 litres milk to give a firm curd in 30 min.
Cutting	Cut to 1–2-cm cubes then leave for 5 min before cutting the curd to rice size pieces (2–3 mm) taking 20 min for cutting. Stir for 10 min.
Scalding	Warm the curds to 36–40°C in 30 min while stirring. Leave to settle the curd.
Pressing	Press the curd with vat press then cut the curd to fit the mould lined with cloth. After half an hour drainage, press for 5 h.
Salting	Immerse the pressed curd in a salt bath (20–24% salt) at 12–14°C for 2 days.
Maturation	Turn the cheese regularly in store for at least 8 weeks at 12–15°C, 75–90% RH. Then store at 5°C, 70–75% RH until required. Manchego fresco – store for 5 days. Manchego curado – store for 20–90 days. Manchego viego – store for over 90 days.
Packing	Preserve the rind with olive oil. Some cheeses are wax coated. Smaller cheeses are packed in film or parchment paper in a carton.

Note: traditional farmhouse cheese use rush plaited linings instead of cloths. These cause a print of basket weave markings on the sides of the cheese.

63. Mascarpone

An acidified cream cheese eaten fresh. Typical size 5–6 cm diameter by 6 cm high and weight 110 g, or up to 500 g for the higher moisture spreadable paste. The yield depends on the fat content of the cream.

Curd	White, from soft curd to paste.
Flavour and aroma	Mild and milky/buttery
Milk	Cream, up to 35% fat, separated from fresh milk or buffalo milk.
Heat treatment	Heated to 90–95°C.
Starter	Nil
Rennet	Nil
Additive	Citric, acetic or tartaric acid, including vinegar or lemon juice.
Coagulation	Hold at 90–95°C in a deep jacketed vat and add the acid, e.g. 2% citric acid solution, over 5–10 min while stirring.
Moulding	Strain off the coagulum and transfer in layers to cloth lined moulds or plastic strainers. After 6–7 h the curd may be blended
Salting	Only if needed, salt cheese on demoulding or blending.
Storage	Store below 4°C and treat as a fresh product.

64. Monterey

A mild flavoured cheese similar to Bel Paese, first made in Monterey, California at the end of the nineteenth century. Moisture should not exceed 44%. Lower fat variants are much harder, suitable for grating. The 'Jack' variant has higher moisture, 44–50%, using a lower scald temperature of 35°C, cooling to 22°C and ripening at 5–10°C.

Milk	Whole milk
Heat treatment	72°C for 15 s and cool to 32°C.
Starter	Add 1% mesophilic culture at 32°C.
Colour	3.3–7.3 ml annatto per 100 litres milk.
Ripening	Allow 10 min for even dispersion of the annatto.
Rennet	19 ml per 100 litres milk at 32°C, well diluted before addition. The curd should be ready to cut in 30–40 min.
Cutting	Cut the curd with 15–16-mm knives then keep curd suspended in the whey for 5 min.
Scalding	Commence heating at 0.2°C per min, raising the temperature to 39°C in 30 min. Maintain temperature and

	stirring for about a further 60–90 min till the whey acidity has increased by 0.02% and curd is firm.
Washing	Drain off whey till 2.5 cm remains above the curd. Add potable cold water till the temperature is reduced to 30°C, leave for 5 min then drain off. Bring curd to the sides of the vat to aid drainage.
Salting	Add salt at 2.5% then dry stir the curds to ensure even dispersion.
Moulding	Put approximately 3-kg lots of salted curd (each sufficient for a cheese) into disinfected muslin cloths, which are then gathered, tied with string and shaped into balls which may then be pressed between boards for 24 h.
Maturation	Remove from the cloths and dry at 15–16°C, 70% RH for 2–5 days with hand brushing. Seal the surface by dipping in hot paraffin wax (118°C) and continue maturation for a further 20–25 days, turning the cheese on alternate days.
Storage	Hold at 8–10°C for a further 2–4 weeks.

65. Mozzarella

Typical cheese of many similar cheeses belonging to the Pasta filata family of cheeses (pulled curd, soft cheese). Shapes various, round cake form, oval or egg shaped, weights from 50 to 500 g. Manufacturing methods variable according to market for cheese. Mozzarella is a rennet curd cheese rather than an acid precipitated curd cheese.

Rind	None. Soft, loose curd under the skin.
Curd	White, soft.
Texture	Soft, elastic, moist, with typical structure of pulled curd cheese. No gas holes.
Flavour and aroma	Fresh, slightly salty and acid taste, pleasant. Buffalo milk cheese more piquant and aromatic than cow's milk cheese.
Milk	Buffalo milk or cow's milk, or mixed. Fresh milk well filtered of extraneous matter is normally used.
Heat treatment	Sometimes none. Heat treatment of the milk may range from 65.6°C to 75°C.
Additives	3 g calcium chloride and 10 g sodium nitrate per 100 litres milk.
Starter	0.05–0.5% thermophilic starter, *Streptococcus thermophilus* and *Lactobacillus delbreuckii* sub-sp. *bulgaricus*. Temperature 31–32°C. Ripen to increase acidity by 0.02%.

Rennet	Temperature 31–32°C but sometimes up to 35°C. Add 10–20 ml rennet extract per 100 litres milk. Curd should be firm to cut in 25–45 min but if cold up to 75 min. Curd should break cleanly.
Cutting	Cut the curd to 1–1.5-cm size (walnut size). Then stir to float curd for 5 min. Leave to settle for 30 min. Keep warm (some cheesemakers warm whey to 40°C and pour back onto the curds) while the acidity of the curd is achieved, for 3–4 h (summertime) and up to 8 h (wintertime)).
Whey off	Drain whey slowly; press curds to make a mat. Cut the curds into 20-cm blocks. Wash in cold water and leave to drain.
Curd ripening	Pack the curd blocks in cloths and place in cold room at 4–5°C. Acid cheese may need covering with broken ice.
Stretching	Test the curd by the stretching test. Place piece of curd (pulling the curd) into hot water at 82°C. When hot the curd should stretch out to 1 m long if the acidity is correct, pH 5.1–5.4. Take the curd out of the cloths and chop the curd into small pieces. Immerse the pieces of curd in hot water sufficient to cover the curd. Temperature 70–82°C in a blender. Leave for a time for the curd to warm. Blend the curd into a smooth long pliable plastic mass. Do not heat the curd to more than 57°C.
Moulding	Fill the hot plastic curd into rigid moulds (stainless steel) or shape the cheese to requirements. Cool in cold water for about 1 h.
Salting	The salt, 0.75%, can be mixed into the plastic curd. Alternatively, immerse the cold, shaped cheese in 16–20% salt brine at 8–10°C for sufficient time (5 min to 24 h) to allow 1.6% of salt in the cheese. Brine strength and the size of the cheese dictate the time in brine. Dry off the cheese for an hour after salting in brine.
Maturation	The cheese is normally eaten fresh but can be stored for a short time at 4°C.
Packaging	The cheeses are often vacuum packed in PVDC film (moisture and air tight) or in laminated pouches.

66. Parenica

Pasta filata type cheese (often smoked) of decorative form. Round curd roll shape, 5–8 cm diameter × 3–5 cm high, weight 200–400 g. Another cheese,

Ostiepok, is a similar curd cheese in which the curd is moulded into egg shapes in wooden moulds with fancy patterns cut into the mould.

Rind	None – colour yellowish to brownish, banded structure.
Curd	White.
Texture	Banded structure, elastic, fibrous, no gas holes.
Flavour and aroma	Mild, slightly acid. Typical flavour of sheep's milk cheese, but smoked flavour distinct.
Milk	Sheep's milk.
Heat treatment	None, raw milk used.
Starter	Usually none.
Rennet	Temperature 34–35°C. 25–30 ml of rennet extract per 100 litres of milk, sufficient to produce firm curd in 30 min (lamb rennet often used).
Cutting	Cut into small cubes (2–5 mm). Stir to float curd then leave to settle.
Curd clumps	Form the curd under the whey into egg-shaped clumps. Acidity 26–30°SH (0.58–0.6% TA).
Ripening	Leave the clumps to form acid in a warm place. Acidity 65–72°SH (1.46–1.62% TA, pH 5.2). Leave for 24 h or up to 3 days to produce acid to pH 4.6.
Cooking	Heat the clumps of curd in cloths to 60–70°C or until soft and pliable. Knead the curds into a plastic mass and then stretch out into long strips 3–5 m long by 5 cm wide.
Moulding	Wrap the strips round and round into circular cheese shape and allow to cool.
Salting	Immerse in salt bath for 5–10 min then dry off.
Smoking	The dry cheese is smoked but the smoking is not necessary and is sometimes omitted. Instead the cheese may be immersed for 30–60 s in 10–12% brine at 72–75°C to plasticize the coat.
Maturation	The cheese matures quickly and should not be stored long.

67. Parmesan

Grating cheese made from a mixture of whole and partly skimmed milk or of settled partially skimmed milk. Cylindrical cheese 35–45 cm diameter × 17–22 cm, weight 30 kg. Moisture 32% in 2 years, 28% in 3 years.

Heat treatment	Raw milk may be used. Standardized to not less than 1.8% fat (average 2.0–2.8%). Varies from 60°C to 70°C to give negative phosphatase result. Cool to 32°C.

Bleaching	Not more than 0.002% benzoyl peroxide may be used for bleaching, and if used then vitamin A must be added to replace that lost. (Note: in some instances masking colour is added in place of bleach).
Starter	1% of mixture of *Lactobacillus delbreuckii* sub-sp. *bulgaricus* and *Streptococcus thermophilus* is added. In some instances the lower temperature *Lactococcus lactis* strains are also added. Ripening time 5–20 min.
Rennet	Setting temperature 32–35°C. Quantity 18–25 ml per 100 litres.
Cutting	Curds should be firm in 20–30 min. Curd knives, steel strip or wire, 6 mm. Curd cut continuously to 3 mm size. Note: too firm curd at commencement of cutting produces loss of fat in the whey ($\leq$0.6%).
Stirring	Stir curd for 10–15 min to float freely in whey.
Scalding	Raise temperature slowly to 42°C and hold for 15 min then continue to scald to 51–54°C taking 30 min (whey acidity should not exceed 0.13%).
Whey off	Push curd away from outlet of vat. Leave until acid whey from curd reaches 0.19%. The whey is removed and the curd is lifted in a cloth into moulds.
Moulding	The curd is pressed into circular steel hoops for 1 h at $\approx$20 kPa (about 250 kg per cheese). Then turn and re-press. The steel hoops are then removed and the cheese dressed in circular steel or wooden bands for 12–24 h at 21–24°C. Turn regularly.
Salting	The cheeses are floated in brine for 14–15 days. Dry salt is rubbed on the surface each day after turning in the brine. Temperature of brine 7–10°C.
Maturation	The curing room should be not more than 10°C and the cheese turned daily at first. to prevent sticking to shelf. The ripe cheese is coated with edible oil to inhibit mould growth. RH below 75%.
Storage	5–10°C for 2–4 years. The olive oil used for coating may be coloured brown (burnt sienna).

68. Peakland

Farmhouse cheese using simple equipment for manufacture. Size various, 2.5 kg using baking tins for moulds, flat, cylindrical.

Rind	Yellow, thin.
Curd	White–yellow.
Texture	Close, elastic.

Flavour and aroma	Clean, pleasant.
Milk	Good quality (3.75% fat), clean, fresh.
Heat treatment	None, but warm to 33–36°C.
Starter	None if using raw milk.
Rennet	3.8 ml in 10 litres milk. Temperature 33°C summer, 36°C winter. Dilute rennet six times with clean water. Leave to set.
Cutting	Curd should be set ready to cut in 35–50 min. Cut to 12-mm cubes. Stir for 25–30 min. Leave to settle for 10–20 min.
Scalding	Stir to float curd and then increase temperature to 37–38°C by using whey taken off the curd and heated to not more than 49°C and then replaced in vat. Stir for 30 min after scald is complete.
Whey off	Take off the whey and stir the curd freely to release whey.
Moulding	Line a hoop with cheese cloth and fill evenly with the curd.
Pressing	Press with convenient weight, 5–6 kg, for half an hour. Turn cheese in hoop and increase weight, 9–10 kg, for half an hour. Press with increased weight, 12–16 kg for 4 h. Take off the cloth; trim rough edges of the cheese and return to hoop.
Salting	Rub 42 g of fine salt onto each cheese surface. Repeat after 8 h. Next morning wash with brine solution (10% salt) and leave to dry.
Storage	Store at ambient temperature, 13°C for 6–10 weeks. Wash with brine occasionally. Turn each day.
Faults	Acid milk or too much acid development gives hard, brittle, bitter cheese. Lack of acid development gives a shapeless, soft, soapy cheese often with gas holes in the curd.

69. Pecorino Romano

Sheep's milk cheese (note: Caprino Romano, goat's milk cheese; Vacchino Romano, cow's milk cheese). Cylindrical cheese, size 20 cm diameter × 30 cm high, weight 5.5–22 kg.

Rind	Smooth, straw coloured, may be coated with oil or with yellow clay paste.
Curd	White to light straw colour.
Texture	Firm but harsh, no gas holes usually.

Flavour and aroma	Piquant and characteristic of the type of milk.
Milk	Fresh sheep's milk, fat 6.8% or higher.
Heat treatment	Usually low temperature, 60–65°C for 30 min, but also HTST, 72°C for 15–20 s. Cool to 38–40°C.
Starter	Several cultures used but mainly contain *Lactobacillus helveticus* and *Streptococcus thermophilus* with acidity up to 0.9% TA. 0.25–1.5% of culture used according to milk.
Rennet	Lamb paste rennet is normally used, 30 g per 100 litres milk. High temperature, 38–40°C gives firm curd in 16–20 min.
Cutting	Curd is cut to hazelnut size (10–12 mm). Stir to float the curds for even cutting. Leave for whey to rise then cut to size of wheat grains.
Scalding	Scald to 45–48°C taking 12–16 min. Stir for further 15–20 min.
Whey off	At 0.22% TA, leave curd to mat in bottom of vat. Take off whey.
Moulding	Fill cloth-lined wooden hoops with curd and stack to drain. Prick curds thoroughly with long metal prongs, to consolidate the curd in the hoop for about 15 min.
Drainage	Leave the curd to drain and then turn the cheese for pressing.
Pressing	Light spring or hydraulic pressure for 24 h.
Salting	Remove cloths from cheese and wash with brine. Then roll cheese in fine dry salt, return into hoop. Temperature 12–14°C. Remove from hoop. Repeat the dry salting for two more days. Temperature 12–14°C and 90–95% RH.
Storage	The cheeses are rubbed with salt during storage a number of times. The rind is kept clean by washing with 10% brine, and the temperature is maintained at 15–18°C and at 75–80% RH. The cheese is mature in approximately 8 months. It has a high salt content (5–7%) and is salty to taste. For sale the rind of the cheese is oiled or coloured chestnut brown or coated with a yellowish clay.
Faults	Cheese mite attack may be serious if not kept down. Bitter tastes develop if the whey is not removed from the curd and the curd is discoloured. Early and late during the ripening blown curds and gas holes may appear in localized patches. Cracks may appear in the rind if the cheese is not firm enough. Reddish spots or mould growth appear on the rind if not kept clean.

70. Pelardon

Family of cheeses made in Languedoc Province, France. Goat's milk cheese made from early summer to autumn in farmsteads. Shape: flat, round, small in size, 6–8 cm diameter × 3 cm; 80–90 g weight.

Rind	Yellowish with overlying blue–grey mould.
Curd	White.
Texture	Smooth, moderately pliable, no gas holes.
Flavour and aroma	Characteristic peppery flavour of goat's milk cheese, mild.
Milk	Fresh milk, not skimmed.
Heat treatment	None.
Starter	Usually none. Evening's milk provides enough acidity.
Colour	None.
Rennet	Use 10–15 ml per 100 litres milk at 20°C to give coagulation in 20 min.
Cutting (ladling)	Coagulum is ladled into moulds.
Drainage	Curd in moulds left to drain for 24 h. Turn once or twice. Then allow to dry off. The moulds (hoops) are removed.
Salting	The dried cheeses are rubbed with dry salt.
Storage	The cheeses are stored at ambient temperature for 7–10 days to ripen.
Faults	Excessive dry chalky curds. Bitterness in high temperature storage. Grainy curds with free fat or moisture.

71. Petit Suisse

A small-dairy or farmhouse cheese, made from whole milk and cream, in proportion 2:1 (e.g. 3 litres of milk and 1.5 litres cream make 16 cheeses). These cheeses are generally eaten fresh but may keep for several days after removal from the mould.

Milk	Good quality milk or mixture of milk plus cream
Heat treatment	Often raw but preferably pasteurized. Warm milk and cream mixed together for 10 min at a temperature of 26–27°C.
Rennet	75 ml per 100 litres of mixture. Temperature 26–27°C. Leave to coagulate.
Drainage	Curd may then be ladled into a coarse cloth held in a wooden box on a drainer. The curd is scraped off the cloth at intervals to assist drainage.

Salting	When the curd is sufficiently firm, 1–2% salt is well mixed into the curd before it is pressed into the moulds with a spatula.
Moulding	The moulds are fixed, 6–12 on a base, lined with absorbent paper and put on a straw mat. Size of usual mould, 2.5–3 cm diameter, 5 cm high.
Packing	Usually in woodchip cartons with overlapping lids. Store if necessary at 3°C but long storage may produce bitter flavours.

72. Pont l'Evêque

Farmhouse or Small Dairy Method: 10 litres milk make three fresh cheeses (i.e. 1 week old) of 350–400 g or three cheeses of about 230 g after ripening for 6 weeks. Pont l'Evêque is now covered by AOC regulations with minimum 45% FDM and three sizes:

Petit Pont l'Evêque 8.5–9.5 cm square, 85 g minimum dry matter per cheese
Demi Pont l'Evêque 10.5–11.5 cm square × 3 cm high
Grand Pont l'Evêque 19–21 cm square, 650–850 g

Characteristics:

Rind	Smooth golden-brown
Curd	Soft consistency.
Flavour and aroma	Taste savoury with pronounced tang but not bitter.
Milk	Pasteurized good quality, or heat treated to 72°C and cooled to 32–34°C in summer, 34–36°C in winter.
Starter	30 ml starter in 100 litres, but if raw milk is used then only 15 ml starter is required.
Colour	(spring or autumn) 3.7 ml annatto per 100 litres.
Rennet	30–50 ml rennet per 100 litres. Temperature 32–36°C.
Cutting (Ladling)	When curd is firm (about 45 min) it is cut vertically with a long knife into 1–2.5-cm squares. When whey is visible, ladle the curd into a coarse calico (nylon) cloth placed over a rack and wooden box. TA of whey 0.11–0.12%.
Drainage	Ambient air temperature 21°C should be maintained. Fold over the sides of the cloth, do not tie. After 10–15 min take up the corners of the cloth and tie. Gradually tighten the cloth over the first 60 min. Cut the curd into 5-cm blocks to assist drainage.

Moulding	When curd is firm and flaky, but not hard or gritty, break with hands into small pieces and place in moulds, pressing firmly at the sides and corners but loosely in the centre. The moulds are set on cane mats on a wooden board and are warmed with boiling water before use. pH 4.9–5.15 at end of draining.
Salting	Three methods of salting are used: (1) Salt is dissolved in the milk (1.3 kg per 100 litres) before rennetting. (2) Salt is mixed in the curd (210 g per 100 litres original milk) at moulding. (3) The cheese is made without salt but is brined in 20% brine at 15–16°C for 4–24 h after removal from the mould and after a short drying off for rind formation.
Maturation	When the cheese has been moulded it is turned frequently (every 10 min) while still warm, and then daily until the cheese shrinks away from the mould. The mould is now removed (3–4 days). Fresh cheese: The cheese is scraped to fill any holes in the sides of the rind. It is ready for sale but will only keep for 7–8 days; packaging is usually greaseproof paper (plastic coated). Mature cheese: If brined, the cheese is dried off. Then it is turned daily for 10 days at a temperature of 18–21°C. Then three to six cheeses are separately wrapped and stacked together on a shelf in a cellar 85–90% RH and 12–16°C. Ripe in 5–6 weeks.
Packaging	Greaseproof paper in woodchip boxes.
Faults	Grainy consistency, hard thick grey rind, leakage of whey, unpleasant farmyard odour.

73. Provolone

Hard cheese of the Pasta filata or spun curd family of cheeses. Made in various shapes. Skittle (ninepin) is traditional but also Flask, Cylindrical, Cubical, Sausage, Pear or Orange shapes and in various weights from 0.5 to 100 kg. Sizes various but cylindrical cheeses are 120 × 25 cm. The cheeses are mainly supported in rope or string harness and hung rather than placed on shelves. Some cheese may be smoked after salting by suspending the cheese in smoke from burning maple, hickory or oak logs until the soot covers the cheese.

Rind	Golden yellow colour – smooth, thin, shiny, sometimes waxed.

Curd	Creamy white to straw colour, close.
Texture	Fairly firm becoming firm, uniform, solid but tending to be slightly crumbly. No gas holes normally but some small holes are accepted.
Flavour and aroma	Mild, pleasant at first but ripe cheese piquant, stronger, sharp and characteristic of the variety especially when kid rennet paste is used.
Milk	Good quality cow's milk (3.2% protein). Fat standardized to 3.5% with skim milk.
Heat treatment	Raw milk was normally used but heat treatment at 72°C for 16–20 s is used now. Cool to 32°C.
Additives	If benzoyl peroxide (0.002%) is used for bleaching, only the cream is bleached and then mixed with the skim milk before starter addition; add vitamin A to replace that destroyed by bleaching.
Starter	1–2% of starter culture containing *Streptococcus thermophilus*, *Lactobacillus casei*, *Lactobacillus delbrueckii* sub-sp. *bulgaricus* (whey cultures have also been used).
Rennet	Rennet pastes (lamb and kid) are used for stronger flavour production, 40–60 g of paste per 100 litres milk, but rennet extracts, 20–25 ml per 100 litres, also used (sometimes with added lipase). Temperature 32°C. In wintertime up to 38°C to give firm curd in 20–50 min.
Cutting	Cut the curd into peanut sized pieces (7–8 mm) with steel knife (9.5-mm blades). Stir to float the curd and then leave to settle.
Scalding	Method 1 (conical type vats) Take off whey down to curd level. Heat the whey to 63–65°C and return slowly to vat. Stir until the curds reach 45–50°C. Stir 10 min more and take off the whey. Method 2 (rectangular vats) Using the vat jacket, heat the curds to 40–44°C. Stir until the temperature is even. Acidity should be approximately 0.19% TA. Take off whey.
Whey off	Take off whey in stages, stir repeatedly.
Texturing	Collect the curd in large pieces on a rack in the vat or drainer. Cut and pile for 4–5 h until the curd reaches pH 5.1 (0.65–0.8% TA).
Stretching	Some cheesemakers soak strips of the curd in hot water at 82°C, to test the curd for stretching (strip should pull out to 1 m long). Allow the curds to stretch out in single layers on the bottom of the vat. Then cut into strips through mechanical mill.

Blending	Put a quantity of milled strips into a hemispherical vat with hot water 79–82°C. Leave to heat up and then stretch the curds (manually or with paddles) repeatedly until a sheen on the smooth plastic body of the curd is developed. The curd should be about 57°C at this stage.
Shaping	Knead the curd into shape by hand or fill and press into the cheese mould. (Note: large cheeses are pulled repeatedly over wooden rollers to obtain the necessary sheen. Then the curd is folded over into the desired shape.) Cool the curd in its shape in cold water.
Salting	Immerse the cheese in 24% salt brine for 1–3 days according to size. Hang the cheese in a rope net to dry.
Maturation	Ripen at 12–13°C for 3 weeks. Store at 4–5°C for 6–12 months for grating cheese.
Packaging	Wash the cheese clean and either oil the rind or wax the cheese. Replace the rope net for hanging the cheese.
Faults	Blown curds giving misshapen cheese, bitter tastes and discoloured curds. Gas holes developed from propionic and clostridial infections. Discoloured curd due to reaction of the proteins during ripening giving pink or salmon tint to the curd.

74. Quark

Unripened concentrated skim milk lactic curd cheese with a typical minimum 18% solids. Normally produced on a large scale. See also Baker's cheese curd. Typical analytical details:

	Fat %	DM %	FDM %	WFFS %
Skim milk curd	0.2–0.4	19–21	1.5–3.5	80.5
Quarter fat curd	2.5–2.9	20–22.5	11–13.5	81
Half fat curd	5.0–6.1	21.5–24	20–26	82
Full fat curd	10–12.5	24–28	43–46	84

1. Manual method for skim milk curds

Milk	Skim milk < 0.1% fat.
Heat treatment	85°C is used to precipitate whey proteins, but also HTST according to equipment, 70–74°C holding time 15–40 s. Cool to 25–27°C.
Starter	Temperature 26–27°C, 1–2% starter of butter type culture with *Lactococcus lactis* sub-spp. *lactis* and *cremoris, L. lactis* sub-sp. *lactis* biovar. *diacetylactis* and *Leuconostoc* spp.

Ripening	Leave for 1–3 h until acidity is 0.2–0.21%.
Rennet	Mix into the milk 1–2 ml rennet per 100 litres (or 0.25–0.5 g powder rennet). Leave until pH is 4.6–4.7 (> 0.5% TA) and a soft coagulum is formed.
Cutting	Cut with knife into 20-cm cubes.
Stirring	Stir gently to form slurry with whey.
Drainage	Pour into cloth (calico) bags and hang to drain. Pile bags to assist drainage.
Pressing	Press curd in steel hoops to dry off and cool to 4°C in cold room.
Flavouring	Add salt (0.06–0.12%) and cream or flavourings, mix well and press curd into steel forms.
Packing	Pack in tubs or wrap in parchment. Keep at not more than 5°C until used.

2. Mechanical concentration method

This method uses large tanks with conical bases, an agitator for the curd and a special Quark separator.

Milk	Skim milk < 0.1% fat.
Heat treatment	Varies from 70 to 85°C, holding time 20–40 s, typically 72°C for 15 s. Cool to 20–30°C according to method used.
Starter	Cottage cheese or butter type culture with *Lactococcus lactis* sub-spp. *lactis* and *cremoris L. lactis* sub-sp. *lactis* biovar. *diacetylactis* and *Leuconostoc* spp. Long incubation method, 1% culture. Short incubation method, 2–2.5% culture.
Ripening (1)	Long method 20–22°C for 16–18 h. Short method 27–30°C for 5–6 h.
Rennet	Add 1–2 ml rennet (or 0.25–0.5 g powder rennet) per 100 litres when acidity is 0.2–0.22%. Mix in well. Temperature according to method.
Ripening (2)	Leave until curd pH is 4.6–4.7, until curds break cleanly.
Breaking	Stir curd into a slurry with high speed until all lumps have broken down then reduce stirrer speed to maintain the suspension.
Concentration	Pump slurry through fine sieve into Quark separator.
Cooling	Curd from separator is pumped through tubular cooler. Cooled to 1–5°C.
Mixing	Cooled curd is mixed with flavour or cream as necessary.
Packing	Curd is pumped to hopper of packaging machine.
Storage	Store at not more than 5°C until used.

Note: The consistency of the Quark is influenced by the heat treatment. Up to 50% of the whey protein may be denatured by high temperature processing, giving a 10% increase in product over the lower temperatures. Low gas producing cultures (cf. Cottage cheese) must be used for the 'thermal' process.

3. Use of ultrafiltration of skim milk to produce Quark

This process gives an 18% greater yield of cheese than the normal method.

Milk	Cow's milk, separated.
Heat treatment	Milk heated to 87°C for 45 s.
Starter	2–5% starter added at 25–27°C.
Incubation 1	Ripened until pH is 5.5–5.9.
Ultrafiltration	Process at 23°C until protein content is 12–13%. The membrane system must be capable of handling viscous media.
Incubation 2	Incubate the retentate until pH is 4.55.
Homogenization	Homogenize the incubated retentate at low pressure to produce a fine, uniform consistency. Cool to below 10°C.
Packing	Pack as required and cold store at not more than 5°C.

Flavoured Quark

Skim milk Quark is a base for adding flavours, i.e. strawberry, cherry, bilberry, whortleberry, banana, etc. Up to 15–17% of fruit mixture is incorporated in the skim milk curd according to trade requirement. Other flavourings, nuts, meats and vegetables (diced) can be mixed in the Quark. Fill into polyethylene lined bags (bulk) or plastic tubs or glass bottles.

Creamed Quark

Cream	18%, up to 35–40% fat, from separation of skim milk.
Homogenization	Preferably double stage, 15 MPa plus 3.5 MPa, reducing the pressure for higher fat levels.
Heat treatment	HTST, cooled to 5°C.
Blending	Batch blending, or more often in line dosing and blending for large scale.

75. Reading Yellow

Washed curd cheese. Square shape, 22 × 22 × 7 cm. Weight 4.5 kg. Characteristics:

Rind	Thin, yellow colour.

Curd	Yellow colour, soft, not spreadable.
Texture	Soft, close with few mechanical holes, waxy, elastic.
Flavour	Mild, slightly acid.
Milk	Good quality (3.75% fat).
Heat treatment	HTST, 72°C for 15 s. Cool to 32°C.
Inhibitory salt	Potassium nitrate 20 g per 100 litres milk.
Starter	1.25–1.75%, mixed culture starter.
Ripening	None.
Rennet	Temperature 32°C, 30 ml of rennet diluted with 200 ml water per 100 litres milk.
Cutting	Cut when the curd is firm, 45–60 min. Cut with 6–10-mm blades. Cut both ways with both knives to reduce particles to pea size. Stir between third and last cut to float curd.
Scalding	Stir until curd floats freely. Scald to 39–39.5°C by the use of hot water as below.
Washing	One quarter of the whey is removed and replaced by equal volumes of hot water at 52°C. Stir for 15 min. One quarter of the whey again removed and replaced by equal volumes of water at 43°C. Stir until curd is firm and tough. Keep the temperature at 39–39.5°C.
Whey removal	Remove whey when curd is tough and plastic. Dry stir for 10–15 min until no free whey remaining.
Moulding	Place the free curd in coarse cloth lined hoops (square).
Pressing	The curd is pressed for 10 min at 50 kPa (about 250 kg per cheese) then turned over in the hoop and re-pressed for 1 h. The cheese is now turned again and pressed overnight at 100 kPa (about 500 kg per cheese) maximum.
Brining	The cheeses are taken out of cloth and placed in brine (20% strength) for 24 h. (Too strong a brine produces a slimy coat.)
Storage	Dry the rind in a warm room at 15–16°C. When sweating has ceased dip the cheese in hot tropical wax or coat with resin plastic fluid. (Note: moisture under the wax causes mould and other growth.) Storage temperature, 12°C. The cheese is edible in 14 days but ripens for 2–3 months in colder conditions.

76. Ricotta

A development from the whey cheese family and found in various forms from the soft Cottage cheese type curds to a hard grating cheese. It is made from whole milk or whey with or without added milk. The proteins are precipitated by acid and heat. Ricottone is made from whey only.

Ricotta cheese (whole milk)

Milk	Raw milk.
Starter	Culture of *Lactobacillus delbrueckii* sub-sp. *bulgaricus*, *Streptococcus thermophilus*. Add sufficient starter (from 1.8 to 2.5%) to produce 0.3% acidity in the milk starter mixture.
Additives	0.05% of stabilizer to prevent foam (as in ice cream), 0.02% sodium chloride.
Heat coagulation	Heat with stirring (including direct steam injection) to 80°C to produce flakes of curd in a clear whey.
Coalescence	Cease stirring and allow proteins to rise to surface. Apply steam to vat jacket to cause proteins to roll towards centre of vat. Let the curds set.
Drainage	Scoop the curd from the surface of the clear whey into perforated tinned steel containers lined with open weave cloth. (Note: a second precipitate can be obtained from the whey by adding citric acid crystals (approximately 1.5% w/w) and further heating. Collect curd and fill into the containers.)
Cooling	Cool the filled containers in cold water and cover with calico and put crushed ice on top.
Packaging	Fill the containers to the top with curd from other containers. Pack in clean parchment and store under ice until required for sale or other use. (Note: the curds may be pressed to remove surplus moisture and then dried out in storage. This forms a grating type cheese for use as a flavouring.)

Ricottone (whey cheese)

Whey	Sweet whey from hard cheese manufacture. Add 5–10% whole milk to the whey and 0.1% salt.
Heat coagulation	Heat to 85°C by direct steam injection. Skim off the foam which forms early in the heating.
Additive	Add 0.05% of citric acid crystals (added as a solution) or 0.4% of white vinegar.
Coalescence	The proteins will rise to the surface. Leave for 5–8 min. Then push curd to the centre of the vat. Leave to become firmer.
Drainage	Scoop curd with a perforated deep scoop into perforated metal cans. Allow to drain for 4–5 h.
Packaging	Pack in containers, cover with parchment and then with ice. Store in cold until required for use.

77. Roquefort

The cheese is made only from sheep's milk (Lacaune breed). Although made mainly in the Aveyron Département, France, cheese has been be collected from Corsica and the Pyrenees, but all cheese must be ripened in the Roquefort Caves in order to carry the 'Appellation d'Origine' mark. Usual size: 19–20 cm diameter × 8–10 cm high, weight 2.5–3 kg. Manufacturing dairies are of fairly small size. Characteristics:

Rind	Thin with white–yellowish smear growth, otherwise dry.
Curd	White creamy with blue, blue–grey or grey–green mould growth.
Texture	Soft, fatty texture, but close curd sometimes short and crumbly.
Flavour and aroma	Strong and aromatic due to fatty acids and methyl ketones, spicy, piquant.
Milk	Sheep's milk, 8% fat and high protein. Resazurin test is used to control hygienic quality of milk. Sheep's milk is white so that no bleach is necessary.
Heat treatment	Raw milk is used, pH 6.5.
Starter	No starter is necessary but some cheesemakers prefer to use a very small amount of starter in cold weather (0.02%). Add *Penicillium roqueforti* culture to milk.
Rennet	Renneting must take place within 48 h of last milking. Lamb rennet extract is used on account of its higher lipase content, 15–17 ml per 100 litres of milk at 30°C, sufficient to give firm curd in 2 h. Calf rennet, 25 ml per 100 litres.
Cutting	The curd is cut with wire or steel blades into small cubes (1–3 cm size) in 2–2.5 h after rennetting.
Stirring	Stir two to five times at intervals to release whey over 40–60 min.
Whey off	Drain off whey to curd depth. Stir as necessary.
Moulding	Soft curds are filled into hoops and the mould culture of *P. roqueforti* is sprinkled over the curds.
Pressing	None, but a plate is put over the cheese in the hoop to give a level top. The cheeses are drained for 2–3 days on drainers.
Salting	The cheeses are dry salted by rubbing coarse salt on the rind as the cheeses are turned for 3 days. Temperature 8–10°C, 96% RH.
Maturation	The salted cheeses (4.5–5% salt content) are ripened at 8–10°C for 18–25 days to cool off and then transferred to the caves at Roquefort for final ripening and blueing at

	5–10°C, 95% RH. The cheeses are pierced by machine from top to bottom for aeration. Final ripening may be at 1°C for 5–10 months.
Packing	Portions are packed in tinfoil and repacked in cartons. Whole cheeses are packed in tinfoil or lacquered aluminium foil.
Faults	Excessive red growths under tinfoil. Excessive rose–red or grey growths on the rind (*Lactobacillus plantarum*, *Sporodonema casei*) which affect flavour and aroma. Mould growth in curd too coarse and waxy. Bitter and sour flavours. Curds too close and irregular mould growth. Antibiotics in the raw milk and large numbers of coliforms in the curd give rise to irregular fermentation. Hard cracked or brittle curds. Soft proteolytic or lipolytic curds.

78. Saint-Maure de Touraine

Farmhouse or small dairy cheese (AOC Touraine département, France). Goat's milk cheese made from early summer to autumn. Tall cylindrical cheese 4–5 cm diameter × 12–17 cm high. Weight 50–220 g. Manufacture is similar to other types of goat's milk cheese.

Rind	White felt of *Penicillium candidum* mould on small dairy cheeses, but with overgrowth of *P. glaucum* mould on farmhouse cheese.
Curd	Whitish
Texture	Close texture in new cheese; older cheeses have more crumbly texture and also slightly brittle curd. No gas holes but spaces between curd particles.
Flavour and aroma	Mild but piquant, typical of goat's milk cheese, but strong mouldy smell in farmhouse cheese and stronger flavour.
Milk	Fresh mixed-herd milks. Acidity developed not more than 0.02%.
Heat treatment	None.
Rennet	Temperature 22°C. Rennet 10–15 ml per 100 litres milk. Coagulum left for 18–24 h.
Moulding	Ladle into moulds and leave to drain for 24 h. Then remove mould and dry for 24 h.
Salting	Dry salt to 1.5% salt.

Inoculation	Spray with culture of *P. candidum* (liquid spray).
Storage	Store in cool room at 12°C, 90% RH for 2–3 weeks for moulds to grow. Mature at 6 weeks.
Packaging	Pack in parchment or paper with trade mark or in woodchip or card cylindrical box.

79. St. Paulin

Derived from Port du Salut cheese made in a monastery near Laval in France. Semi-hard mild cheese. Flat, round shape, size 20 cm diameter × 5 cm high, weight 2 kg.

Rind	Smooth rind with cloth marks, yellow to orange in colour.
Curd	Creamy white.
Texture	Greasy, soft but firm to cut. No gas holes but some small curd spaces.
Flavour and aroma	Mild flavour, slightly acid with delicate aromatic aroma.
Milk	Cow's milk, clean and fresh. Sometimes standardized to 2.4% fat for 40% FDM cheese.
Heat treatment	Flash heated at 74–80°C, but also HTST when equipment is available at 70–75°C for 30–60 s.
Starter	1.5–2% mesophilic lactic starter and 1% of an aroma-producing bacterial culture (e.g. *Leuconostoc mesenteroides* sub-sp. *cremoris*).
Additive	0.003–0.08% calcium chloride in heat treated milk. Vegetable colour (annatto) in winter, 1–2 ml per 100 litres milk.
Rennet	30 ml rennet extract per 100 litres milk. Temperature 30–32°C. Acidity 0.22%, pH 6.5. Curd should be firm in 30 min.
Cutting	Cut to 5-mm cubes with wire frame or bladed knife, taking about 20 min. Stir to float curd and leave to settle.
Whey off	Take off 50% of the whey, i.e. down to curd level.
Washing	Replace whey with water at 30°C. Use sodium nitrate and sodium chloride in the wash water (0.02%).
Scalding	Warm the curds to 36–39°C for 15 min with stirring to firm up the curd to wheat grain size. Settle the curd. Vat pressing. Press the curd with metal plates for 10–15 min then cut pieces of curd to fill the moulds (i.e. steel or plastic hoops on wire or nylon mats).

Pressing	Cheeses are pressed for 2–4 h. Turn regularly and increase pressure from 15 to 34 kPa (about 100–200 kg per cheese). Cool the cheese while in the press to 28°C. Acidity 0.105%.
Salting	Immerse cheese in brine (22% salt) at 10–15°C for 15–18 h.
Ripening	Ripen at 12–13°C. Mainly mature in 1 month.
Packing	Wash the rind for 3–4 days with anti-mould solution (potassium sorbate or natamycin). Dry off the coat and colour the cheese in annatto solution or give plastic or resin coat then market 2 days later. Alternatively pack in portions in metal foil. Sometimes in clear polystyrene box.
Faults	Infection with coliform bacteria or yeasts. Blown curds and large gas holes. Too much wash water giving a soft curd and lack of acidity and wet curds. Two colours in the curds and some unripe curds in the body of the cheese.

80. Sage Derby (or Derbyshire)

Made as for Derbyshire cheese up to the moulding of the curd. Ground up sage leaves are mixed with one-third of the curd. Chlorophyll extracted from green leaves of wheat, kale or spinach is poured over the sage and curd and mixed well. The curds are moulded in layers, white and green. The cheese is then pressed as for white Derbyshire cheese.

For sale, the cheese is decorated on the coat with country scenes made from sage leaves stuck on the curd.

81. Samsoe

Curd similar to Emmental cheese. Firm-bodied cheese sold when young. Mature cheeses have stronger flavour and may be used for processed cheese. Shape flat, round or block form. Size 44 cm diameter × 10 cm, weight 14 kg.

Rind	Yellow to brownish, dry, smooth, even. Sometimes waxed.
Curd	White to yellow colour.
Texture	Firm, with few gas holes about pea size.
Flavour and aroma	Nutty, mild, buttery when young. Stronger, even pungent, when more mature.
Milk	Good hygienic quality. Standardized to 3.4% fat.

Heat treatment	HTST, 68–72°C for 15–20 s. Cooled to 30°C.
Additive	Sodium nitrate, 0.02% to control gas-producing bacteria. Rarely calcium chloride; 0.02% if used.
Starter	Culture added at 0.7–1%. Sometimes pre-ripened milk is used. Normally ripened for 10–15 min.
Rennet	28–35 ml of rennet per 100 litres of milk. Temperature 30–31°C. Curd should be firm in 30 min.
Cutting	Cut to 5-mm cubes with steel bladed knives (6 mm spacing). Stir to float curds for even cutting. Continue stirring for 30–40 min after cutting.
Scalding	Settle curd for 5 min and run off 30% of whey. Stir to float curd and add hot water at 65°C. Stir more rapidly to raise temperature to 37°C. Stir for 50 min and add 20 g salt per 100 litres.
Whey off	Draw off whey, channel the curd to give good drainage. (Note: stir in cumin seeds if required.)
Moulding	Pre-press curd in vat then cut into blocks and fill hoops (steel or plastic).
Pressing	Press cheese for 30 min (200 kg per cheese). Turn cheese into a cloth, replace in hoop and re-press for 3–4 h at 400 kg per cheese.
Salting	Cool the cheese to 14°C for 16–18 h and then immerse in 24% brine at 12°C for 3 days. Drain and dry off the rind for 24 h at 14°C.
Maturation	Mature the cheese at 17°C and 85% RH for 4 weeks. Reduce the temperature to 11°C and 80% RH for 7–8 days. Final maturation at 10°C and 75–80% RH for 5–6 months.
Packaging	The cheese may be waxed when 1–2 weeks old during the first ripening period if intended for early sale.

82. Sbrinz

Hard cheese for grating when 2–3 years old. Often made in high mountain pasture lands. Size 50–70 cm diameter × 10–15 cm high, weight 20–45 kg. Cheeses not intended to ripen are sold fresh and called Spalen. Characteristics:

Coat	Grey, oiled coat.
Texture	Hard and grainy when broken
Flavour and aroma	Piquant
Milk	Whole milk (3.75–3.9% fat).

Heat treatment	None. Temperature adjusted to 24–27°C.
Starter	Sour whey is used for starter in which predominant organisms are lactococci. Clear whey from the cheese is allowed to sour for several days. Since the whey contains rennet, care is needed regarding inoculation rate, 0.3–0.5%.
Rennet	25–30 ml of standard rennet per 100 litres milk but home-made calf vell extract may be used. The curd should be ready to cut in 35–45 min.
Cutting	The cutting takes 10–15 min to produce 2–3-mm cubes. Harp or knife is used for cutting but either the harp or wooden hay rake is used for stirring.
Scalding	Scalding takes 45 min to reach 57°C and stirring is continued for 40 min.
Washing	When the final scald temperature is reached, approximately 4% by volume of cold water is added to the whey. This reduces the temperature by 2°C. The curd is allowed to pitch for 30–35 min before the whey is removed.
Moulding	The curd is transferred to wooden or metal moulds (hooping) lined with a cloth.
Pressing	The cheeses are pressed over 2–3 days with a final pressure of 45–55 kPa, but the type of press and pressures vary from farm to farm.
Salting	The cheeses are immersed in brine for 9–10 days. Some cheesemakers prefer to dry salt before brining.
Maturation	The cheese is dried off and kept at 18–20°C for 4–6 weeks. During maturation, the cheese is wiped clean periodically with cloth soaked in brine, treated with linseed oil and then stored on edge. At 4–5 months the cheeses are sold to a cheese factor where they are kept for 2–3 years to achieve full maturity.

83. Single Gloucester

Rarely made today. Farmhouse cheese made from low fat morning milk. Made in spring and early summer only when the milk contains less fat. The acidity is lower and the curd quicker ripening. Size 38 cm diameter × 7.6 cm high, weight 6.8–8 kg. (Compare with Caerphilly cheese.) Characteristics:

Curd	White to yellow.
Texture	Open, soft body.
Flavour	Milky

Milk	Farmhouse only: 68–82 litres per cheese. Morning's milk and skimmed evening's milk (3.0–3.2% fat).
Heat treatment	None.
Starter	0.125–0.5% mixed culture starter added at 24–26°C. Ripen for 45–75 min to acidity 0.18%.
Colour	None.
Rennet	Temperature 30°C. Quantity 28–35 ml per 100 litres. Setting time 45–50 min.
Cutting	Cut curd to size of maize, 3–6 mm. Stir to float curd.
Scalding	Not more than 31°C. Stir until firm.
Whey off	Take whey off at acidity 0.13–0.14%.
Texturing	Break curds into 10-cm squares. Disturb curds and break frequently until dry and firm. Curds the size of large walnuts finally.
Moulding	Curds are mixed with 1.7% salt and put into hoops. (No milling.) Line hoops with coarse cloth (hessian).
Pressing	Press at 22 kPa (about 250 kg per cheese) for one day. Increase pressure to 66 kPa (750 kg per cheese) on second day after changing cloth to a fine weave. Turn into new cloth and re-press on following day or until the last cloth is dry.
Marketing	The cheese is greased with whey butter or lard. Clean off mouldy coat at 1–2 months for marketing.
Faults	Hard acid curd. Chalky body. Bitter or off-flavours. Discoloured curds. Mouldy cracked coats due to draughts.

84. Steppenkäse

Example of the family of cheeses of this type (55–60% water in fat-free substances). Similar curds are Stepnoj Syr (USSR), Danbo (Denmark) and Edam (The Netherlands). Block cheese 24 cm square, 9 cm high, weight 5–6 kg. Characteristics:

Rind	Smooth, even, dried-up smear coat.
Curd	Colour yellow.
Texture	Fatty, close, with pea-sized gas holes.
Flavour and aroma	Piquant, but milder than some similar cheeses (i.e. Tilsiter).
Milk	Partly skimmed (2.45–3.0% fat).
Heat treatment	Pasteurized at 72°C for 15 s or flash at 75°C. Cool to 30–31°C.
Colour	None.

Starter	0.5%, i.e. 500 ml per 100 litres milk. Short ripening period.
Additives	0.02% calcium chloride, 0.02% sodium nitrate.
Rennet	30 ml per 100 litres.Coagulation in 30–40 min.
Cutting	Cut into 6–8-mm cubes (9.5-mm knives). Stir to float curds.
Scalding	Remove 25% whey and add equivalent water at 55–65°C to give a final temperature of 40–42°C. Stir for a further 30–40 min.
Whey off	Remove whey slowly when pH is 5.6–5.8.
Moulding	Cut curd to fit hoops lined with cloth. Alternatively, fill hoops with curd and turn cheese into cloth. Replace in the hoop.
Pressing	Press for some hours with 22 kg per cheese. (Note: whey from press should be pH 5.1–5.2.)
Maturation	Store at 12–13°C. When mature, the pH of the cheese should have increased to pH 5.3–5.4.

85. Stilton

Premier English blue cheese but also a white variant originally made from poorer quality milk (in spring). Cheeses unlikely to blue are culled as white cheese. Characteristics:

Rind	Thin but complete – no cracks. (A moist coat is preferred by some makers.) Well developed coat on blue cheese.
Body	White cheese: mellow, soft, creamy–white, open texture. Blue cheese: as above but with green–blue veins throughout.
Flavour and aroma	White: clean flavour, slightly acid. Blue cheese: piquant, peppery flavour, no bitter aftertang, mellow.

Farmhouse method

This method has been used for single cheeses, and also for up to 9000 litres of milk daily, but has not been used since 1970 in factories. 63.5 litres are needed for one standard cheese; but smaller cheeses of 2.5, 3.1 and 4.5 kg have been common.

Milk	Mainly raw, but pasteurized on some farms. Good quality milk (3.8% fat) for blue cheese.
Starter	Temperature 21°C, for pasteurized milk 16 ml per 100 litres, for raw milk 8 ml per 100 litres. For larger

quantities of milk, smaller quantities of starter may be used, less than 1 ml per 100 litres.

Ripening None

Rennet 37 ml per 100 litres at 30°C. Coagulation in about 1 h. When curd firm enough, ladle out into cloths placed in the Stilton sink. These cloths are arranged by means of laths to form bags and the curd is ladled evenly into each bag to ensure uniform drainage. Four bags of 15.5 litres are in each sink. The cloths are then folded over and the laths removed. Whey is allowed to collect round the bundles until an acidity of about 0.12% has developed. The bundles are then drawn loosely together by the four corners and secured by a Stilton knot. The whey is allowed to escape. At intervals of about 1 h the knots are tightened, care being taken not to break the curd. At about 0.15% acidity the bundles are turned onto the knot. When firm enough the curd is turned out of the bundles into the sink. It is then left in the whey to mellow until an acidity of 0.5% is reached, when the curd may be cut into 10-cm blocks and turned. The turning is continued at intervals of half an hour and the curd may be placed on a rack to facilitate drainage.

Salting At not less than 0.75% acidity the curd is broken into pieces about walnut size, salted at the rate of 23 g per kg of curd weighed out on scales. The salt is then mixed into the curd before moulding in tall Stilton moulds.

Moulding The moulds are placed on boards covered by calico squares and the curd is placed loosely without pressure. About half an hour after moulding each cheese is turned onto a fresh dry board and cloth. The following morning this process is repeated, then daily for 4–6 days. The mould will then be loose enough to slip off (often at this stage after 3–4 days) and the surface of the cheese should be greasy and have the smell of ripe pears. The surface of the cheese is then scraped and a calico bandage is pinned around. The mould is then replaced. The following morning the bandage is removed, a second scraping given and a clean dry bandage pinned on. The cheeses are then replaced on the shelves and turned daily until the bandage appears dry in places. This indicates that it may be removed. When a good coat has been formed,

	the cheeses are removed to the ripening room and turned daily. At this stage cheeses to be sold white are segregated from cheeses for blueing.
Blueing	As below.

Throughout manufacture the curd should be soft and mellow and when rubbed between the fingers should feel buttery.

Factory method

Milk	High quality milk, 3.8% fat.
Heat treatment	Pasteurized at 72°C for 15 s and cooled to 30°C.
Starter	1.6 ml up to 13 ml per 100 litres milk at 21°C. *Penicillium roqueforti* spores may be added to the milk or sprayed over the curd. No ripening period.
Rennet	25–35 ml rennet extract per 100 litres milk. Temperature 30°C.
Cutting	Cut at 60–90 min using a vertical knife only (12.5 mm blade) mainly lengthways. Crossways only sufficient to release whey. No horizontal cutting.
Pitching	No stirring or scalding. Whey is run off when curd and whey are equal depth. The curd is then shovelled (with a square and deep-sided shovel) into sinks lined with drain trays and left until evening. Alternatively, the curd vat may be higher than the draining tray with a perforated base in the shallow sink. The whey curd mixture can be run from vat to drainer. In the evening the curds are cut into 15–20-cm blocks and turned over. The following morning the curd is cut into 10-cm blocks to finish drainage. (Acidity 1.2%.) Repeat breaking or cutting curd to assist drainage if necessary.
Salting	At the rate of 2.5% salt, 2.5–3% for white Stilton.
Moulding	Well mixed salt curd is weighed (10–11.3 kg) into moulds set on calico cloth on a square board. Cheeses are turned daily and the surface of each cheese is rubbed with palette knife after 3–4 days.

Factory and Farmhouse methods of maturation

Blueing	At 5–9 weeks old the curd will be mellowed to the stage when blueing will occur if the curd is aerated. The aeration is accomplished by piercing the curd with wire (stainless steel) spears 3 mm diameter, 10–12

	holes down each of four sides. This is repeated weekly until blueing commences. (The factory method with mechanical piercing uses a number of wire spears fastened together.)
Maturation	6–15 months. Temperature 18–21°C for drying off, 7°C after blueing has commenced.

86. Surti

Also called Surati cheese. Small round cheese, weight 110 g.

Milk	High fat buffalo milk (6% fat).
Heat treatment	Temperature 80°C flash (or HTST 15–20 s holding). Cool to 30°C.
Starter	Active lactic starter culture, *Lactococcus lactis* spp. mixture. Add 0.03–0.2% starter, ripen milk as necessary.
Rennet	5 ml rennet per 100 litres milk. Dilute ten times. (Tablet or powder rennet may be used as per instructions for strength.) The curd is firm in 60–90 min and should break cleanly with clear whey. Note: vegetable rennets should be used where there are religious reasons why animal rennet cannot be used.
Moulding	The hoop is a freshly woven basket of bamboo washed with hot water and sprinkled with salt. The curd is ladled into the basket in layers, each layer being sprinkled with salt. (Note: salt addition 0.5 kg to cheese from 100 kg milk.)
Drainage	Turn cheese in basket after 1 h to make a ball shape. Collect whey during drainage and immerse balls of curd in the whey after a further hour.
Maturation	Mature the cheese in whey for 1–2 days, then dry off. Alternatively keep the cheese in whey in earthenware crocks for 1–2 weeks.

87. Svecia

Cheese of the type made in various provinces of Sweden for many years. A hard pressed cheese made in various shapes and sizes. Typical flat round shape 35 cm diameter × 15 cm high, weight 13–16 kg. Typical block shape 35 cm × 33 cm × 15 cm high, weight 22–24 kg. Manufacturing methods resemble those for Gouda cheese. Svecia cheese is sometimes flavoured with cloves or cumin. Characteristics:

Rind	Dry, tough and hard, but soft and elastic when ripened in film.
Curd	Pale yellow colour deepening with age.
Texture	The body is firm but pliable with irregular gas holes and some spaces between curd particles.
Flavour and aroma	When new the flavour is slightly acid but mild. When ripe the flavour is strong, aromatic and sometimes bitter. The cheese is made in four qualities, 30, 40, 50 and 55% FDM. One of the most serious defects in low fat cheese is a hard dry curd with bitter flavour.
Milk	Cow's milk; fat adjusted from 2.7% to whole milk according to type of cheese.
Heat treatment	Normally HTST, 72°C for 15 s then cooled to 30°C.
Additives	10 g calcium chloride and 15 g sodium nitrate per 100 litres milk.
Starter	0.5–1.0% of starter culture (*Lactococcus lactis* sub-spp. *lactis* and *cremoris*, *Leuconostoc mesenteroides* sub-sp. *cremoris*).
Rennet	Temperature 30–32°C, 25–30 ml rennet extract per 100 litres milk. Setting time 30–35 min.
Cutting	Cut the coagulum into 1-cm cubes. Stir to float the curds for 20 min.
Scalding	Scald the curds to 40–42°C taking 20–40 min. Stirring in total, 60–90 min.
Whey off	The whey is removed without any pressing of the curd and up to 2.5 kg coarse salt per 1000 litres of original milk is added into the free curds. Mix the salt into the curd.
Moulding	Fill the curd into stainless steel moulds.
Pressing	Press for 24 h at 18°C. Then dry off in store for 2 days.
Salting	Immerse in brine bath for 2–3 days (salt content 20%). Temperature 12–15°C. Dry off in store then transfer to ripening store.
Maturation	Mature for 10–12 days at 12°C, 65–70% RH, then at 16°C for 21–28 days at 70% RH. Finally hold at 12°C. Clean the rind, pack in film or dip in wax bath as necessary for maturation, especially the block cheese. Mature in 3–6 months.

88. Telemea

Brine pickled cheese made in the Danube area. Made from sheep's, buffalo or cow's milk. Mainly cubical shape, or 10 cm square × 8–9 cm high, weight 1 kg. Characteristics:

Rind	None, colour porcelain white but may be yellowish if made from cow's milk. Covered with seeds of the black cumin (*Nigella satura*).
Curd	White or yellowish.
Texture	Soft, almost spreadable.
Flavour and aroma	Pleasant, slightly salty and acid when made from sheep's milk, piquant.
Milk	Cow's milk (3.5% fat), sheep's milk (7.5% fat) or buffalo milk (7% fat).
Heat treatment	LTLT at 60–68°C for 20–30 min.
Starter	0.5 litres culture per 100 litres milk (includes mesophilic lactococci and *Lactobacillus casei*). Temperature 28–31°C in summer, 3°C higher in winter.
Additives	50 ml of 40% calcium chloride solution and 20 g sodium nitrate per 100 litres milk.
Rennet	25–30 ml of rennet extract per 100 litres milk. Temperature 30°C summer, 34°C winter. Setting time 1 h.
Ladling	Coagulum ladled into cloths in layers. Cut the curd crosswise with a long knife. Take up the corners of the cloth together and press the curds for two or more hours with weight of 1.5 kg for each kg of curd.
Cutting	Cut the curd into blocks 10–11 cm square and 10 cm thick. Cool the cheese blocks with cold water.
Salting	Immerse in salt bath (18–22% salt) at 15–16°C for 15–16 h.
Pickling	Layer the curds in casks, barrels, steel or plastic cans. Fill the container with acid whey (1.6–2.0% TA) with 4–5% salt.
Maturation	Cheese mature in 1 month. Store at 5–10°C.

89. Tilsiter

Normally flat, round shaped cheese made in several qualities in respect of fat (i.e. 30, 40, 45, 50% FDM). Caraway seeds are often incorporated in the lower fat cheese. Various sizes and shapes are produced. Round: 25 cm diameter × 10 cm high, weight 4.5–5 kg. Rectangular: 30 × 12 × 10 cm, weight 3.5 kg. Characteristics:

Rind	Thin rind with dried smear coat or washed thin rind.
Curd	Ivory white to yellow colour, firm and fatty.
Texture	Firm, close, not short or crumbly with slits or oval or round gas holes, barley corn size.
Flavour and aroma	Slight acid or sour taste, piquant, but not strongly acid.

Milk	Cow's milk whole or with part skim milk for low fat cheese. Mixed evening and morning milk (0.17% TA).
Heat treatment	HTST 72°C for 16–20 s or flash 75°C. Cool to 28–33°C. (Lower temperature for low fat cheese.)
Starter	0.2–1.0% starter culture.
Colour	Vegetable colour (annatto/carotene) 2 ml per 100 litres milk in wintertime.
Additives	Calcium chloride 0.01–0.015% (added in dilute solution), also 0.02% of sodium nitrate or potassium nitrate.
Rennet	30 ml of rennet extract per 100 litres milk (or 2–5 g rennet powder) to give a firm curd in 30–40 min, temperature 28–33°C (2°C warmer in wintertime).
Cutting	Cut with bladed knife or harp to hazelnut or pea size (10 mm), take at least 40 min. Stir to float curds between the cutting to obtain even size.
Scalding	Warm the whey to 36–42°C, the lower temperature for low fat cheese, and higher temperatures in winter. Stir thoroughly.
Whey off	Take off the whey (30–40%) and continue to stir the curds for 40–60 min, then remove the whey.
Hooping	Fill the metal hoops with curd when firm enough (pressed together in the hand, curds should not stick together). Turn the cheese into cloth and replace in hoop. Turn frequently while the whey drains freely. Temperature 26–30°C.
Salting	Salt bath, 18–20% salt and 0.2–0.4% acid. Temperature 12°C in summer, 14°C in winter. Immerse cheese for 2–3 days. Dry off on shelves.
Maturation	Mature cheese, summer 12–14°C, winter 15–16°C, humidity 90–95% RH. Turn frequently and wipe coat of cheese with cloth dipped in brine. Smear coat develops during maturation. Mature in 5 months or longer.
Storage	Store cheese at 2–3°C for 3–5 months as required.
Packing	Pack in parchment paper and plastic film as required.
Faults	Main faults arise from bacteriological growths and size or regularity of gas holes. Normal salt content 2.0% but lower salt allows clostridia to grow and produce bitter rancid flavour. Short textured curds or chalky curds are caused by high acid development.

90. Trappist

Semi-hard cheese (washed-curd type) made in Bosnia, Hungary, Germany, etc. A similar method to that used for Port du Salut (St. Paulin). Flat, round shape, also block form. Size 18 cm diameter × 6 cm high, weight 1.5–2 kg. Characteristics:

Rind	Yellow, thin, elastic, smooth, wax-coated for sale.
Curd	Ivory to yellowish.
Texture	Tender, pliable, with a few gas holes the size of peas.
Flavour and aroma	Clean, mild, pleasant aromatic flavour, stronger if smear coat developed.
Milk	Cow's milk (3.6–4.0% fat).
Heat treatment	Flash 75°C, or HTST 72°C for 16–20 s. Cool to 27°C.
Colour	None in summer, 1–1.5 ml per 100 litres milk in winter.
Additive	Sometimes 0.02% calcium chloride.
Starter	0.25% of mixed culture starter (mesophilic bacteria). Ripen not more than 30 min.
Rennet	Temperature 30–31°C. Rennet extract 18–25 ml per 100 litres milk. Sufficient to give firm curd in 40–45 min.
Cutting	Cut to 12-mm cubes with multi-bladed knives. Stir to float curd.
Scalding	Stir gently at first and heat to 34–35°C taking 45 min. Allow curd to settle.
Washing	Take off whey to curd level. Fill with water at 35°C to original level with stirring to float curds. Stir for 10–15 min.
Moulding	Fill metal perforated hoops placed on a drainage mat. (Each hoop contains approximately 2 kg curd). Turn immediately and repeat turning in 4 h. Leave overnight to cool off.
Salting	Remove hoops, immerse cheese in brine (20% salt) at 15–16°C for 6–8 h.
Maturation	Mature for 2–3 weeks at 15–18°C, wash with brine to remove any mould growth (90% RH). Some cheese develops a smear coat if ripened at 12°C.
Packing	Dry cheese in humidity 75% RH. Dip in hot wax 63°C.
Storage	Store at 2–5°C until sold.
Faults	Blown curds (i.e. large holes in cheese) due to coliform bacteria. Bitter flavours. Blown curds due to growth of *Clostridium butyricum*. Unpleasant flavour is produced.

91. Vacherin

Small flat round cheese made in autumn. Size 15–28 cm diameter × 3–4 cm high, weight 0.5–2.0 kg. Characteristics:

Rind	Thin with dark red smear coat.
Curd	Creamy white, soft.
Texture	Very soft when ripe, almost liquid, not chalky or rough. No gas holes or open curd spaces.
Flavour and aroma	Milk aromatic with typical flavour with savoury after-taste, almost like the Munster cheese.
Milk	Fresh cow's milk of good quality (3.2–3.4% fat). Usually end of lactation milk (0.15–0.16% TA) with 3.0–3.3% protein.
Heat treatment	Mainly raw milk used, sometimes heated to 65°C for 1–2 min.
Starter	0.5–1.0% culture added at 20–25°C and ripened to pH 6.5–6.6.
Rennet	Sometimes 3–4% water added to the milk, then 15–18 ml rennet extract per 100 litres. Setting to a firm curd in 40–45 min.
Cutting	Cut the curd with harp or knife to 15-mm cubes (hazel-nut size) and stir slowly for 15 min.
Whey off	Allow the curd to settle then drain off most of the whey.
Moulding	Ladle the curd into the hoops and drain on cloth covered shelves in a warm room. Whey acidity 11°SH (0.24% TA), pH 4.9–5.1. Weights may be used to assist drainage. After 18 h transfer the drained cheese to a cool room for 24 h.
Salting	Dry salt with coarse salt or immerse in a salt bath for 20–60 min. Dry off and transfer to cellars at 13–15°C and 90% RH for 4–5 weeks.
Maturation	Brush the cheese coat and allow smear coat to develop (the pH rises to pH 5.5–6.0 and even pH 7.0 under the smear coat following the growth of *Brevibacterium linens*).
Packing	Pack in greaseproof paper in pine or spruce woodchip boxes.
Faults	Growth of *Penicillium glaucum*, *Oidium* and *Mucor* spp. as well as coliform bacteria cause off-flavours in the cheese.

92. White pickled cheese

Sheep's milk cheese (brine pickled). Made in Bulgaria and adjacent areas with simple equipment (in shepherd's huts, etc.).

Milk	Sheep's milk, occasionally cow's milk may be added.
Heat treatment	None but may be flash heated to 70°C, or at 68°C for 10 min.
Colour	None.
Starter	Sour milk may be used. Normally a mixed culture of *Lactococcus lactis* sub-spp. *lactis* and *cremoris* is added at 0.5–1%. Thermophilic starters including *Lactobacillus casei*, *Lactobacillus delbrueckii* sub-sp. *bulgaricus* and *Streptococcus thermophilus* are also used. Ripening period short or none.
Rennet	25–30 ml of standard rennet per 100 litres milk (lamb rennet extract is also used). Temperature 30–31°C. Coagulation complete in 60–70 min.
Drainage	The curds are poured into a fine weave cloth (cotton, nylon or hessian) laid over a large wooden frame placed over a drainer. The curds are compressed during drainage by tightening the cloth over them. Finally the curd is pressed by weights (approximately 2 kg) on the curds. The drainage is assisted by the acidity (0.9–1.2%).
Salting	The curds are cut into 10-cm squares and immersed in saturated brine at 10–12°C for 12–14 h (usually overnight).
Packing	The brined cheeses are packed in barrels, casks or drums or plastic cartons and the container filled with whey brine (whey at 0.35% TA with 8–12% salt).
Maturation	Maturation proceeds for 30 days – movement of the containers assists even maturation. Store at 5–8°C.
Faults	Coliforms cause most storage spoilage in spite of the high acidity (1.25% TA) of the curd.

93. Wensleydale

White or blue cheese from 1.5 kg to 5 kg, the larger size similar to that of Stilton. Blue Wensleydale cheese is difficult to manufacture satisfactorily and is therefore not produced widely. Small white cheeses are similar to Cheshire cheese. Factory-made white Wensleydale cheese for 0.45- and 0.9-kg sized cheeses have been made without the use of cloths using a process similar to the Cheshire recipe. Characteristics:

Coat	Banded contours due to stitching.
Texture	Blue cheese soft, and open with blue veins, white cheese as for Cheshire.

Flavour and aroma	Typical peppery flavour of blue cheese, but more mellow than some of the more quickly ripened blue cheeses. White cheese clean, mild flavour.
Milk	Fresh milk, raw for farmhouse but pasteurized for creamery production. Higher fat (3.85%) milk needed for blue cheese. Low fat (spring) milk satisfactory for white cheese.
Starter	0.2–0.4% mixed culture starter added to milk at 21°C.
Ripening	Temperature 30°C. Ripen for 45 min or until acidity rises by 0.01%, i.e. from 0.185 to 0.195%.
Rennet	Temperature 29–30°C according to ambient temperature. Quantity 30 ml per 100 litres. Dilute six to eight times with clean water.
Cutting	Curd firm with clean break in 35–45 min. Cut both ways with vertical knife, 13-mm blades, then cut with horizontal knife to size of large beans, acidity 0.13–0.145%.
Stirring	Stir for 5–10 min to float curd. Settle in whey for 15–20 min and then refloat curd to scald.
Scalding	Scald to 1°C above rennetting temperature and stir for 20 min.
Pitching	Pitch until acidity is 0.35–0.155% if curd is free, but stir to float if curd mats together.
Drainage	Draw whey when acidity is 0.14–0.15%. Ladle curd onto a coarse (hessian) cloth on rack in drainer. Tie in bundle. Open cut curd and disturb then retie in bundle.
Texturing	Disturb the curd in the bun every 10 min until acidity is 0.15–0.16%. Then cut blocks and turn over. Repeat process until acidity is 0.27–0.35% for blue cheese (0.39–0.45% in cold weather or for white cheese).
Milling	Broken by hand or coarse peg mill.
Salting	1.6% salt (1.8% for wet cheese or blue cheese). Mix salt in thoroughly.
Moulding	Mould small pieces tightly in the corners of the mould (to give firm corners of the cheese). Fill remainder of curd loosely, but place small pieces on the top and corners. Leave overnight at 21°C.
Pressing	Turn cheese into coarse hessian cloth, stand for 2 h then press overnight at maximum 80 kPa (about 250 kg per cheese if using a 200 mm diameter mould) for white cheese, but at 5–10 kPa only for blue cheese. Bandage with sewn-on bandages (gather top and bottom, stitch at intervals down side). Re-press for 3 h. Dry off at 18°C.

Storing	Store at 12–13°C in humid atmosphere (90% RH) for blue cheese, but drier for white cheese. Blue cheeses are ripe in 6–12 months, white in 2–3 weeks. Turn regularly in store.
Faults	Due to the low acidity of the curd, gas-producing and taint-producing bacteria may occur, especially in raw milk cheese. Wet curd at milling gives cracked coats due to evaporation in store. Cracked coats become mouldy. Ripening in a high humidity causes excessive mould growth on the coat of the cheese. Waxing of the cheese too early results in fermentation under the coat and off-flavours developing.

94. Whey curd cheese

Whey protein produced as a pressed curd (15% moisture) or as a paste (80% water) used for baking purposes.

Whey	Separated, free from fat.
Additives	Neutralized with calcium phosphate or calcium hydroxide.
Heat treatment	Use indirect heating, heat to 91–96°C.
Whey off	Allow whey to cool slowly then decant liquid whey (lactose solution).
Pressing	Collect whey protein curd in cloth bags and press in cheese press for 4 h, turn bags and re-press.
Packaging	Pack in film, waxed paper, waxed card or a crock.
Storage	Cold store below 5°C until required and protect against mould growth.
Alternatives	Air dry to crumb as required then mix into paste for spreading purpose or further dilute to a fluid for baking purposes.

95. Whiteknights

Soft cheese based on Cambridge and Coulommiers types, designed for small-scale production.

Milk	Whole milk, 3.6–4.2% fat.
Heat treatment	Pasteurize at 72°C for 15 s, cool to 31–33°C. Fill into pairs of stainless buckets in 10-litre lots.
Starter	1 ml per litre of mesophilic butter type starter, no ripening period.

Colour	Add 0.2 ml annatto solution per litre to one of the pair of buckets.
Rennet	Add 0.2 ml rennet solution per litre diluted five-fold to each bucket, stir for 1 min then leave for 60 min to form a curd.
Cutting	Cut vertically in both directions at 20–25-mm intervals with a knife then leave for 20–30 min.
Ladling	Ladle the uncoloured curd into plastic hoops 110–130 mm diameter × 120 mm high on a plastic grid, initially to give a layer 25 mm deep then 20–30 min later a second ladling to two-thirds fill the moulds. After a further 30 min ladle coloured curd into the moulds
Drainage	Leave to drain for at least 5 h with the room temperature over 20°C, preferably 25°C, then turn and leave overnight then turn again.
Salting	Dry salt the surfaces at each turn.
Packing	When cheese adequately drained (30–50 h) wrap in greaseproof paper.
Storage	Store at 2–4°C for up to 7 days.
Faults	Excessively moist curd if draining temperature too low. Bitter off-flavour in old cheese.

96. Yorkshire cheese curds

Used for bakery purposes – cake fillings with egg, jam or sweet fruit filling.

Milk	Skim milk.
Heat treatment	None.
Starter	None.
Additives	Sodium sulphate (Glauber's salts) 1 kg or magnesium sulphate (Epsom salts) 1 kg added to wheat flour paste to produce a creamy liquid.
Precipitation	Heat the skim milk to 100°C then add sufficient paste to flocculate the protein and give greenish clear whey.
Whey off	Leave for 40–60 min and pour off the whey.
Drainage	Pour the curd into hessian bags and press out the free moisture.
Marketing	Dry off the curd with wheat flour and pack in moisture-proof bags.
Storage	Store at 1–2°C until required for use.

Index